THE DESIGNER'S HANDBOOK OF PRESSURE-SENSING DEVICES

THE DESIGNER'S HANDBOOK OF PRESSURE-SENSING DEVICES

JERRY L. LYONS, P.E.
Essex Cryogenics Industries, Inc.
Fluid Controls Division
St. Louis, Missouri

VNR VAN NOSTRAND REINHOLD COMPANY
NEW YORK CINCINNATI ATLANTA DALLAS SAN FRANCISCO
LONDON TORONTO MELBOURNE

Van Nostrand Reinhold Company Regional Offices:
New York Cincinnati Atlanta Dallas San Francisco

Van Nostrand Reinhold Company International Offices:
London Toronto Melbourne

Library of Congress Catalog Card Number: 79-19259
ISBN: 0-442-24964-0

Manufactured in the United States of America

Published by Van Nostrand Reinhold Company
135 West 50th Street, New York, N.Y. 10020

Published simultaneously in Canada by Van Nostrand Reinhold Ltd.

15 14 13 12 11 10 9 8 7 6 5 4 3 2 1

Library of Congress Cataloging in Publication Data

Lyons, Jerry L.
The designers' handbook of pressure-sensing devices.

Includes index.
1. Pressure transducers. I. Title.
TJ223.T7L96 629.8 79-19259
ISBN 0-442-24964-0

DEDICATION

To my friend Carl L. Askland, P.E. for all of his help and encouragement in the making of this book.

Preface

This book is written to benefit users and designers of pressure switches and related devices, and of systems incorporating such devices. Users include design engineers in such fields as aerospace, industrial process control, fluid power systems, and so on, as well as plant and industrial engineers, maintenance engineers, instrumentation engineers, researchers, and others who use pressure switches.

As the monitoring of any system today is increasingly complex, the proper selection of the pressure switch has become more necessary. With the aid of this book, the systems engineer may for the first time select the proper switch for his or her application. An especially useful aid to the pressure switch designer, this book is the first of its type to be published.

Part One is devoted to a general description of basic types of pressure switches. Its purpose is to give an introduction to the pressure devices more commonly used throughout the industry. In general, it explains the main function of these basic types. Because of the numerous pressure switch companies throughout the world, many types of basic pressure switches have been developed that may differ in style and function, and many of the descriptions and terminologies may differ from one company to the next. I have decided to take the middle ground and attempt to give a generalized view of the industry as a whole, and to try to avoid the specialty area where new techniques are still on the research and development drawing boards and are closely guarded by individual firms.

Part Two is the heart of the book. In this section every attempt has been made to put at the fingertips of the design engineer a guide that will help him or her resolve the many problems arising from each requirement or specification.

Part Three gives the design engineer or user additional information or alternatives in the selection of devices that may be used in monitoring or measuring various processes, experiments, and so on.

Part Four, which should be of particular interest to application engineers, designers, consultants, and users, is the applications section of the book.

Finally, the Appendix contains several helpful sections: A glossary of terminology gives definitions of the basic pressure switch devices. Standards and specifications are also listed, with formats illustrating pressure-sensing requirements. Also, there are charts for metric conversions, which are becoming increasingly important in our society.

The author is deeply indebted to Mrs. Kathy Lyons, Mr. Dennis Mertz, and Mr. D. S. Beckham and to the following manufacturing companies and engineering societies for materials and/or advice and comments contributed during the preparation of this book:

Allen-Bradley, Industrial Control Division, Milwaukee, Wisconsin

American Society of Mechanical Engineers, New York, New York

Ametek Controls Division, Feasterville, Pennsylvania

Automatic Switch Company, Florham Park, New Jersey

Barton ITT, Process Instruments and Controls, Monterey Park, California

Belfab Division, Daytona Beach, Florida

Bellofram Corporation, Burlington, Massachusetts

BLH Electronics, Incorporated, Waltham, Massachusetts

Bourns, Incorporated, Instrument Division, Riverside, California

Bristol Division of Acco, Waterbury, Connecticut

Burling Instrument Company, Chatham, New Jersey

Canadian Standards Association (CSA), Rexdale, Ontario, Canada

Celesco Industries, Incorporated, Environmental & Industrial Products, Canoga Park, California

Chemical Engineering, New York, New York

Chemiquip Products Company, Incorporated, New York, New York

Compac Engineering, Incorporated, San Jose, California

Consolidated Controls Corporation, Bethel, Connecticut

Control Products, Incorporated, East Hanover, New Jersey

Cook Electric, Morton Grove, Illinois

Curtis Industries, Incorporated, Milwaukee, Wisconsin

Custom Component Switches, Incorporated, Chatsworth, California

Datametrics, Wilmington, Massachusetts

Delaval Turbine, Incorporated, Barksdale Controls Division, Los Angeles, California

R. B. Dension, Incorporated, Bedford, Ohio

Design News, Boston, Massachusetts

Dresser Industries, Industrial Valve and Instrument Division, Stratford, Connecticut

Dwyer Instruments, Incorporated, Michigan City, Indiana

Electroid Company, Springfield, New Jersey

Electronics Corporation of America, Photoswitch Division, Cambridge, Massachusetts

Essex Cryogenics Industries, Incorporated, St. Louis, Missouri

Fenwal Incorporated, Ashland, Massachusetts

Fluid Components, Incorporated, Canoga Park, California

Fluid Products, Incorporated, Eden Prairie, Minnesota

The Gas Spring, Montgomeryville, Pennsylvania

Gould, Incorporated, Control and Systems Division, Wilmington, Massachusetts

Richard Greene Company, St. Louis, Missouri

Gulton Industries, Incorporated, Costa Mesa, California

Handy and Harman Tube Company, Norristown, Pennsylvania

Harwil Company, Santa Monica, California

Haskel Engineering and Supply Company, Burbank, California

Hydra-Electric Company, Burbank, California

Hydraulics and Pneumatics, Cleveland, Ohio

Instrumentation Technology, Pittsburgh, Pennsylvania

Instruments and Control Systems, Radnor, Pennsylvania

ISA (Instrument Society of America), Pittsburgh, Pennsylvania

J. P. Semiconductors, Incorporated, Arcadia, California

Kavlico Corporation, Chatsworth, California

Kratos Pressure Devices, Pasadena, California

Leslie Company, Parsippany, New Jersey

Machine Design, Cleveland, Ohio

Material Control, Incorporated, Aurora, Illinois

McDonnell and Miller, Chicago, Illinois

The Mercoid Corporation, Chicago, Illinois

Metal Bellows Company, Sharon, Massachusetts

Micro Switch, A Division of Honeywell, Freeport, Illinois

Mid-West Instrument, Troy, Michigan

Edward W. Mooney Company, Incorporated, St. Louis, Missouri

Namco Controls, Cleveland, Ohio

National Sonics Division, Envirotech Corporation, Hauppauge, New York

NEMA (National Electrical Manufacturers Association), New York, New York

Neo-Dyn, Incorporated, Glendale, California

NFPA (National Fluid Power Association, Incorporated), Milwaukee, Wisconsin

The Norson Company, Clarkston, Michigan

Oil-Rite Corporation, Manitowoc, Wisconsin

Orange Research, Incorporated, Orange, Connecticut

Paul-Munroe Hydraulics, Incorporated, Pico Rivera, California

PCS Piezotronics, Incorporated, Buffalo, New York

Penton/IPC, Publishers of *Automation,* Cleveland, Ohio

Pressure Controls, Incorporated, Belleville, New Jersey

Princo Instruments, Incorporated, Southampton, Pennsylvania

Product Engineering, New York, New York

Qualitrol Corporation, Fairport, New York

Sealol, Incorporated, Warwick, Rhode Island

Servometer Corporation, Clifton, New Jersey

Sethco Manufacturing Corporation, Freeport, New York

Setra Systems, Incorporated, Natick, Massachusetts

Sirco Controls Company, Seattle, Washington

The Solon Manufacturing Company, Chardon, Ohio

Square D Company, Middletown, Ohio

Static "O" Ring, Pressure Switch Company, Olathe, Kansas

Stewart-Warner Corporation, Springfield, Illinois

Texas Instruments, Incorporated, Control Products Division, Attleboro, Massachusetts

Tyco Instrument Division, Lexington, Maine

UL (Underwriters Laboratories, Incorporated), Chicago, Illinois

United Electric Controls Company, Watertown, Massachusetts

Validyne Engineering Corporation, Northridge, California

Whitman General, Terryville, Connecticut

Contents

THE DESIGNER'S HANDBOOK OF PRESSURE-SENSING DEVICES

PART ONE

SELECTION, SPECIFICATION AND USAGE OF PRESSURE SWITCHES

I An Introduction to Pressure Switches

A pressure switch may be defined in several essentially equivalent ways. One common definition defines it as the link between a pressure source and an electrical circuit.[1] Another considers it as a bistable pressure transducer with on and off modes.[2] For purposes of this book we define pressure switch as follows: A pressure switch is a process control device that upon sensing a change of pressure in a medium causes an external circuit to switch from one mode to a second mode.

Most often the circuit to be switched is an electrical circuit, but it may be of another type, such as hydraulic or pneumatic. Normally, the circuit will switch between on and off modes, although in some cases it may switch between two different "on" states.

Pressure switches are available in a wide variety of pressure ranges, electrical ratings, construction, materials, accuracies, and, of course, cost. They range from fairly unsophisticated commercial switches to high-quality industrial control instruments, to extremely critical devices for aerospace. They may be found on products ranging from automobiles and washing machines to manned spacecraft. They may sense pressures ranging from high vacuum to 50,000 psi or higher and may have to operate with fluids ranging from water or air to very corrosive liquids or gases, to highly radioactive materials.

In spite of the enormous variety of pressure switches available, we shall see that they have many features in common and may be grouped into a few basic categories based on their use or on their construction.

SENSING PRESSURE CHANGE

A change in pressure is usually sensed or detected in one of two basic ways—the motion or displacement of an object under an increasing force such as the travel of a piston in a cylinder, or a change in some physical property such as the electrical resistance of a material due to increasing internal stress caused by the increasing applied force. In most commercial and industrial pressure switches the movement of a member such as a piston or bourdon tube is utilized.

The change in internal stress under change in pressure is the basis of most types of strain gage pressure transducers, which are becoming of increasing importance in the so-called high technology industries such as aerospace. In some cases they are made to serve as the element in an on-off pressure switch.

Other possibilities include the use of the piezoelectric effect, change of index of refraction with pressure, and similar physical phenomena. These effects are employed relatively infrequently in pressure switches, as the cost is usually very high, and such devices are quite limited in their range of response.

TYPES OF PRESSURE SWITCHES

Nearly all pressure switches will fall into one of three categories, depending on the nature of the pressure to be sensed. All pressure switches respond to a difference between the pressure to be sensed and a reference pressure. In this respect all pressure switches are differential switches.

Absolute Pressure Switch. An absolute pressure switch is one that is designed to sense a change in absolute pressure independent of the prevailing atmospheric pressure. One side of the sensing element is exposed to the pressure source, and the other side is exposed to a fairly high vacuum, such as in an aneroid capsule or sealed bellows. Hence the reference pressure in this case is (essentially) zero. (See Figure 1-1a,b,c.)

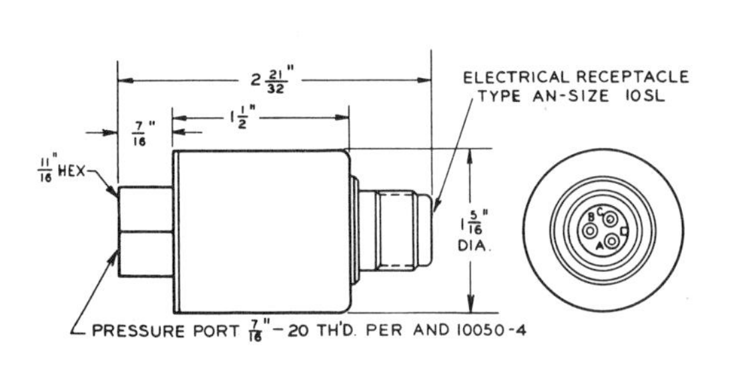

PHYSICAL CHARACTERISTICS

Electrical Connection
Hermetically sealed receptacle, size 12S shell. Mates with MS connectors.

Pressure Connection
7/16″ —20 UNF —3B, per AND 10050–4.

Maximum Weight
3.2 ounces.

Operating Media
All fluids compatible with Ni-Span-C.

Temperature Range
—65°C to +125°C.

Vibration Limits
MIL STD 202C, Procedure 204A, Condition B, 15G, 10–2000 Hz.

Military Specification
MIL-E-5272C (applicable portions).

OPERATING CHARACTERISTICS (Typical Models)

Model 506104	–26–5	–27–3	–28–1
Calibration Range	8.0–100.0 psia	8.0–100.0 psia	1.25–25.0 psia
Operating Differential, Typical*	1.4/7.5 psi	1.4/7.5 psi	.4/1.5 psi
Setting Accuracy	±1.25 psi	±1.25 psi	±0.25 psi
Overall Accuracy	±2.5 psi	±2.5 psi	±0.5 psi
Proof Pressure	125 psia	125 psia	30 psia
Contact Arrangement	SPST-NC	SPST-NO	SPDT

Fig. 1-1a. Typical absolute pressure switch operating characteristic. *(Courtesy of Acco Bristol Div.)*

Fig. 1-1b. Capsule. *(Courtesy of Acco Bristol Div.)*

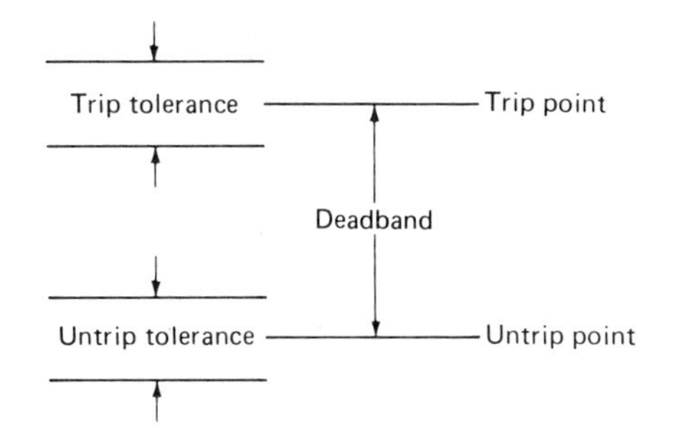

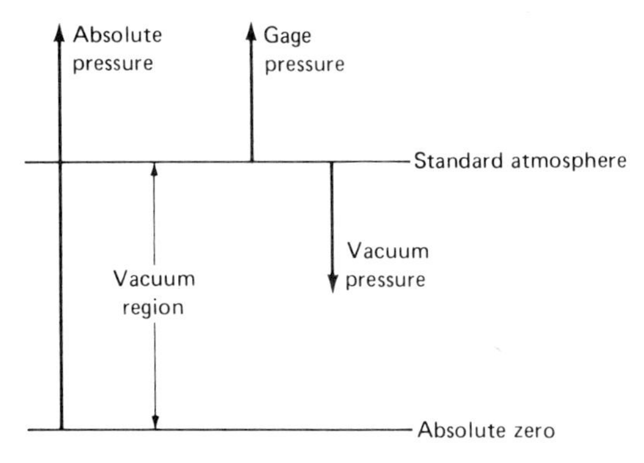

Fig. 1-1c.

An absolute pressure switch may be needed when the pressure to be sensed is less than atmospheric or when the reference pressure could change, such as with a change in altitude.

Gage Pressure Switch. A gage pressure switch is designed to sense pressures above the prevailing atmospheric pressure. One side of the sensing element is exposed to the pressure source, and the other side is open to the atmosphere, i.e., in this case, the reference pressure is that of the atmosphere. The gage pressure switch is probably the most common type. (See Figures 1-2 and 1-1c.)

There is one type of switch, usually called a vacuum switch, in which the role of the ports is reversed, i.e., the atmospheric side is the high pressure side. In this case, the set point is measured down from atmospheric pressure.

Differential Pressure Switch. As its name implies this switch responds to the difference of two pressures. In this case, the reference pressure is actually one of the variable pressures. Physically this switch may be direct acting, or it may be a combination of two absolute or gage pressure switches with coupled outputs. (See Figure 1-3.)

This terminology has also been applied to pressure switches in which a prescribed pressure differential or deadband exists between trip and untrip points. More will be said about this type of switch in later sections.

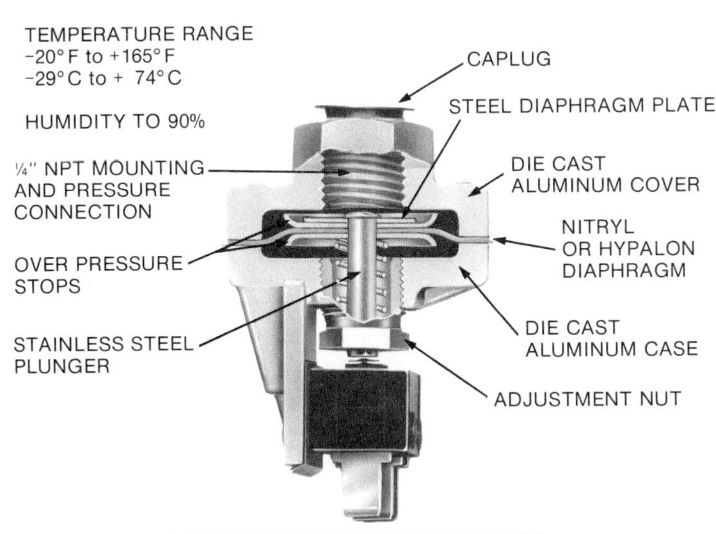

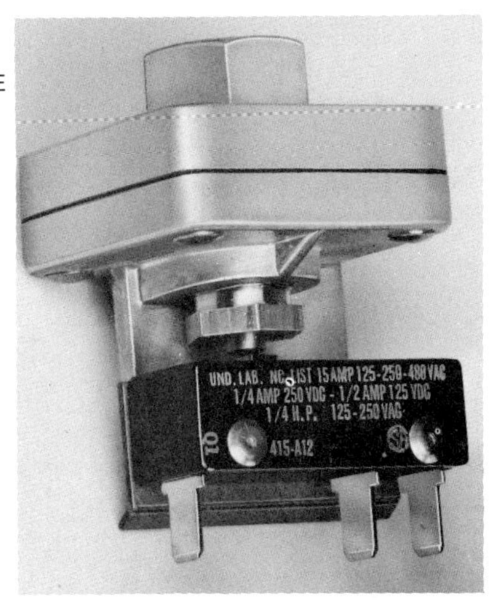

Fig. 1-2. Gage pressure switch (also vacuum switch). *(Courtesy of Qualitrol Corp.)*

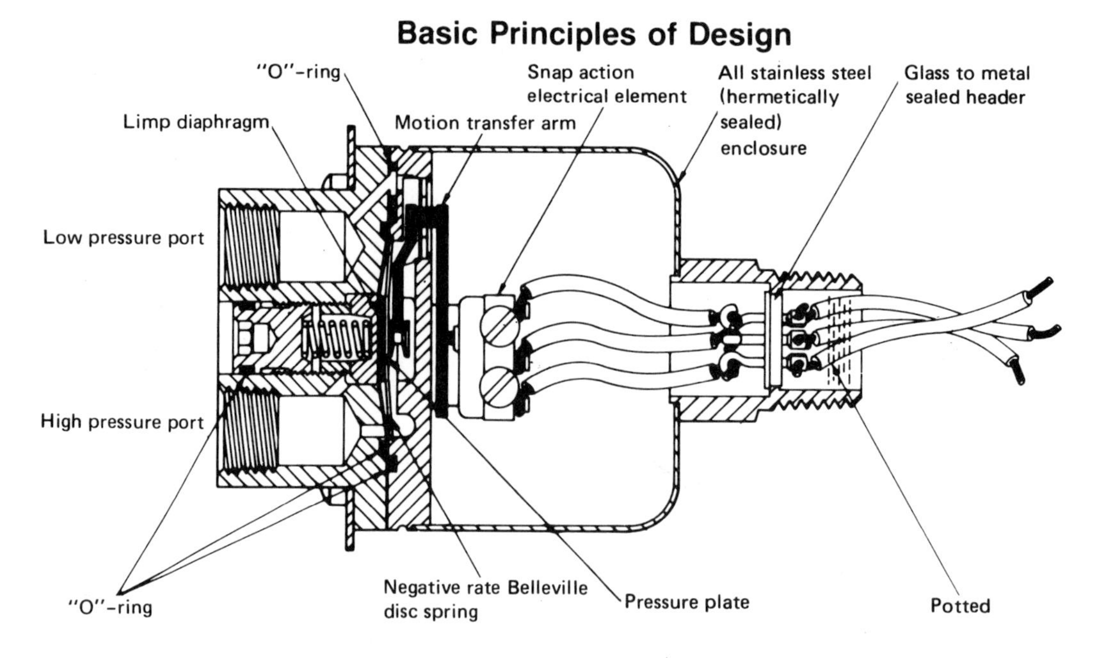

Fig. 1-3. Differential pressure switch. *(Courtesy of Neo-Dyn, Inc.)*

CONSTRUCTION

All pressure switches can be said to be composed of three basic elements, a pressure sensing device such as a piston or diaphragm, a loading spring which may be physically separate or may be inherent in the pressure sensor, and the mechanism to be switched—generally an electrical switch. (See Figure 1-4.) Some authors consider a fourth basic element to be the pressure transmitter, which serves to transmit the "signal" from the sensor to the electrical switch. This may be the pressure sensor itself, e.g., a piston, or it may be composed of mechanical linkages, or in more exotic cases such means as photocoupling may be employed. (See Figure 5-3.)

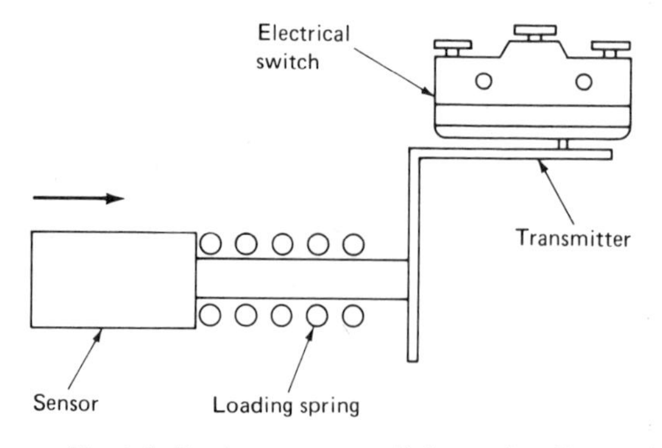

Fig. 1-4. Basic pressure switch construction.

Other physical features include plumbing to the pressure source, electrical connectors, mechanisms for adjusting the pressure settings, and housings, mounting hardware, etc.

Pressure switches are available in physical sizes ranging from very tiny devices for aerospace applications where size and weight are at a premium to very large industrial switches that may contain several high current electrical switches.

While most pressure switches are discrete components, they may be combined into other components such as valves and flow controllers, automatic recording equipment, etc.

SOME TYPICAL USES

Pressure switches have been used for many different functions in commercial applications, industrial processing and control, aerospace applications, and laboratory measurements and testing. In each instance they are intended to switch one or more circuits when a given pressure or differential is reached.

Perhaps the most important usage in terms of the numbers of switches used is in the industrial control field. Pressure switches may be used to control system pressures in piping systems by sending signals to solenoid valves and other controllers; they may be used to control the automatic filling or emptying of fluid vessels; and they may be used as sequencing mechanisms in various types of machinery.

A very important use of pressure switches in many different fields is as a safety switch. At dangerous pressure levels they may be made to shut down a processing system, turn on other safety equipment, such as automatic fire sprinklers or exhaust fans, shut off a malfunctioning machine, sound an alarm, etc. (See Figure 1-5.)

Another important use of a pressure switch is as a precision measuring instrument. In many situations it is necessary to know fairly accurately when a specified pressure has been reached, but it is not feasible to monitor it continuously as with a pressure gage. As one example, many aircraft use differential pressure switches connected to pitotstatic tubes to indicate when a specified air speed has been reached.

Finally, in some instances a pressure switch may be used to perform the functions of a limit switch or a proximity switch.

SOME TYPICAL SWITCHES

In this section we shall look very briefly at a few examples of the various pressure switches available today in terms of function and physical construction. Later chapters will cover much of this material in greater detail. While the examples chosen are arbitrary and in no sense cover the entire industry, they will serve to illustrate the wide variation in the pressure switch field as well as to illustrate a few of the features discussed in previous sections.

Low Pressure Switch. The switch shown in Figure 1-6a,b is a diaphragm actuated type designed for use with pressures in the range of a few inches of water. The large-diameter diaphragm transforms a small pressure change into the relatively large force necessary to trip a large electrical switch.

Miniature Pressure Switch. Figure 1-7a,b shows a miniature bellows actuated pressure switch. It is fairly typical of instrumentation and laboratory switches in size and construction. This type of pressure switch is available with set points ranging from a few

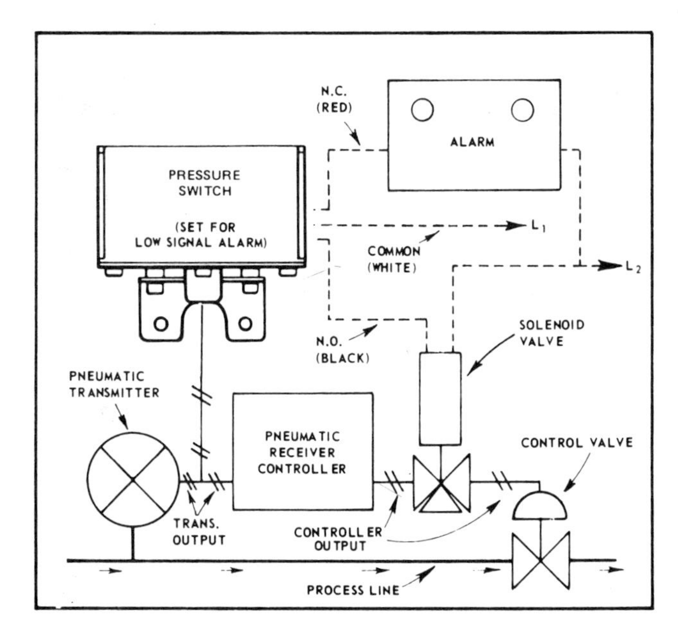

SCHEMATIC OF TYPICAL
INSTRUMENT AIR SWITCH APPLICATION

Fig. 1-5. Schematic diagram of typical instrument air switch application. *(Courtesy of Static-O-Ring Pressure Switch Co.)*

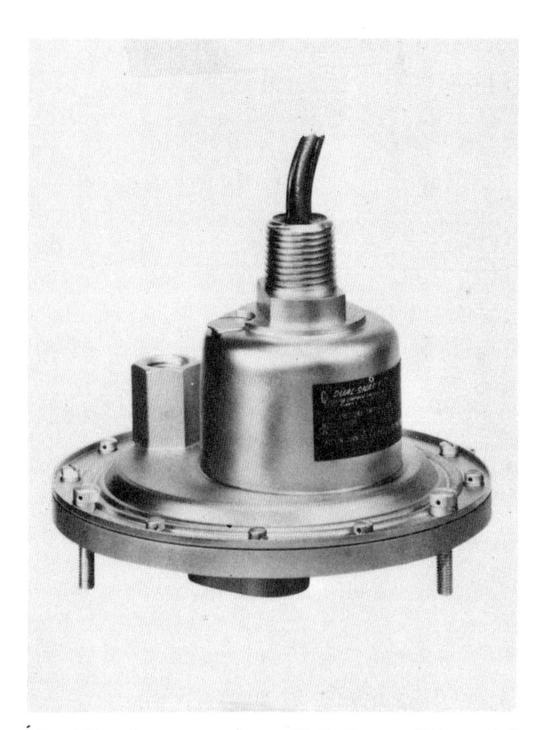

Fig. 1-6a. Low pressure switch. Some differential pressure switches are built exclusively for low and very low pressure differentials ranging from .01" water column (.0004 psig) to 20 psig, as shown in the figure. *(Courtesy of Custom Component Switches, Inc.)*

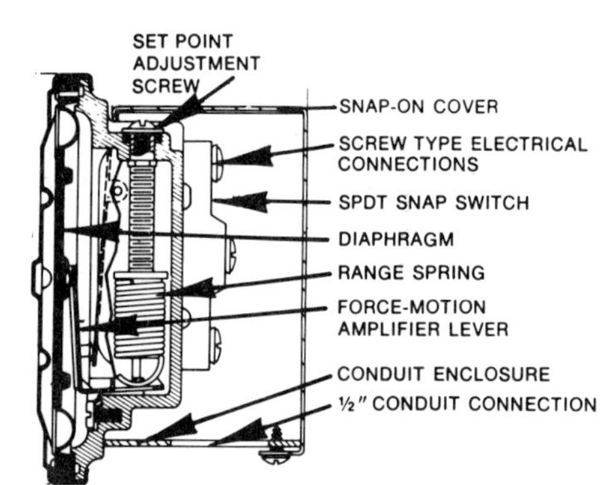

Fig. 1-6b. Low pressure switch. *(Courtesy of Dwyer Instruments, Inc.)*

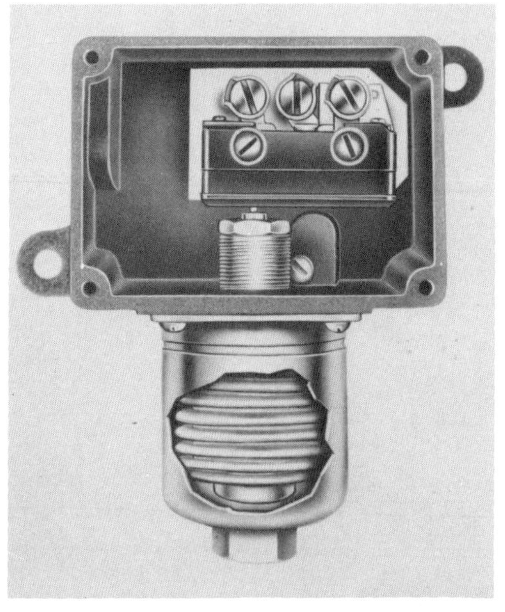

Fig. 1-7a. Miniature bellows pressure switch. *(Courtesy of United Electric Controls Co., Watertown, Massachusetts.)*

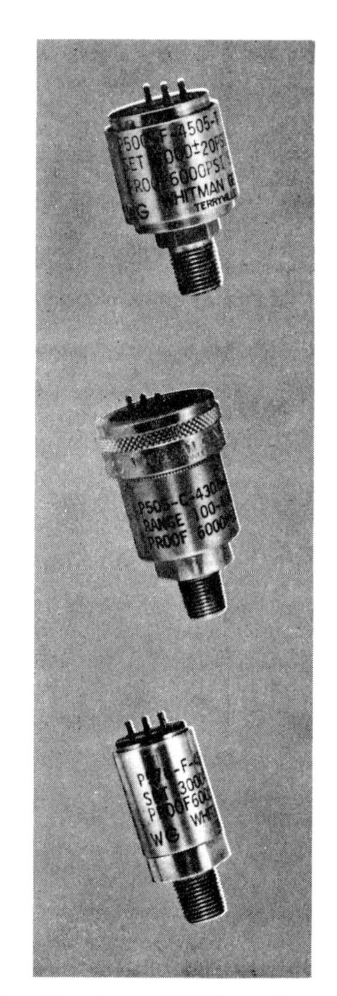

Fig. 1-7b. *(Courtesy of Whitman General, Terryville, Connecticut.)*

hundredths of an inch of water to many psi. Miniature pressure switches are available with a wide range of actuator configurations and electrical ratings. This type of switch is of great importance in the aerospace industry where space available and weight are of critical importance.

Field Adjustable Switch. In many industrial applications it is necessary to be able to change the pressure setting of a switch without having to remove it from an installation or to disassemble it. Figure 1-8 shows a typical field adjustable switch provided with an adjustment wheel that enables adjustments in pressure setting to be made under normal operating conditions. This feature is also useful for calibration of instrument and laboratory switches. The range of adjustment is usually fairly small, but in special cases it may be quite large.

High Pressure Switch. Figure 1-9 shows a bourdon tube actuated pressure switch designed for high pressure usage. This switch is rated for 18,000 psig working pressure. Such switches are needed in many high pressure industrial processes and in many types of hydraulic-powered machinery. Most switches of this nature are large in size and housed in heavy metal cases for safety. The bourdon tubes for high pressures are normally made of type 316 stainless steel.

Gage/Pressure Switch. This device (Figure 1-10) is a combination of pressure gage and pressure switch for low pressure use. A sensitive diaphragm controls a shutter in a photocell. By eliminating the mechanical coupling to the switch, effects such as switch actuation force and hysteresis are eliminated, thus decreasing the diaphragm size necessary. Also, by use of relays large current may be switched.

Explosion-Proof Switch. In some instances pressure switches must be installed in areas where there may be hazardous or explosive vapors. Figure 1-11 shows a typical explosion-proof housing designed to keep the vapors out of the electrical switch. Such switches usually carry Underwriters Laboratory approval or CSA Testing Laboratories approval for particular hazardous locations.

Fig. 1-8. Field adjustable pressure switch. *(Courtesy of Custom Component Switches, Inc., Chatsworth, California.)*

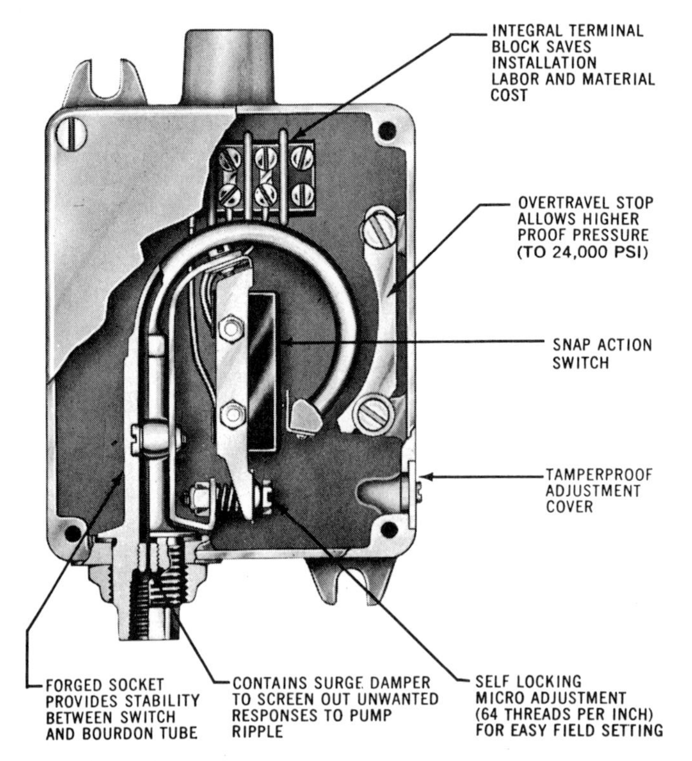

Fig. 1-9. High-pressure pressure switch. *(Courtesy of Delaval Turbine, Inc., Barksdale Controls Division, Los Angeles, California.)*

Fig. 1-10. Gage pressure switch. *(Courtesy of Dwyer Instruments, Inc., Michigan City, Indiana.)*

Fig. 1-11. Explosion-proof switch. *(Courtesy of Custom Control Sensors, Inc.)*

Manual Reset Switch. The pressure switch shown in Figure 1-12 is a manual reset type. As pressure increases, the switch will trip at the actuation point. As pressure decreases from the actuation point to the reset point, the switch does not automatically reset but instead must be reset manually after the pressure has fallen to the reset point. Such a switch is useful as a safety switch where it is desirable to maintain the action obtained at set pressure (as in alarms, machine shutdown, etc.) until the problem can be studied and a decision made to override or restart the system.

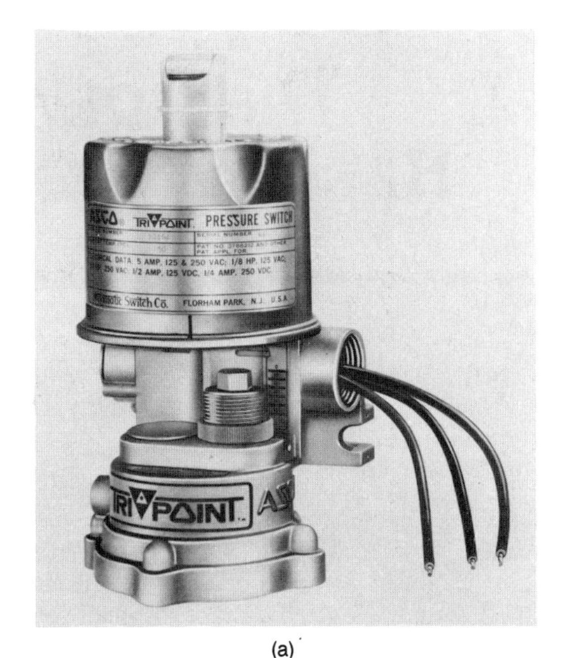

(a)

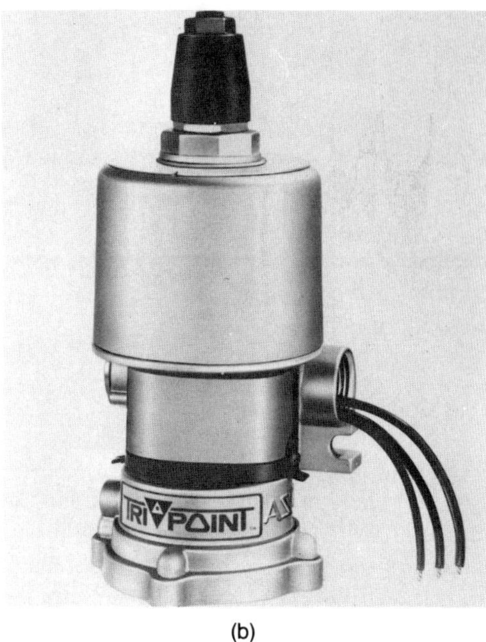

(b)

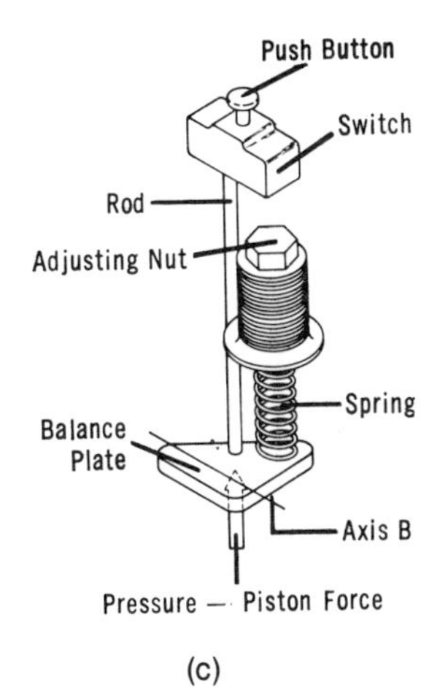

(c)

Fig. 1-12. Manual reset pressure switch. The operation of the manual reset pressure switch is similar to that of the fixed deadband pressure switch except that reactuation does not automatically occur. The arrangement of parts is shown in diagram C.

The actuation point is adjustable over the full scale pressure range. The switch automatically operates when the pressure reaches the set point. This occurs when the balance plate rotates about axis "B." The switch doesn't automatically reactuate when the pressure returns to allow the balance to rotate about axis "B" to its original position. However, after this occurs the electrical switch may be manually reset by depressing a push button. *(Courtesy of Automatic Switch Co., Florham Park, New Jersey.)*

REFERENCES

1. Brown, James O., How to Select a Fluid Pressure Switch, *Product Engineering*, May 14, 1962.
2. Askland, Carl L., An Introduction to Pressure Switches, *Design News*, October 6, 1975.

II The Four Basic Types

The overwhelming majority of the common pressure switches commercially available today utilize one or more of the four basic pressure sensing elements. These four are the piston, bellows, bourdon tube, and diaphragm sensors. Each type may, in theory, be used in an almost limitless variety of pressure switches, although for each type there are limits of feasibility insofar as cost and complexity are concerned. In this chapter we shall consider the typical construction, theory of operation, and some typical applications of each type, as well as consider in a general way their various advantages and disadvantages. Later chapters will consider some special types as well as some additional sensing elements.

Each of the four types to be discussed in this chapter detects a change in the pressure applied to the sensor element by dynamic means, i.e., the sensor responds by moving under the pressure change. The switching mechanism may detect either displacement, velocity, or acceleration, although in practice only displacement is normally considered. Examples will be given in later sections.

PISTON TYPE

There are two basic types of piston actuated pressure switches—the unsealed type, which requires a line to the reservoir for the fluid that leaks past the piston and hence necessitates a return line, and the sealed type, which uses an "O"-ring or similar seal to prevent leakage. Most of the piston type pressure switches manufactured today are of the sealed variety. Modern pressure switches can readily be made to be absolutely leak-tight around the piston, although usually at a higher cost and at some sacrifice in their range and speed of response. A large variety of seal elements are available, including such things as elastomeric o-rings, plain and filled Teflon and similar plastic rings, rolling diaphragms, limp diaphragms, and metallic diaphragms. We shall consider diaphragm seals in a later chapter on combined sensor elements. In this section we shall restrict the discussion to ordinary rubber "O"-ring seals. (See Figure 2-1a.)

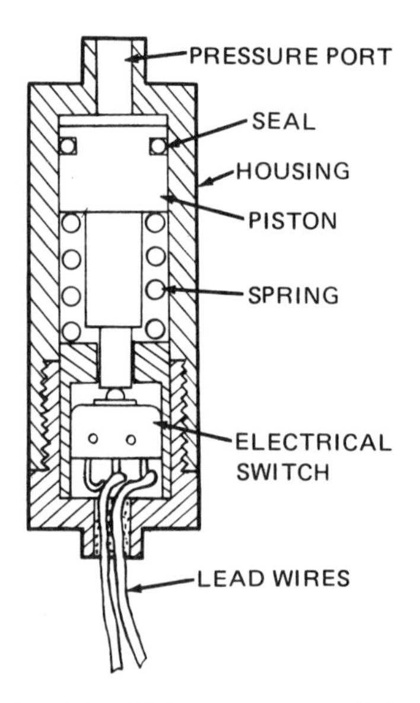

Fig. 2-1a. Piston pressure switch.

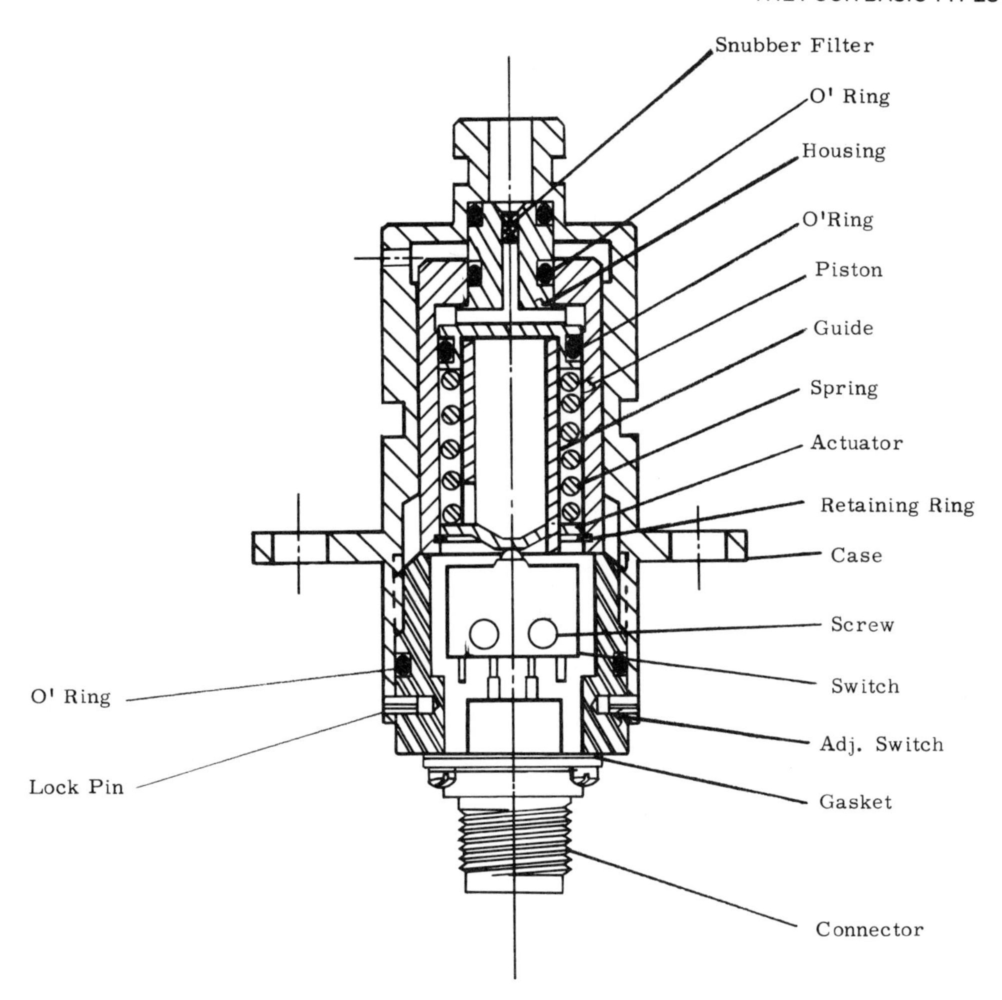

Fig. 2-1b. Reverse acting piston pressure switch. *(Courtesy of Essex Cryogenics Ind., Inc., St. Louis, Missouri.)*

Operation

The basic principle of operation of a piston actuated pressure switch, as with the other types, is the transformation of a pressure change in a confined fluid into a movement of the sensor sufficient to actuate the electrical switch. Normally, this is required to occur at a specific pressure called the set pressure (or set point).

At the set point the force acting on the piston is given by

$$F_1 = P_1 \cdot A$$

where symbols have the meanings and units given in Table 2-1. The force of the pressurized fluid on the piston is transmitted either directly or through mechanical linkages to the actuator button of the electrical switch.

Table 2-1. Piston Pressure Switch Force Symbols

A = Area (in.^{2})
F = Force (lb)
F_s = Spring force (lb) F_b = Bellows force (lb)
P = Pressure (psi)
1 = Set point value
2 = Release point value

Since a force of any magnitude could cause the piston to move (neglecting friction), there must be a force equal in magnitude to F_1 but oppositely directed to prevent movement until the set point is reached. This force may in theory come from any source, but in practice a preload F_s is usually applied to the piston by a loading spring of an appropriate size. (See Figure 2-2.) This is normally a helical compression wire spring, but it may be of another form such as a Belleville or column spring. In the installed position with no fluid pressure applied to the piston the spring exerts a force F_s on the piston directed away from the electrical switch. When the force due to fluid pressure equals the spring force in magnitude, the piston is just ready to begin moving and thus actuate the electrical switch. A typical basic electrical switch requires a force ranging from a few hundredths to a few tenths of a pound to actuate it. Usually this is negligible compared to the force acting on the piston at the set point, i.e., a very slight rise in pressure above the set point will trip the switch. In a theoretical switch with no friction, no additional force would be needed, and the switch would always trip at the set pressure.

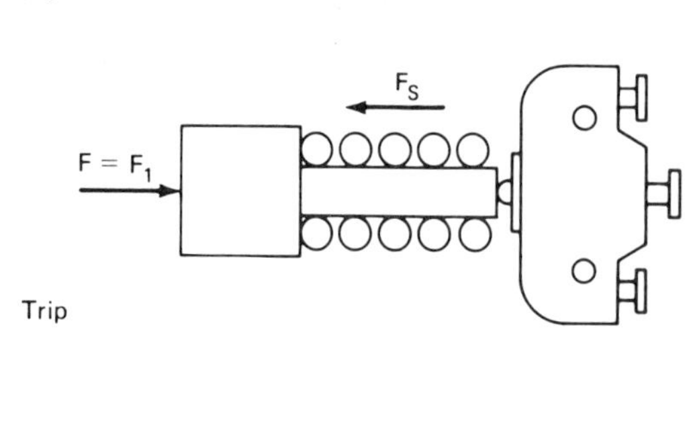

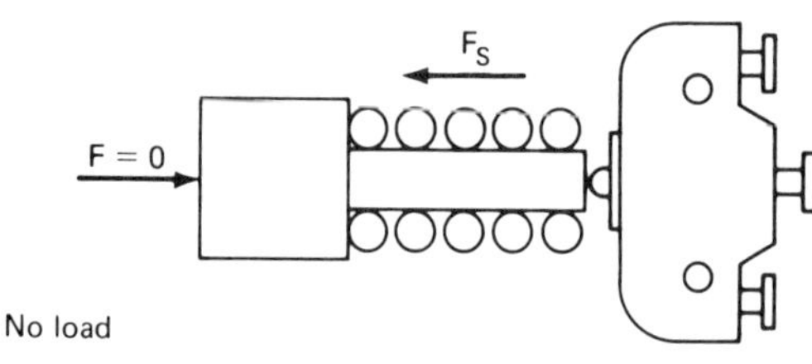

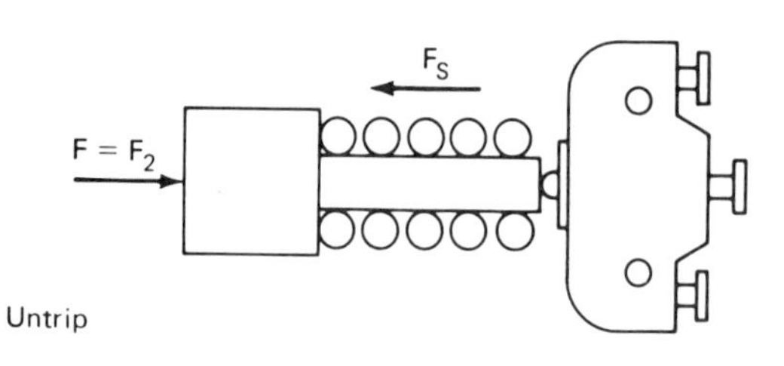

Fig. 2-2. Piston pressure switch force schematic

In real pressure switches the effect of seal friction becomes significant and must be considered. In this instance it is not feasible to wait until the set pressure is reached for the piston to begin moving, as the breakout friction can be great enough to move the actual trip point beyond the allowable tolerance band.

The actual motions and forces involved will be discussed in detail in Chapter 7, which is concerned with the design of piston actuated pressure switches.

There may be cases such as very high or very low set pressures where it is not desirable to have the force on the piston equal to the force necessary to trip the electrical switch. For example, it may not be possible to size the required loading spring to fit the space available. In such cases various linkage mechanisms may be used to increase or decrease the force transmitted to the electrical switch.

As the applied pressure decreases, the electrical switch must untrip or reset to its original state. This is accomplished by the loading spring pushing the piston away from the electrical switch. The pressure at which this occurs, called the untrip pressure, is a finite distance away from the trip point because of the effect of friction. Often this distance (deadband) is a specified requirement for the switch.

Advantages

A piston actuated pressure switch, like the other types to be considered later, has several advantages and disadvantages compared to other types, which are dependent on the application in question. A few of its strong points are given here.

First of all, a piston actuator is relatively immune to small pressure surges and ripple because of the friction between the piston seal and the cylinder wall and also because there is a relatively large volume change as the piston moves. Hence small surges will be "soaked up" before the electrical switch is actuated, thus eliminating false actuation and switch chatter from this cause. In addition, most piston type pressure switches are provided with an absolute mechanical stop, thus making them insensitive to pressure surges up to the point of material failure, or they are made reverse acting, as shown in Figure 2-1b. In this arrangement, force is away from the electrical switch. Note, however, that large surges may actuate the electrical switch in direct acting pressure switches.

Piston type pressure switches are generally less costly than the other basic types except for those in the low pressure range (to approximately 500 psi). Generally, the manufacturing tolerances are more generous, and the operations involved in manufacture and assembly are not as stringent as for other types of pressure switches.

With its simple but rugged construction a piston type pressure switch normally has a longer life expectancy than other types, often exceeding several million cycles, primarily because the stress load is distributed over almost the entire length of the loading spring instead of being concentrated in one or a few locations, such as in bellows convolutions.

Most piston type pressure switches are also relatively insensitive to moderate shock and vibration levels, such as encountered in aerospace and similar applications, and also to other severe environmental conditions, such as high and low temperatures, high humidity, and similar conditions.

Among their other advantages, piston type pressure switches offer ease of adjustment of the set pressure by providing for adjustment of the installed load on the loading spring, and they are generally able to respond to relatively high cycle rates (up to about 75) without overstressing or fatigue failure.

Piston actuated pressure switches may be fabricated from a wide variety of materials. Most pistons are metallic, but others are of plastic, composites, elastomers, etc. Loading springs are generally wound of metal wire.

Finally, a very important practical advantage to the systems engineer, plant engineer, and others is the tremendous variety of commercially available pressure switches. For many applications all one need do is select the appropriate switch from among those available. One of the intended purposes of this chapter is to enable the purchaser to obtain the proper switch by consideration of its strong and weak points.

Disadvantages

One of the most important disadvantages of piston actuated pressure switches in certain types of applications is their lack of repeatability compared to other types because of the effects of friction between the seal and the cylinder wall. As a general but very rough guideline, piston actuated pressure switches are accurate to plus or minus 2% of set point value, while most other types are accurate to plus or minus 1%. This fact may eliminate consideration of this type of switch for many instrument applications where high accuracy is required. Accuracy is improved with unsealed pistons, but the need for return lines usually outweighs this benefit.

In the case of pistons without seals, not only are return lines needed to carry any leakage back to a reservoir, but it is also usually necessary to have lapped finishes on the piston and cylinder walls, thus increasing their cost considerably over the sealed type. Also, in this case the pressurized fluid must be kept considerably cleaner than in the case of sealed piston types, as contamination could cause the piston to jam up.

A very serious problem with sealed piston type switches in some instances is the tendency of the seal to stick if the switch has been sitting unused for a long period of time. In these cases the set point is usually raised a considerable amount where the switch is first actuated again, a situation that could prove disastrous to the intended function of the switch. Whenever this problem can be reasonably anticipated, a different style of switch should be considered; or else provision should be made for "dry" cycling of the switch after it has sat unused for a period of time. This effect must be considered particularly in cases where the switch is to be used as a safety switch.

Generally speaking, a piston actuated switch is only suitable for sensing gage pressure, as it is very difficult to hold a high vacuum or other reference pressure behind this type of element in order to have an absolute or differential pressure switch, and the cost of doing so is generally prohibitive.

In some cases a piston actuated pressure switch should not be considered for use in gaseous systems, as the seal may leak slightly or may be permeable to the gas. For reasons of safety they are generally unsuited for use with toxic or otherwise dangerous gases unless special sealing methods are employed.

Finally, piston actuated switches have a long response time compared to the other types, and hence are not as sensitive to rapid pressure changes. In fact, some switches of this type may require over a full second to respond to a pressure change. This puts them at a disadvantage in certain types of instrument usage.

BELLOWS TYPE

A bellows is a thin-walled, flexible, circularly corrugated cylinder that can expand or contract axially under a change in pressure. Bellows used in pressure switches normally have an integral end cap. A bellows may be made of almost any material that is sufficiently elastic; however, in practice, most bellows are made of metal such as one of the stainless steels, beryllium, copper, nickel, bronze, and similar metals. Some bellows are made of synthetic rubber or various types of polymer compounds.

There exists in practice several different forms of bellows, which differ in configuration and/or method of manufacture. Typical forms include single-ply, two-ply, and multiple-ply bellows. They are formed by such processes as welding, brazing, soldering, electroplating, vacuum deposition, and hydroforming. Most of the bellows used for pressure switch applications are of single-ply construction and are either hydroformed or welded. (See Figure 2-3.)

Bellows actuated pressure switches are normally used for low pressure applications and other critical uses. They are available commercially in a fairly wide variety of bellows configurations and sizes, and may be that do not require very high reliability, but special designs find applications in aerospace obtained on special order in just about any desired specifications. The bellows is especially well suited for miniature pressure switches. The bellows actuated switch is also useful in applications where a relatively long actuator stroke is required but where the sensor element must be absolutely leaktight.

Operation

A bellows actuated pressure switch utilizes the accordionlike expansion or contraction of the bellows convolutions to provide the displacement and thus the force necessary to actuate the electrical switch. As in the case of the piston actuated switch, a pressure acting on a cross-sectional area provides a force that is transmitted either directly or through a mechanical linkage to the actuator button of the electrical switch.

Unlike the free piston actuator, the bellows itself can act as a loading spring to provide an opposing force for the pressure to work against. In most cases, however, a secondary loading spring, usually of the helical compression type, is used to assist the bellows.

When installed in the pressure switch, the bellows and loading spring exert a force F_b directed away from the electrical switch (Figure 2-4). When the force due to the fluid pressure reaches this value, the bellows can begin to move and cause the electrical switch to actuate. This is, then, the set point pressure.

A bellows is normally installed so that the applied pressure acts over its external surface, as a bellows is much less subject to buckling, squirming, etc., than other types. If the pressure must be applied internally, the bellows should be supported laterally.

A detailed discussion of bellows operation and design will be found in Chapter 8, on the design of bellows actuated pressure switches.

Advantages

Depending on the application, the bellows pressure sensor may have several advantages

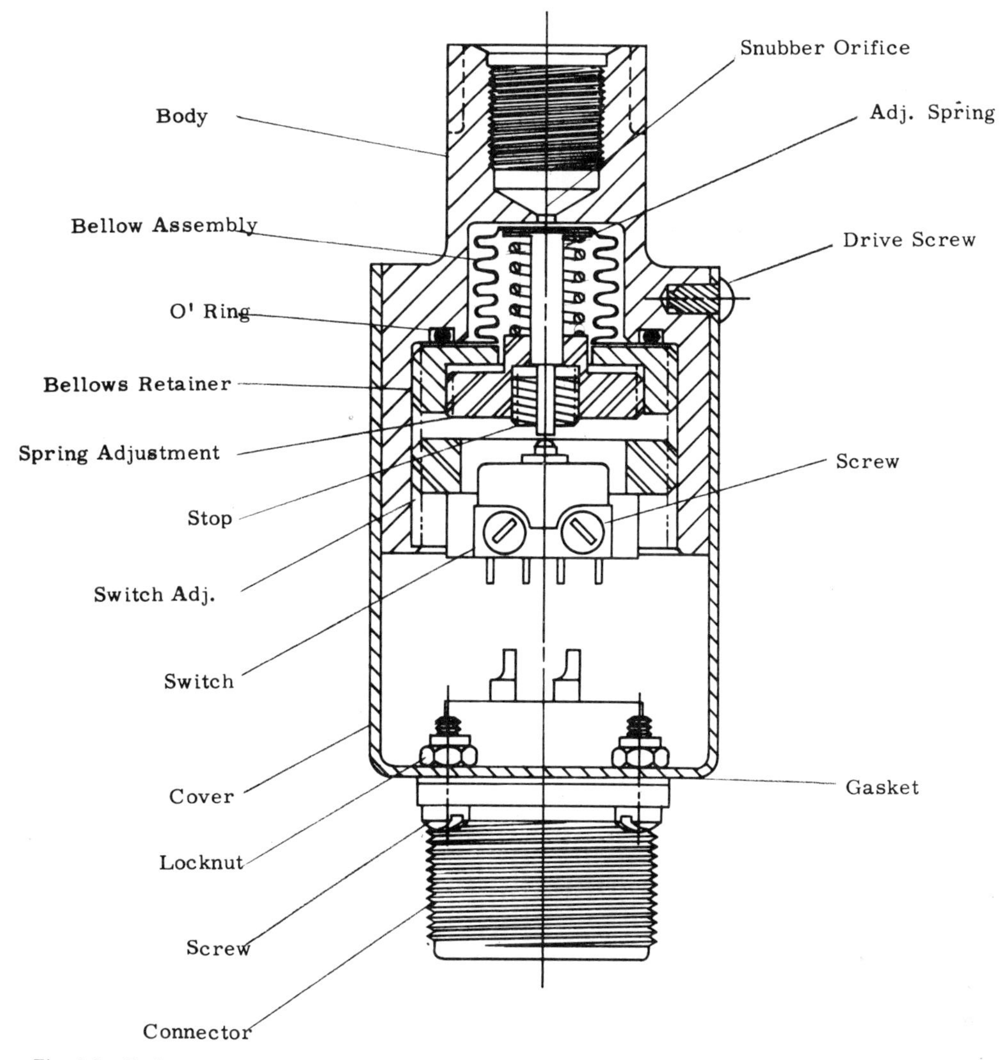

Fig. 2-3. Bellows pressure switch. *(Courtesy of Essex Cryogenics Ind., Inc., St. Louis, Missouri.)*

over the piston sensor and the other types to be considered later.

First, a bellows type pressure switch is generally smaller than the other types for a specified set of operating parameters. This is due to the physical configuration of the bellows, which gives it a fairly long stroke under a relatively small force increase. Thus, these pressure switches are of importance in the aerospace industry where weight and size are critical.

Second, some styles and construction of bellows may be less expensive to manufacture than the diaphragm or bourdon tube types. The hydroformed bellows, in particular, is relatively inexpensive in many cases.

A high-quality bellows will have a high accuracy comparable to a high-quality diaphragm. Typical repeatabilities for quality bellows are of the order of 1/2 to 1% of set point value. Cheaper bellows are not usually as accurate. Bellows are often used in instrument switches because of the high accuracy obtainable.

Bellows are available over a fairly wide range of operating pressures, although not as high as piston or bourdon tubes except in special designs. They are suitable for use from moderately high vacuum up to approximately 1500 psi, and when sealed or evacuated may be used for absolute pressure sensing (Figure 2-5), although generally they

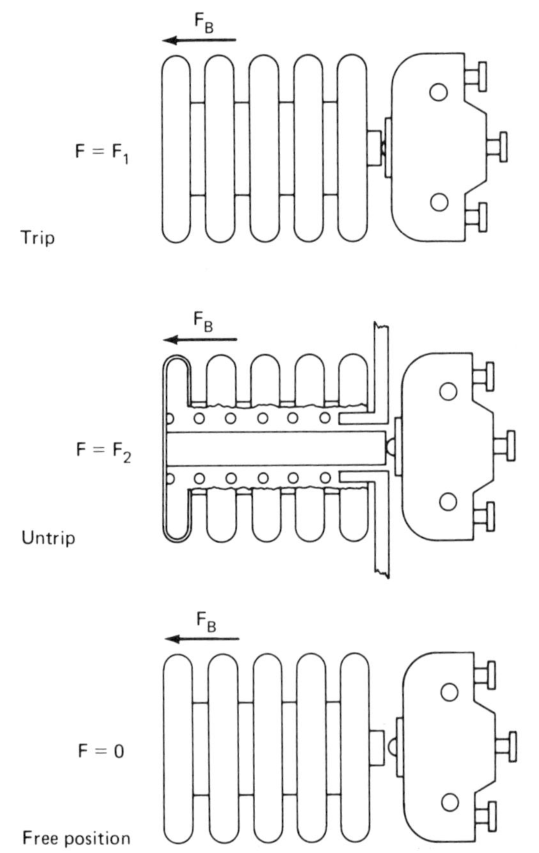

Fig. 2-4. Bellows pressure switch force schematic diagram.

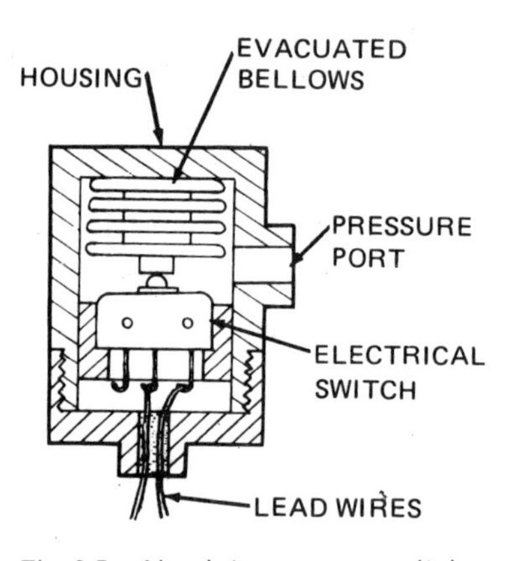

Fig. 2-5. Absolute pressure switch.

are not used above a few hundred psi.

Because of their construction and the range of available materials, they are useful in switches exposed to corrosive or otherwise dangerous fluids. Since they are leaktight and normally impermeable to gases, they are suited for many harsh environments in industrial plants.

A quality bellows sensor has a small differential or deadband, which may be advantageous in certain applications. This is not generally true, however, if a secondary loading spring (as shown in Figure 2-6) is used with the bellows. Normally the spring rate of the bellows, i.e., the force/deflection is quite linear over a fairly large pressure range.

Bellows have a very fast response to pressure change, usually in the range of milliseconds, which may be an advantage, depending on the intended use.

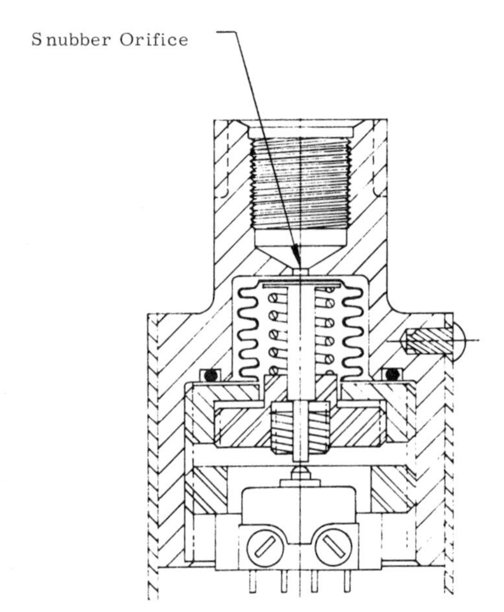

Fig. 2-6. Bellows pressure switch with snubber orifice. *(Courtesy of Essex Cryogenics Inds., Inc., St. Louis, Missouri.)*

Disadvantages

A bellows sensor pressure switch, like the other types, has several features that may be disadvantageous, depending on the application to which it will be put.

First of all, a bellows is highly susceptible to vibration. Because of its physical structure it is a free-standing unit without rigid support and thus is relatively free to move under shock or vibration. Because of the unpredictable effects of friction, it is not usually feasible to provide external support in order to damp out vibration. In some special designs a viscous fluid may be used to provide vibration damping.

Some types of bellows are quite costly to manufacture inasmuch as wall thicknesses must be held to very close tolerances to prevent fatigue and premature failure. The quality of the material used in manufacture must also be very high, thus raising the cost.

Because operation of the bellows involves flexing of the convolutions and the flexing is unequally distributed over the convolutions, a bellows has a limited life span. There is no known way to predict the life of a bellows with good accuracy.

A further drawback to bellows actuated switches in some cases is their fast response time, which may make them sensitive to surge pressures and ripple in pumping systems. If a bellows must be used in such cases, the surges must be reduced by means of a snubber, surge reservoir, or similar means. (See Figure 2-6.)

BOURDON TUBE TYPE

There are two basic styles of bourdon tube commonly used as sensor elements in pressure switches—the plane single-loop or "C" type shown in Figure 2-7 and the coiled or spiral tube shown in Figure 2-8. In addition, two other styles of bourdon tube seldom used in pressure switches are the helical and axially twisted configurations. Bourdon tubes are usually elliptical or oval in cross section and are nearly always of seamless construction. They may be fabricated from a variety of materials; however, most of the commercially available tubes are made of type 316 stainless steel, beryllium copper, phosphor bronze, or similar metals, which have good spring characteristics as well as corrosion resistance.

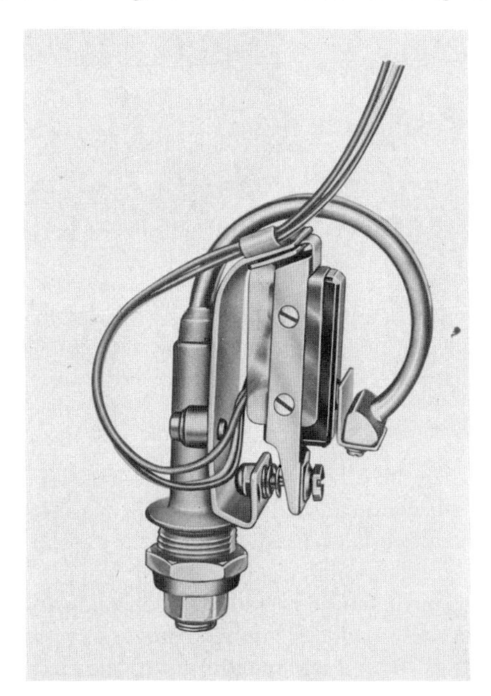

Fig. 2-7. Bourdon tube "C" type. *(Courtesty of Delaval Turbine, Inc., Barksdale Controls Division, Los Angeles, California.)*

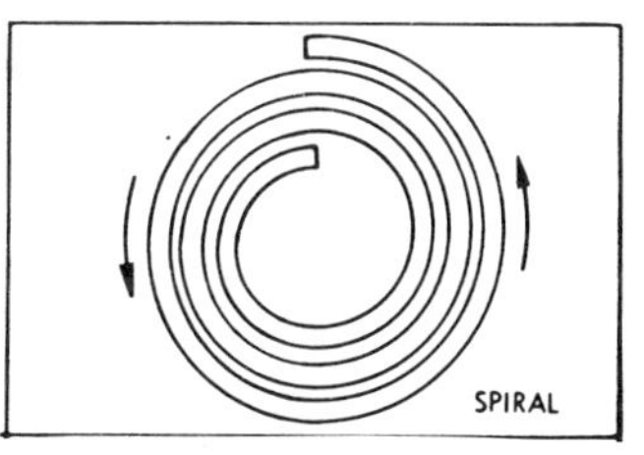

Fig. 2-8. Bourdon tube "coil" type. The spiral tube is similar to the plain tube except that its radius of curvature changes along the length of the tube, creating a greater length of tube for a given circumference. The number of turns in the spiral varies from three to six. Owing to the larger active length of tube, the arc length of stroke is greater. A straight-line pickoff from this larger angle of sweep causes a larger nonlinearity than with a plain bourdon tube.

The plane single-loop type is the one most often used in pressure switches, as relatively little travel of the tip is needed to actuate an electrical switch. The coiled or spiral type, which is not used as often in switches, is generally used in conjunction with large electrical switches that require sizable movement to actuate, or in conjunction with mechanical linkages where a large movement of the tip is desirable.

Operation

A bourdon tube has one end sealed off and the other end open to the pressure source so that it is completely filled with the pressurized fluid. An increase in the pressure of the fluid causes the walls of the tube to bulge slightly in the direction of their original circular cross section, thereby causing the tube to try to straighten out, thus resulting in movement of the tip in an outward direction

(Figure 2-9). The amount of travel of the free end is directly proportional to the pressure or vacuum applied up to the elastic limit of the material used, or more often until it comes to rest against a mechanical stop.

The bourdon tube itself acts as its own loading spring, thereby eliminating the need for an external compression spring. The applied pressure produces high stresses in the tube, thus setting up restoring forces in the tube walls, which attempt to return the tube to its original unstressed configuration.

The free end of the bourdon tube usually presses directly on the actuator button of the electrical switch, but it may be made to act on various types of mechanical linkage, which, in turn, act on the electrical switch. In most of the common bourdon tube pressure switches the tip of the tube holds the actuator button of the electrical switch fully depressed in the no-load configuration. As the pressure in the tube is raised to the set point, the force builds up until at the set point the tip moves away and allows the electrical switch to actuate. It is usually much easier to control the set point in this case as compared to having the tip move up to the electrical switch button under increasing pressure to actuate it.

The calculation of the forces acting on the bourdon tube and electrical switch will be given in detail in Chapter 9, on the design of bourdon tube pressure switches.

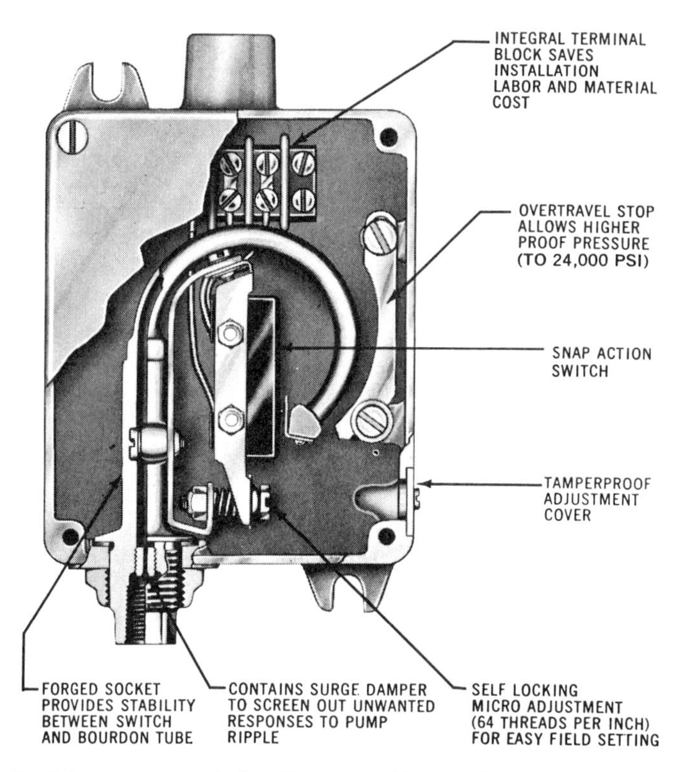

Fig. 2-9. Bourdon tube pressure switch. *(Courtesy of Delaval Turbine, Inc., Barksdale Controls Division, Los Angeles, California.)*

Advantages

Like the other pressure sensors considered in this chapter, the bourdon tube has several advantages and disadvantages, depending on the application in question.

Bourdon tube pressure switches are commercially available in a wide range of pressure ratings from high vacuum up to the vicinity of 15,000 psi, and in a wide variety of sizes and configurations.

A bourdon tube has the fastest response of the four basic types of pressure switch to changes in pressure up to a frequency of about 10 cycles per second, owing to the fact that there is little or no volumetric change in the tube aside from that due to the compressibility of the pressurized media. For liquids, of course, this is very small. However, the fast response conceivably could be lost in cases where the bourdon tube acts on external linkage rather than directly on the electrical switch actuator.

The accuracy of the bourdon tube is very high, with typical repeatability being in the range of 1/2 to 1% of set point. Thus, it compares very favorably with bellows or diaphragm sensors in this respect.

Bourdon tubes are usually quite rigid and hence relatively immune to vibration and shock loads, so that they are very useful in severe environments such as near vibrating machinery or piping and in aerospace applications.

Because of their relatively tough construction and the fact that they can be provided with mechanical stops, bourdon tubes can carry a very high proof pressure rating. Thus, while it is sensitive to surge pressure and may actuate, the high pressure will not in general damage the switch.

Disadvantages

Bourdon tubes are very sensitive to surge pressures and ripple, since there is little or no effective volume change to absorb them. Hence, measures must be taken to protect against these effects, such as by installing snubbers or surge suppressors in the inlet line. When close differential electrical switches are used with bourdon tubes, there can be a tendency to "chatter" when surges or ripple is present.

A bourdon tube pressure switch is normally larger in size and weight than most of the other types with a similar rating. This can be a severe disadvantage in aerospace applications where size and weight are critical.

A bourdon tube is usually costly, as very tight control of the wall thickness, shape, homogeneity, and heat treatment is required to ensure high accuracy and long life. This often means that specially selected materials and special processes must be used, both of which raise the manufacturing costs.

DIAPHRAGM TYPE

There are two basic styles of diaphragms used as sensor elements in pressure switches —the flat type and the corrugated or convoluted type, shown in Figures 2-10 and 2-11, respectively. Other types of diaphragms, such as the rolling diaphragm and limp diaphragms used for sealing purposes, will be discussed in later chapters.

The flat diaphragm sensor is normally a circular disc of the diameter necessary to give the required operating force and of the order of a few hundredths of an inch thick. Diaphragms are usually made from a spring material, such as beryllium copper, stainless steel, nickel, and similar metals. The flat diaphragm is generally used where a very short stroke is required, usually with moderately high set points and/or very small differentials, since it cannot flex very far without exceeding the elastic limit of the material.

Corrugated or convoluted diaphragms may be made in many different forms. A few examples will be discussed in Chapter 10, on the design of diaphragm actuated pressure switches. A convoluted diaphragm is useful where a longer stroke is needed (compared to that obtainable from a flat diaphragm), as it is considerably more flexible than a plane diaphragm. Convoluted diaphragms may be made of many of the same materials that flat diaphragms are made of. Convoluted diaphragms are nearly always made by die stamping methods, although some are made by electroplating and vacuum deposition.

Closely allied with and functioning in much the same way as diaphragms are aneroid capsules consisting of two diaphragm plates sealed together on the circumference and with the enclosed region highly evacuated. They are normally used in absolute pressure switches, whereas the diaphragm is normally restricted to gage or differential pressures.

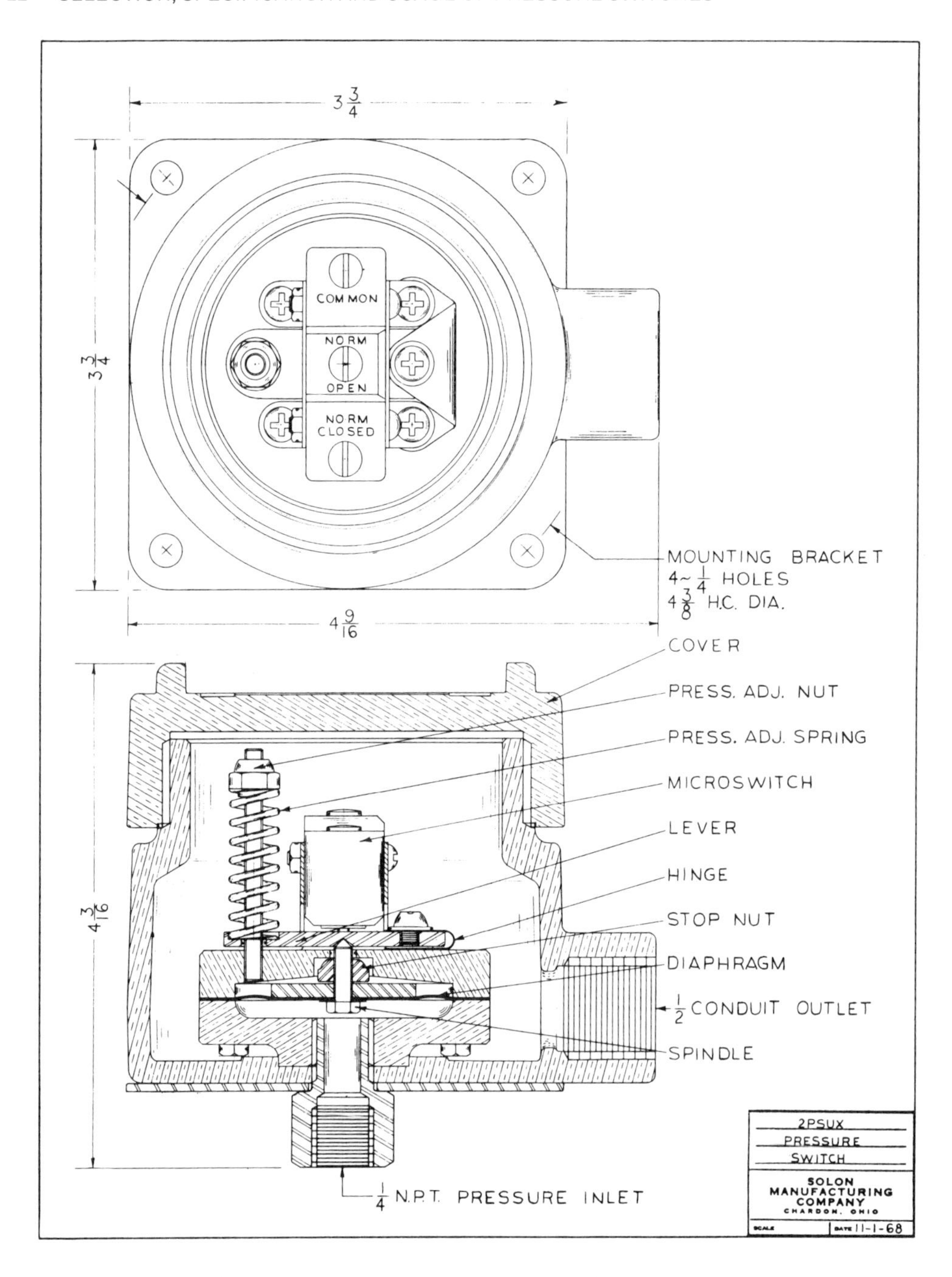

Fig. 2-10. Flat diaphragm pressure switch. *(Courtesy of The Solon Mfg. Co., Chardon, Ohio.)*

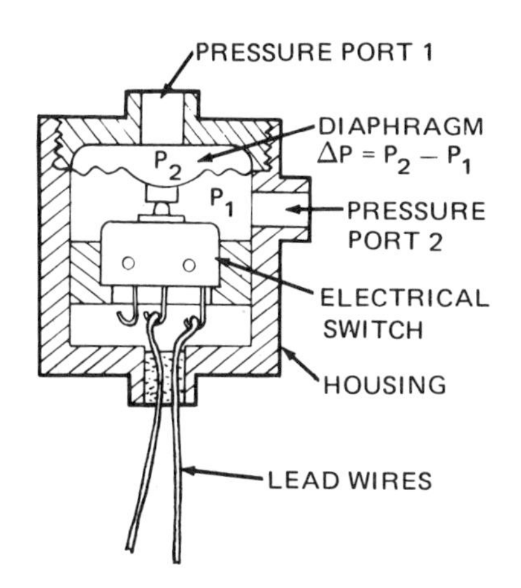

Operation

The diaphragm sensor functions very much like a bellows or piston in that a pressure acting on its surface area yields a force directed to the actuator button of an electrical switch. The area in question may be the actual exposed surface area in the case of a plane diaphragm, or it may be an "effective" area in the case of a convoluted diaphragm. The calculation of effective areas will be illustrated in Chapter 10.

In many instances the diaphragm acts as its own loading spring in the same way as the bellows or bourdon tube sensors. This is particularly true for pressure switches with low set points. In other cases it is necessary to back up the diaphragm with a helical compression spring. In this case the action of the diaphragm is somewhat similar to that of the piston except that the circumference is fixed in position, and the stroke at the center is usually much less.

Diaphragms may be set to depress the electrical switch button to actuate, or they may sometimes be used to release the depressed button to actuate the switch.

Advantages

Foremost among the advantages of a diaphragm is its relatively large area, which enables it to respond to very low pressures. A diaphragm is generally the best choice for use with pressures below a few psi. Most of the common diaphragm sensors are in the range of 2 to 4 in. in diameter.

Diaphragms are also very fast in response to pressure change, with reaction times of a few milliseconds being typical. Only the bourdon tube has a faster response on the average. A typical diaphragm has a movement of about 0.03 in. and thus a very low volume change, but one slightly larger than that of the bourdon tube.

A diaphragm, especially the flat style, has high accuracy, since the thickness and composition of the material can be closely controlled. Convoluted diaphragms are only slightly less accurate.

Most diaphragms are relatively insensitive to shock and vibration, both because of their being supported circumferentially and because of the rigidity imparted to them by the convolutions.

In most applications the diaphragm has a back-up plate to protect it from overtravel. This gives it the capability to withstand high proof pressures.

Diaphragms have some immunity to surge or ripple pressure due to their natural rigidity and the slight change in volume. Fortunately, however, most diaphragms see little or none of this pressure in the range of pressures in which they operate.

Diaphragms, like bellows or bourdon tubes, have the ability to provide an absolutely leaktight seal between the pressure source and the electrical switch. For this they are useful as elements in safety switches.

Disadvantages

Perhaps the biggest disadvantage of the diaphragm sensor is its limited range. Because of its large exposed area a diaphragm is normally limited to pressures below about 400 psi. Of course, special designs may function at much higher pressures, but at a much higher price.

The differential or deadband is somewhat larger than that of a bellows, as its deflection rate is about half that of a bellows. This may restrict its usage in certain types of differential pressure switches.

Diaphragms of the convoluted type may also suffer most of the disadvantages of the bellows sensors in that stresses, nonuniformities, etc., in the corrguations may cause fatigue failure, loss of accuracy, and similar problems.

Because of their fast response diaphragms are sensitive to surge pressures and hence must be protected from them.

Table 2-2 compares the performance of various types of pressure switches, including two diaphragm types.

Table 2-2. Comparative Pressure-Switch Performance

	BELLOWS	SLACK DIAPHRAGM	SPRING DIAPHRAGM	BOURDON	PISTON
Repeatability	Excellent	Excellent	Excellent	Good	Good
Ruggedness to:					
Pressure surge	Fair	Good	Fair	Fair	Excellent
Vibration	Poor	Good	Fair	Poor	Excellent
Temperature	Excellent	Excellent	Excellent	Good	Good
High overpressure	Fair	Good	Fair	Good	Excellent
Corrosive media	Excellent	Excellent	Excellent	Excellent	Poor
High cycle life	Fair	Good	Fair	Good	Excellent
Low settings (less than 100 psi)	Excellent	Excellent	Excellent	Good	Fair
Medium settings (75 to 750 psi)	Good	Excellent	Good	Excellent	Good
High settings (500 to 5,000 psi)	Fair	Good	Poor	Good	Excellent

G = good, recommended for normal usage; F = fair, better choices available; P = poor, not normally recommended.

Ratings are intended as guidelines only and refer only to "typical" units. Special designs may have entirely different properties.

REFERENCES

1. Askland, C. L., Piston Actuated Pressure Switches, *Machine Design,* March, 1976.
2. Coleman, J. N., Selecting a Pressure Switch, *Fluid Power International,* May, 1963.
3. Griffith, D., Sorting out Pressure Switches, *Machine Design,* August 10, 1972.
4. Jennings, F. B., Theories on Bourdon Tubes, *Transactions of the ASME,* January, 1956.
5. Mason, H. L., Sensitivity and Life Data on Bourdon Tubes, *Transactions of the ASME,* January, 1956.
6. Myer, J. L. and Bitzer, E. B., Bourdon Tube Selection and Spring Design, Unpublished.

Characteristics of Pressure Switches

This chapter is concerned with some of the important operating characteristics or parameters of pressure switches and with the selection of the proper switch for a given application based upon consideration of these parameters. Due consideration of these properties should enable the selection, specification, or design of a suitable pressure switch for a particular application to be made with a minimum of difficulty and with reasonable assurance of proper operation when it is installed.

One or more of the four basic types of pressure switch discussed in Chapter 2 may have a natural advantage over the others for a particular characteristic. In other instances all four basic types may be equally good, and the choice is dependent on such things as cost and availability. Reference should be made to Table 2-2 of Chapter 2, which summarizes some of the advantages and disadvantages of the four basic types.

In a few applications none of the four basic types of pressure switch is suitable, and special designs, perhaps combining two of the basic sensor elements, are necessary. Some of the characteristics given may not directly apply to such devices and should be treated with due caution. However, in almost all applications it will be found that one of the basic types is suitable, and the discussions given will apply.

TYPE

The first thing to be determined is what type of pressure is to be sensed—absolute, gage, or differential—as this tells immediately the type of switch required. It is, of course, possible to use an absolute pressure switch in place of a gage pressure switch in some cases, but it is usually more costly to do so. Similarly, while it is possible to couple the outputs of two gage pressure switches to act as a differential pressure switch in certain instances, it is nearly always better to use a differential pressure switch directly.

OPERATING PRESSURE

The operating pressure is usually considered to be the normal set point or actuation pressure of the switch. If an adjustable switch is required, the operating pressure will be the highest set pressure necessary. For operating pressures in the vacuum region consider bellows, bourdon tube, or diaphragm actuated switches; for pressures up to a few hundred psi any of the four basic types may be suitable; for higher pressure consider piston or bourdon tube types; and for very high pressures the piston actuated switch is usually the best choice.

PROOF PRESSURE

The proof pressure rating of a pressure switch is the pressure representing the extreme condition that the switch may safely withstand in use without permanent change in the action point. It should be higher than any peak surge pressure in the system or any other expected transient pressure. At pressures above the proof rating, the loading spring and/or sensor element may be stretched beyond the elastic limit and take a permanent set, thereby changing the actuation point.

For design purposes in the aerospace industry, proof pressure rating is often taken as some multiple of the operating pressure, such as 1.5 times or 2 times operating pressure. In the commercial field, most pressure switches carry a proof pressure rating established by the manufacturer, usually from extensive testing.

As a general guideline, if expected surge pressures exceed twice the normal operating pressure, a piston actuator should be considered first. In most cases, whatever type of sensor element is chosen, the switch should be protected against surge pressures. The best way is to incorporate a positive mechanical stop to limit the travel of the actuator.

ACCURACY

The accuracy or repeatability of a pressure switch is a measure of how close to the desired set point the actuation comes with repeated cycles. Because of hysteresis it will not always trip at exactly the same pressure each time. The accuracy required in a pressure switch depends on the application to which it will be put. Since the cost of most pressure switches goes up significantly with increased accuracy, careful consideration of actual need should be made. Nothing is gained by calling for a 1/2% rating on the switch when a 2% rating is sufficient for the job. Conversely, in critical applications such as in instrument switches the highest accuracy obtainable should be sought. Another good rule, and one often ignored, is to not call for a tightly controlled differential if only a trip or untrip point is specified. Usually sufficient deadband to prevent switch chatter due to ripple pressure is necessary, but the tolerance on deadband should be quite generous.

The accuracy of the pressure switch is a function of the amount of hysteresis in the loading spring and to a lesser extent of the electrical switch employed. In most cases the bellows, bourdon tube, and diaphragm sensors are more accurate than the piston type.

When specifying the required accuracy of a pressure switch, allow as large a tolerance band as possible for the intended application; and in those situations where a tolerance band is required for both trip and untrip pressures, try to make each as large as possible, make each the same percent, and try to keep the two bands from overlapping.

RESPONSE TIME

In many instrument applications and some safety or warning switches it is necessary for the pressure sensor to actuate very rapidly in response to a sudden change in the system pressure. The time required for a pressure switch to respond to this change in pressure may vary from a few milliseconds in the case of bellows, thin diaphragms, and bourdon tubes to nearly a full second for some piston actuated switches. The response time of a pressure switch is, among other things, a function of the amount of friction in the sensor and electrical switch mechanisms.

CYCLE RATE

The frequency of pressure pulses or changes to which a pressure switch can respond is clearly related to the response time discussed above and to the required accuracy of the switch. Generally speaking, no pressure switch is highly responsive to a pulse rate above ten cycles per minute unless specially designed. A high cycling rate can lead to premature failure of the loading spring, bellows joints, etc., especially when the stress is also very high. For frequencies above approximately ten cycles per minute it is usually advisable to select a piston actuated switch, if possible. If both a high cycling rate and high accuracy are required, a bourdon tube, bellows, or diaphragm in the proper working range may be used.

FATIGUE LIFE

The number of cycles of operation obtainable from a pressure switch depends on several factors, such as the cycle rate discussed above, the stress on the spring element, and the environment to which it is exposed. In many cases it is possible to estimate an expected life mathematically, while in others, particularly the case of the bellows, it is very difficult to predict with any accuracy. Under the same operating conditions pistons, bourdon tubes, and diaphragms tend to outlive bellows. If the required service life is over one million cycles, a piston switch should be given first consideration. When available, actual data on life expectancy for the switch under consideration should be consulted.

Commercially available pressure switches commonly have life expectancies in the range of ten thousand to two million cycles under normal temperature conditions. Switches used in aerospace applications, which are usually under more severe environmental stresses, have lifetimes commonly in the order of five to fifty thousand cycles.

OPERATING MEDIA

The physical and chemical nature of the pressurized fluid will dictate the type of material to be used in the construction of the sensor element to be utilized and the nature of any seal elements employed. Any of the four basic types of pressure switch can be used with corrosive or otherwise dangerous fluids if special sealing methods are used. However, because of possible seal wear and subsequent leakage it is best to avoid the piston actuated switch and instead employ a bourdon tube, diaphragm, or bellows of a suitable resistant material such as stainless steel. Piston switches should also be avoided in pneumatic systems where leakage past or permeation through the seals could occur, particularly where hazardous gases are used. In the case of very viscous fluids or slurries, it is usually best to use a diaphragm actuated switch if possible. Bellows joints and piston seals could be subjected to severe abrasion from some types of slurries, and the drag from viscous fluids could result in additional friction in the switch, which could be objectionable in some applications.

TEMPERATURE

Many pressure switches are required to operate over a fairly wide range of temperature. For example, aircraft switches must usually function at temperatures ranging from $-65\,^{\circ}\mathrm{F}$ to $+250\,^{\circ}\mathrm{F}$. This generally means that methods must be employed to correct for temperature effects such as expansion or contraction, change in spring rate, and change in friction forces.

The biggest problems brought about by temperature changes usually occur in switches having a specified differential between trip and untrip points. Temperature change usually changes the spring rate of the loading spring and tends to widen or narrow the differential. Operation at continual high temperatures leads to a marked reduction in the life of the switch.

Any of the four basic pressure switches can be designed to operate over a wide temperature range; however, it is usually cheaper and easier to consider bellows or diaphragm types for low temperatures and piston or bourdon tube types for high temperatures.

SPRING RATE

The rate of the loading spring defined as force per unit displacement will determine how far the sensor element will travel under a specified change in the applied pressure. In simple switches where only a trip point is specified, the spring rate may be completely arbitrary, since the switch may untrip at any point. In the case of the differential pressure switch where a specified deadband is required, the spring rate must be tightly controlled, as the amount of deadband depends directly on it. In some cases, such as bellows and diaphragms, the geometry of the sensor element determines its spring rate. Calculation methods for spring rate will be presented in later chapters.

ELECTRICAL RATING

In most applications of pressure switches the electrical parameters such as current, voltage, and type of load in the circuit to be switched are already known, and the problem is one of choosing an appropriate electrical switch with the required rating. In other applications the electrical load can be chosen to suit the electrical switch available. In many commercially available pressure switches a variety of basic electrical switches are available, and the problem is reduced simply to specifying the proper one. In most custom designs the pressure switch is built around a particular electrical switch.

There are three or four types of electrical switch elements in common use in pressure switches, of which the basic snap action switch is the most widely used. Another possibility is to use direct acting contacts of a size and type needed to carry the expected current. For small currents, as in electronic circuits or instrument applications, the Hall Effect switch or photoelectric switch is a possibility. The nature and properties of the various switches will be discussed in detail in Chapter 5.

STRESS

In designing or modifying a pressure switch to operate under heavy load conditions, particular attention must be paid to possible stresses set up in the loading spring and/or sensor element. If the stress becomes too high, bellows may collapse, diaphragms deform, and compression springs fail. The actual calculation of these stresses may be very difficult or even impossible, but fortunately in most cases it is possible to arrive easily at a "ball park" approximation or to compare the device in question to actual data for similar switches. In most cases manufacturers of pressure switches and sensors such as bellows and diaphragms often can supply data on allowable pressures under prescribed conditions.

In the case of purchased pressure switches, stress is seldom a problem if the switch carries an appropriate proof pressure and life cycle rating and is operated under moderate environmental conditions.

Some calculation methods for estimating stress will be given in the chapters on pressure switch design.

WORKING RANGE

The term working range refers to the range in pressure that a pressure switch may encounter under normal operating conditions. Normally, the working range is the adjustable range of the switch.

The accuracy and cycle life of a pressure switch depend strongly on the location of the set point within the working range. Greater accuracy is obtainable when the set point is in the lower 65% of the working range. In most applications the set point is located in the middle of the working range to give the optimum combination of accuracy and cycle life. (See Figure 3-1.)

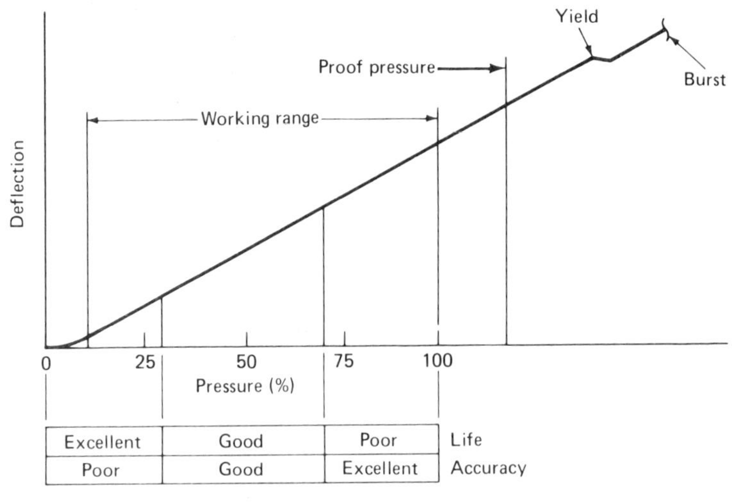

Fig. 3-1. Working range of pressure switch.

IV Installation and Maintenance

The proper functioning of a pressure switch depends not only on the inherent quality of the switch itself but also on how it is installed and used in the system and how well it is maintained during use. This chapter will consider several important factors related to installation and usage of the switch and present several ideas for compensating for environmental conditions.

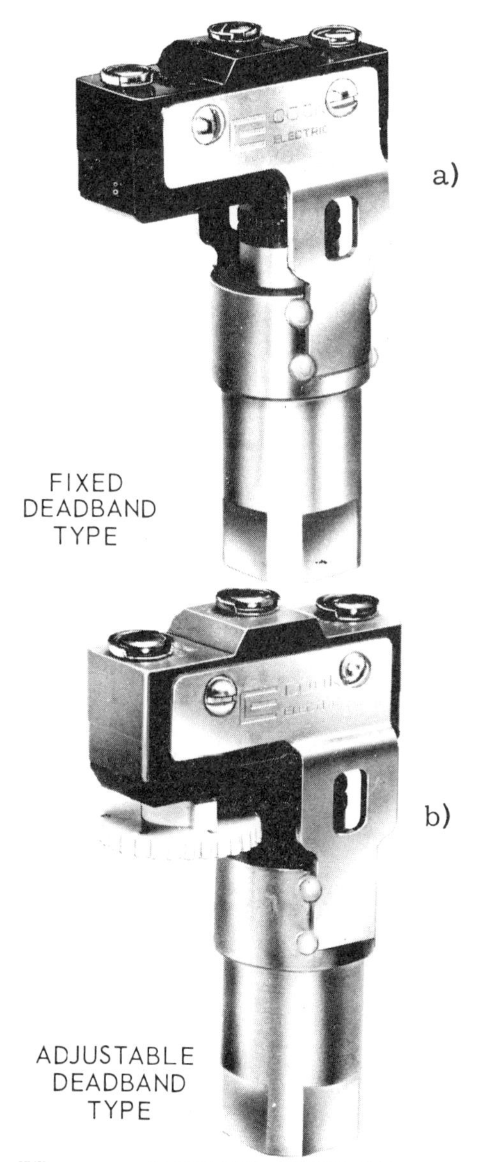

Fig. 4-1a,1b. *(Courtesy of Cook Electric, Morton Grove, Illinois.)*

HOUSING

Of prime importance to the user of a pressure switch is the nature of the housing. The housing must provide protection against the anticipated environment, a means for mounting the switch, and in many cases access to the internal components of the pressure switch.

The various types of housing are discussed below:

Stripped Switch. The stripped pressure switch shown in Figure 4-1a,b,c is normally sold to OEM (original equipment manufacturing) users for installation in custom housing or other uses. In a few instances, such as in very clean environments and where the electrical rating is low, the stripped switch may be used without a housing.

Terminal Block. This, like the stripped switch discussed above, is normally sold to be installed in the user-supplied housing. In some instances the switch is inside a housing with the terminal block on the outside. This allows easy wiring into the circuit and can eliminate

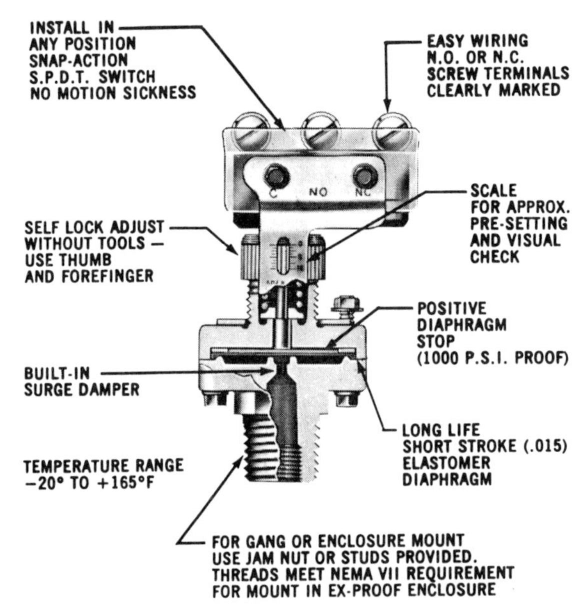

Fig. 4-1c. *(Courtesy of Delaval Turbine, Inc., Barksdale Controls Div., Los Angeles, California.)*

the need for external junction boxes. In some applications no cover is needed. (See Figure 4-2a,b,c,d.)

Standard Housing. This term refers to a housing that completely encloses the switch and all wiring connections to provide protection against certain environments and to protect against electrical shock hazards from exposed connections. There are various categories of enclosures, such as dusttight, watertight, etc. Normally, they are metal containers, but they may be made of plastic or other materials. Several of them are described further in the NEMA standards given in Appendix B.

Explosion-Proof. An explosion-proof housing is designed to isolate the electrical switch from explosive atmospheres, such as fuel vapors and various other gas mixtures. Such housings are generally of heavy gage metal

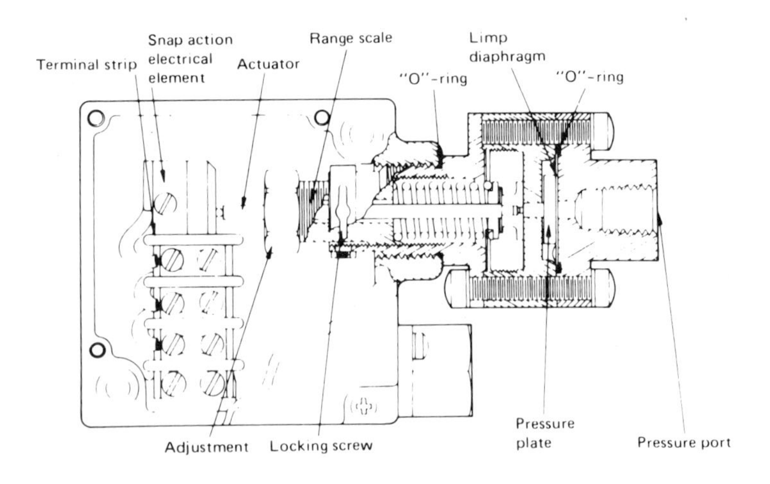

Fig. 4-2a. Pressure switch with terminal block. *(Courtesy of Neo-Dyn, Inc., Glendale, California.)*

Fig. 4-2b. *(Courtesy of Delaval Turbine, Inc., Barksdale Controls Division, Los Angeles, California.)*

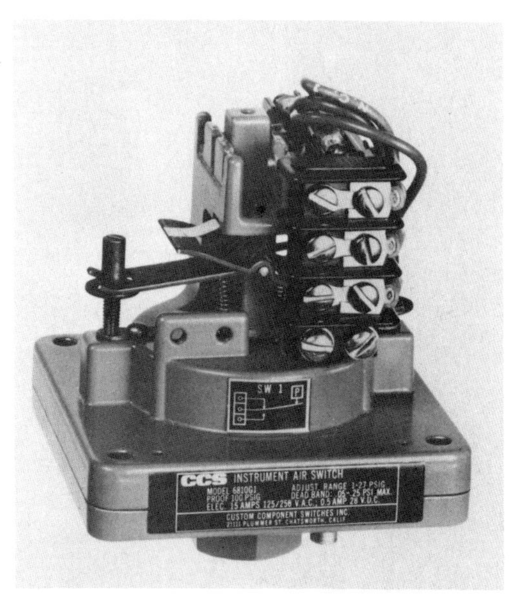

Fig. 4-2d. *(Courtesy of Custom Components Switches, Inc., Chatsworth, California.)*

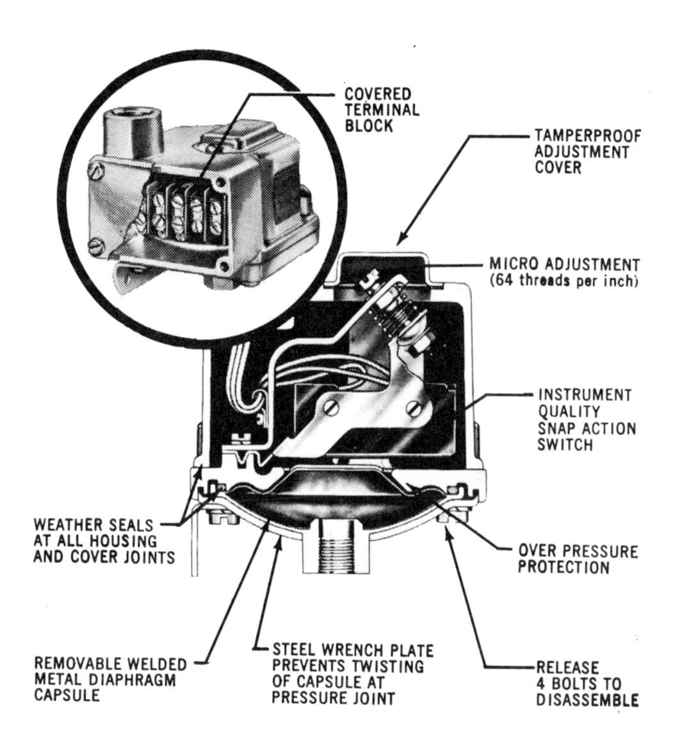

Fig. 4-2c. *(Courtesy of Static "O" Ring Pressure Switch Company, Olathe, Kansas.)*

construction with provision for hermetic sealing of the electrical connections to keep out vapors. External provisions for adjustments are often made. (See Figure 4-3, as well as Figure 1-11, in Chapter 1.)

Explosion-proof housings complying with applicable NEMA, CSA, or other standards are required whenever a switch is to be used in a hazardous location where contact arcing could cause an explosion.

Fig. 4-3. Pressure switch with explosion-proof housing. *(Courtesy of Custom Component Switches, Inc., Chatsworth, California.)*

ELECTRICAL CONNECTION

There are several ways in which the electrical leads may be connected to the pressure switch. The choice depends on many factors, such as the type of electrical switch used, type of housing used, accessibility, environment, frequency of connection and disconnection for repair or other reasons, cost, applicable safety codes, and others.

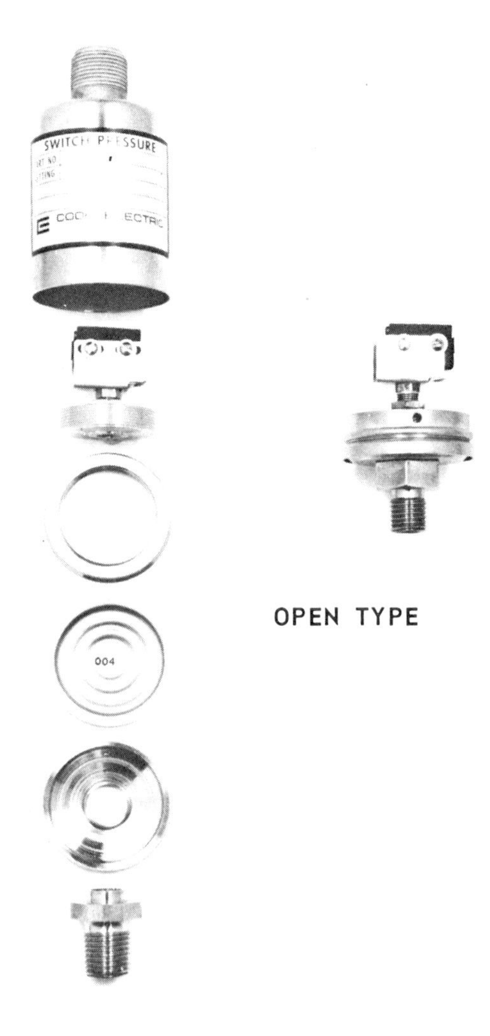

Fig. 4-4. Pressure switch with electrical connector. *(Courtesy of Cook Electric, Morton Grove, Illinois.)*

Possible types of connections in approximate order of increasing cost and complexity are screw terminal, push-in or clip terminal, solder terminal, and spot welding. These connections are usually used for connecting directly to the electrical switch and are usually considered for permanent connections.

Often the electrical switch is connected to some type of connector (see Figure 4-4) mounted on the switch housing, and connection is made from there to the electrical circuit. Any of the above methods may be used, and in addition several types of pin and socket type connectors are available. The latter are especially useful in aerospace and similar applications where switches may be removed periodically.

PLUMBING

Connection of the pressure switch to the pressurized media is generally determined by the type of plumbing in the system. In the case of industrial pressure switches the connection is generally by use of pipe threads, either male or female form. The connection is usually permanent barring unforeseen failure and need for repair. In aerospace applications the plumbing is typically tubing with straight thread fittings. Pressure switches usually have male or female fittings of the military standard (MS) type. Other possibilities include flange fittings, welding fittings, brazing or solder fittings, quick disconnect fittings, etc. Material chosen should be rated for expected pressures, type of media, etc.

LOCATION

The location of the pressure switch in a system is usually determined by the layout of the pressure system. Whenever there is a choice of location, the switch should be mounted in the least hazardous area (which can obviate the need for explosion-proof housings) and in a location where the environment (temperature, cleanliness, etc.) is fairly constant. This can eliminate unwanted effects of temperature change and similar problems. It is also highly desirable from the standpoint of economy to locate the switch where it can be easily serviced or replaced without interrupting or having to dismantle other equipment. Where possible, consistent with the intended function, the switch should be mounted within arm's reach on the outside of pressure vessels or on walls in a stable environment. For convenience several pressure switches may be grouped together on a panel if such arrangement does not mean excessive plumbing or affect conditions of operation.

MOUNTING

Generally, pressure switches are mounted with bolts or screws to a mounting surface. (See Figure 4-5.) In some cases adhesives, welding, peel-off mounting, and other methods may be convenient. Normally a pressure switch should not be held by plumbing and electrical fixtures alone. Mounting should be sufficiently firm to resist movement due to shock, vibration, pressure surges, and similar phenomena that can affect the operation of the device.

Most pressure switches in common use can be mounted in any orientation. However, a few unusually sensitive instrument switches must be mounted with the proper surface up, as the force of gravity may cause relative movement of component parts and throw off settings.

In cases of severe environment or system conditions, special methods must be employed in locating and mounting the switch. One fairly common problem is that of a high varying temperature in the pressurized media. Temperature changes can affect the spring rates of the loading spring and thereby change the set point and/or differential. Also the plumbing to the switch may expand and cause structural problems. To overcome this the switch should be mounted in a location where the ambient temperature is under approximately 100 °F, and the inlet line should be coiled three or four turns. This will normally result in a temperature at the switch of under 100 °F for system temperatures up to approximately 700 °F and will allow for thermal expansion of the line. Figure 4-6 shows a typical unit.

BACK MOUNTING

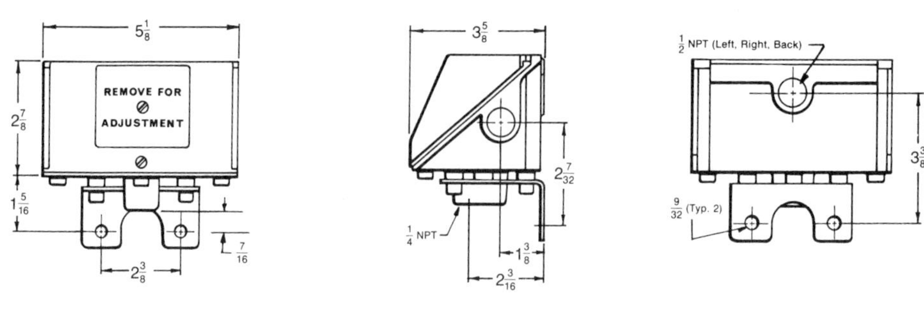

SIDE MOUNTING

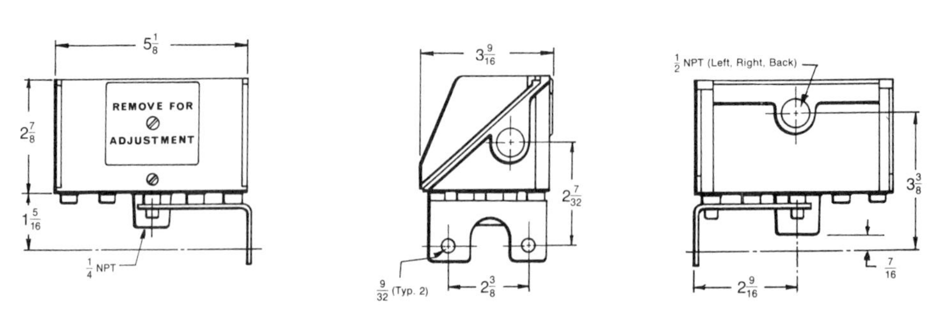

Fig. 4-5. Pressure switch with mounting bracket. *(Courtesy of Static "O" Ring Pressure Switch Co., Olathe, Kansas.)*

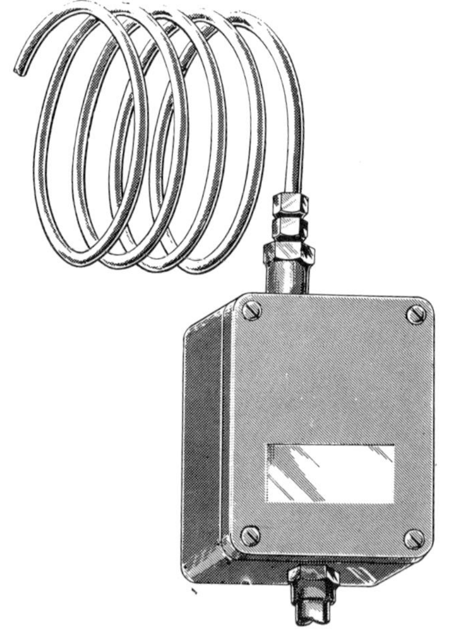

Fig. 4-6. Thermal expansion line for pressure switch. *(Courtesy of Delaval Turbine, Inc., Barksdale Controls Division, Los Angeles, California.)*

If the ambient temperature where the switch must be mounted will vary widely, and it is not feasible to insulate the switch from it, the switch should be raised to the highest expected temperature before setting it in order to "season" it. This gives the switch components a chance to expand to their fullest and stabilize, and should eliminate problems of binding and set point change due to stress relaxation.

When the temperature of the pressurized fluid is very low (below −65 °F), many pressure switches will not function properly. To alleviate this problem, the normal solution is to run a length of uninsulated tubing through a warmer ambient region to the switch. In the case of liquid media this acts as a vapor generator. The tube should run horizontally a distance of about one to two feet and then up to the switch. The switch must be mounted higher than the highest liquid level in the system (Figure 4-7).

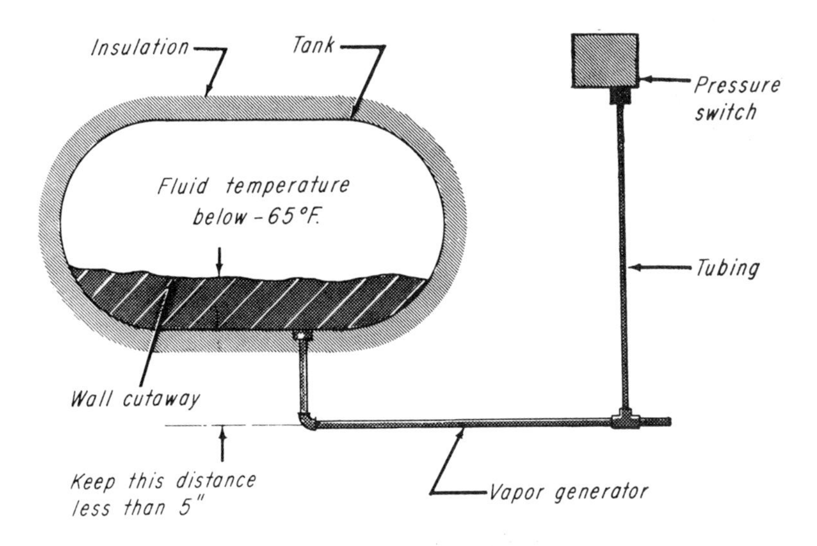

Fig. 4-7. Typical pressure switch set-up for low temperatures. *(Courtesy of Delaval Turbine, Inc., Barksdale Controls Division, Los Angeles, California.)*

Mounting is also important in minimizing the effects of pressure surges or ripple. There are several methods used to dampen out surges, of which the two most common are to install a snubber or surge reservoir (accumulator) in the line ahead of the switch (see Figure 4-8a,b,c) or to put several loops or coils in the inlet tubing as described above for high temperature systems. The loops will respond to the pressure surges in the manner of bourdon tubes.

The installation of the pressure switch can also be important when the operating media are very viscous, or contain solid particles. The switch should be located as close as possible to the pressurized system, and the line to the switch should be as large in diameter as possible—at least 1/2 in. Also, in the case of abrasive media any changes in direction or diameter should be smooth to minimize abrasive scrubbing by the media, and the number of such changes should be held to a minimum.

In situations where pressure switches must be mounted on vibrating machinery or in similar environments, steps must be taken to isolate the sensor element from the vibration. This may be accomplished in some cases by the design of the pressure switch, by building in controlled friction or damping. In other cases the pressure switch must be shock-mounted using rubber mounting pads, springs, or similar devices to soak up the vibration. Usually only trial and error will tell what will work in any given situation.

MAINTENANCE

The amount of maintenance required by a pressure switch will depend on the nature of its environment, its function, and the quality of the device. Most industrial or aerospace quality pressure switches are designed for replacement of component parts and recalibration of the switch. Most commercial or appliance pressure switches are of "throw away" type, since repair costs usually would be a sizable fraction of or even exceed replacement costs.

Most pressure switches do not require periodic maintenance other than perhaps a periodic recalibration in critical applications. Occasionally periodic replacement of seals and similar parts may be called for. Otherwise maintenance usually means repair or replacement of a failed unit.

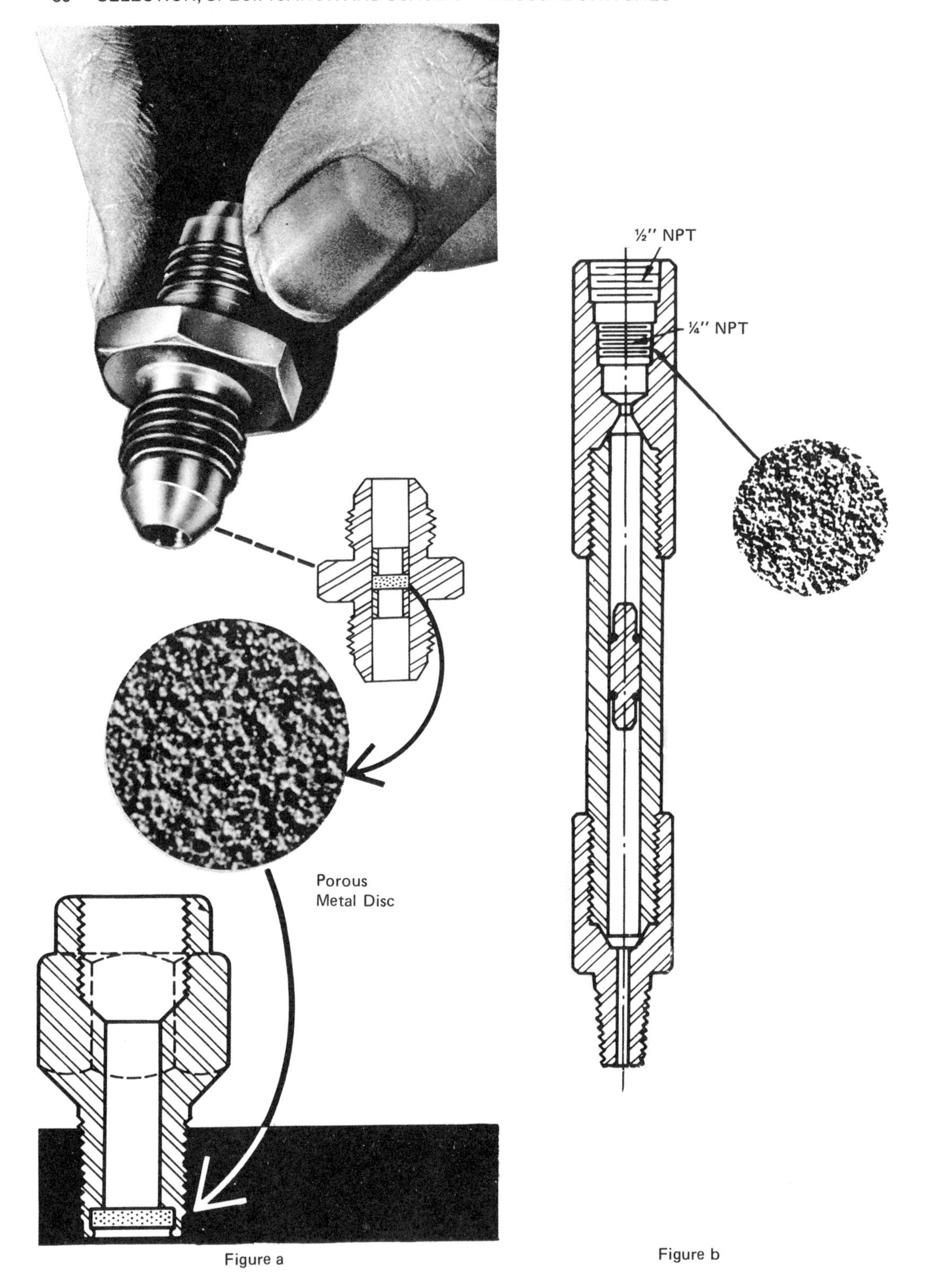

Fig. 4-8a,b. Snubbers. *(Courtesy of Chemiquip Products Co., Inc., New York.)*

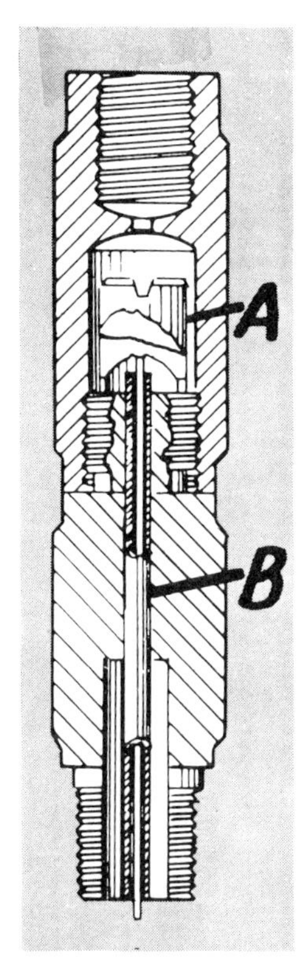

Fig. 4-8c. Snubber. *(Courtesy of Operating and Maintenance Specialties.)*

How It Works

Pressure shocks and pulsations are absorbed in the doughnut-shaped orifice formed by piston *A* in tube *B*. Piston *A* rises and falls with the pulsation, automatically kicking out any sediment or pipe scale that would clog a simple orifice, needle, filter, or porous disc.

PART TWO

DESIGN OF PRESSURE SWITCHES

V Electrical Switches

Most commercially available pressure switches come with specific electrical ratings and incorporate a basic snap action electrical switch in the design. Often a series of electrical switches can be used interchangeably to produce a family of pressure switches. The choice of pressure switch will depend on the nature of the electrical circuit being switched.

In the design of a pressure switch it is necessary to specify the nature of the electrical switch to be employed. In addition to its electrical ratings, often its physical nature must be specified. This chapter will deal with the selection and specification of electrical switches for use in pressure switches, the first step of which is to write down a list of requirements a switch must meet. Subsequent paragraphs will point out features that may be required in various applications. In other applications, there may be no requirement, and such things as cost and availability will determine the switch to be used.

TYPE OF CONTACTS

There are three categories of electrical switch based on the nature of the contacts and their method of operation.

Most pressure switches incorporate a basic electrical switch having snap action contacts. Figure 5-1 shows the construction of a typical basic snap action switch. Increasing force on

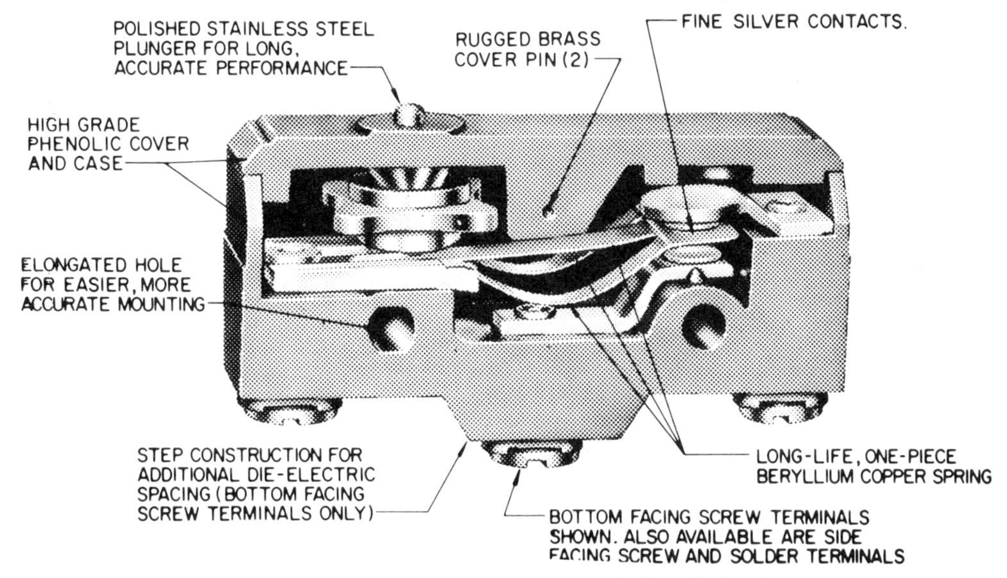

Fig. 5-1. Basic switch. *(Courtesy of Micro Switch.)*

the plunger causes the spring to snap over center and make contact. (See Chapter 6 on springs for a discussion of overcentering action of springs.) Switches of this type are available in a wide range of sizes, current ratings, and contact materials. They may also be obtained in a variety of contact arrangements, as shown in Figure 5-2. These arrangements may be needed for switching multiple circuits and similar functions. The prime virtue of a snap action contact is the rapid extinguishing of any arcs that form as contacts open and close, thus greatly reducing burning and pitting of the contacts.

A second form of contact configuration is called direct acting, since there is no overcentering or snap over of the contacts. As may be anticipated, direct acting contacts are more susceptible to arcing and associated problems. In addition, they are often subject to vibration and false contact closure, which may mean trouble when they are used in a pressure switch. Also included in the category of direct acting switches are dry and wet reed switches, mercury switches, etc. Reed switches, which are magnetically actuated, have the advantage of not requiring mechanical contact between the pressure sensor and the electrical switch and the disadvantage of being highly susceptible to shock and vibration. In the case of pressure switches, direct acting contacts are generally used when space or other restrictions do not allow the incorporation of a purchased snap action electrical switch. Direct acting contacts may also be used with negative rate springs (Chapter 6) to effect a snap action switch.

The third and newest type of electrical switch, which is finding increasing use in pressure switches for aerospace and other specialized applications, is the contactless solid state switch. There are two basic varieties of this switch available—the electro-optical coupler (Figure 5-3) and the Hall Effect switch (Figure 5-4). In the electro-optical device a moving shutter interrupts a light beam which couples a light emitting diode to a photocell, thus interrupting the current in the output circuit. In the Hall Effect switch a moving magnet and the Hall Effect are used to produce a current in the output circuit. These devices eliminate the problems found with moving contacts, such as arcing, contact bounce, wear, etc., but introduce problems of their own, chief among which is the necessity of input power to the light source or Hall Effect circuit. Also, they are sensitive to temperature change and overloading. At the present time the output currents are limited to very low values (typically 10 milliamperes)

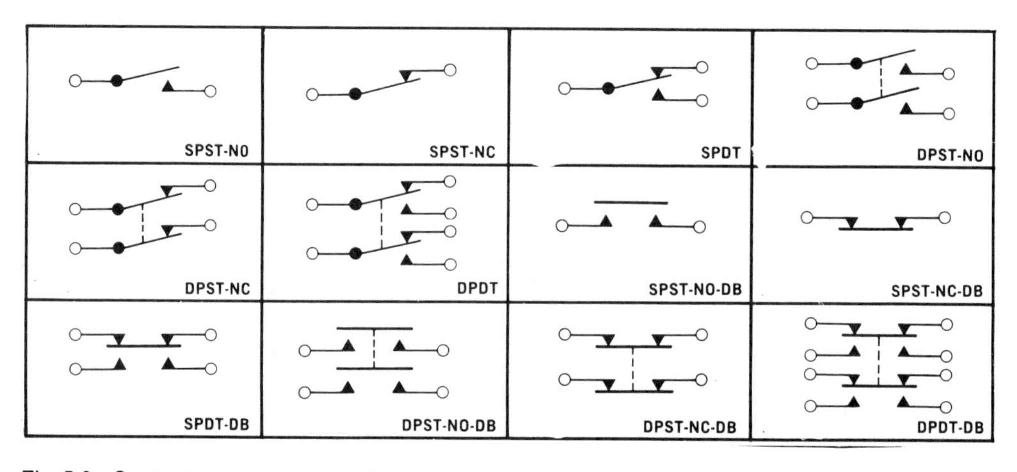

Fig. 5-2. Contact arrangements. *(Courtesy of Licon, A Div. Illinois Tool Works Inc., Chicago, Illinois.)*

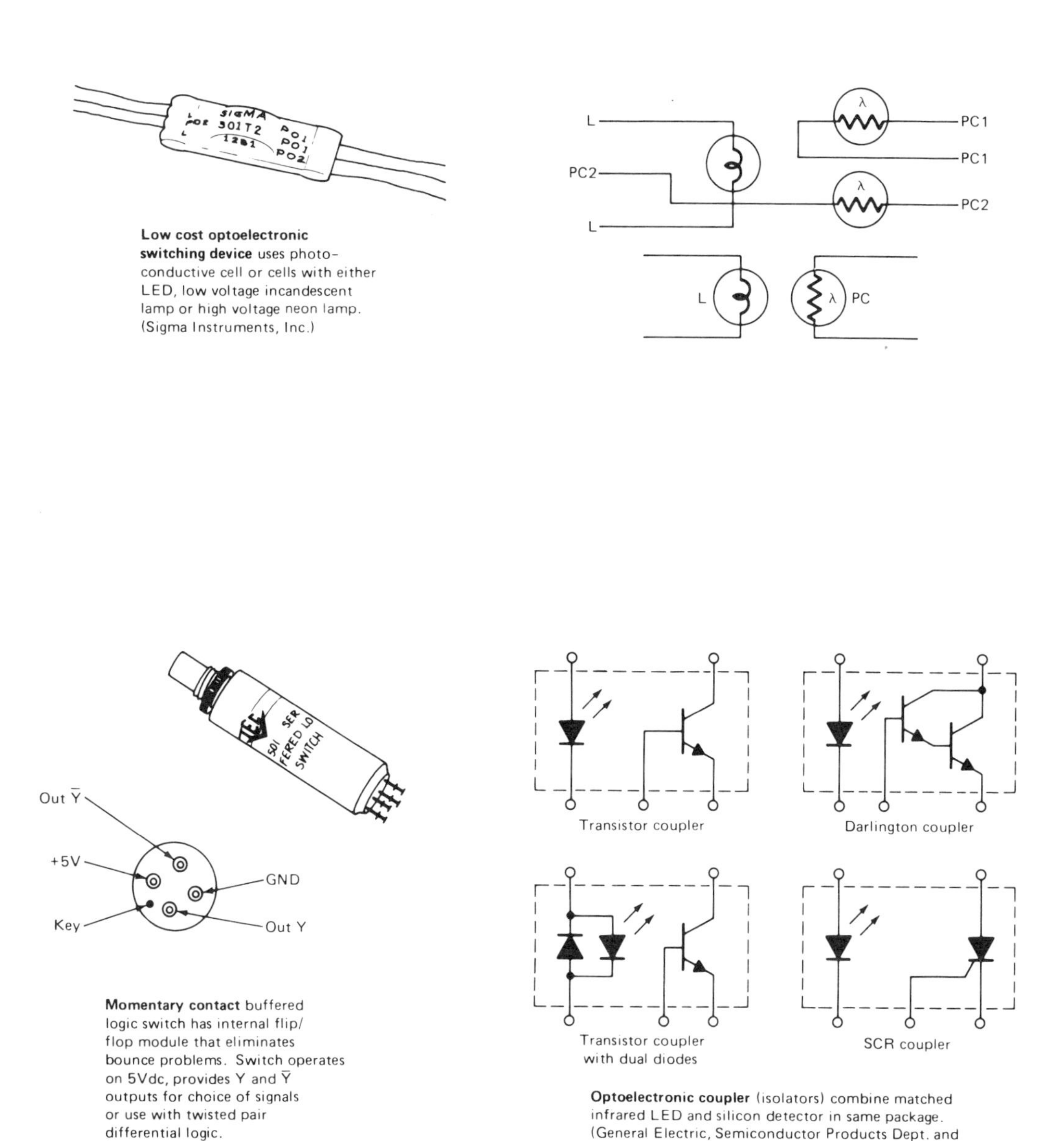

Fig. 5-3. Electro-optical switch (solid state). *(Courtesy of* Design News Magazine.*)*

unless relays or amplifiers, etc., follow the switch. Hall Effect and electro-optical switches are useful in pressure switches used in various types of instrumentation and controllers that send their output signals to a computer. Many of these devices can be directly interfaced with Transistor-Transistor Logic and other forms of computer circuitry.

MECHANICALLY OPERATED SOLID STATE SWITCH

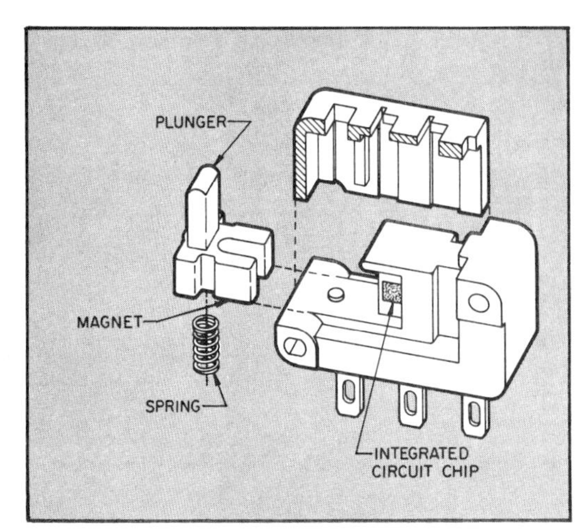

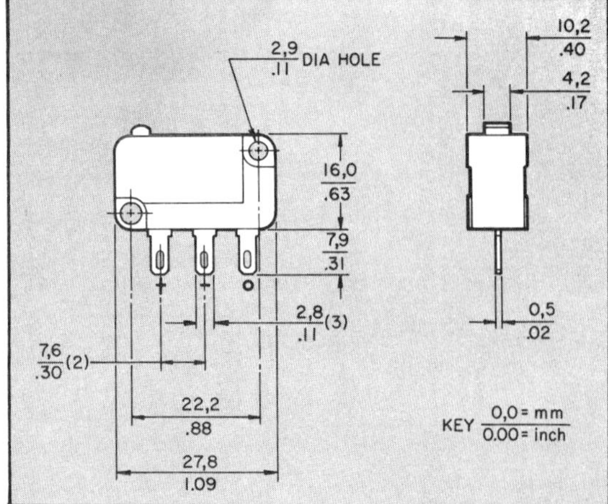

SINKING

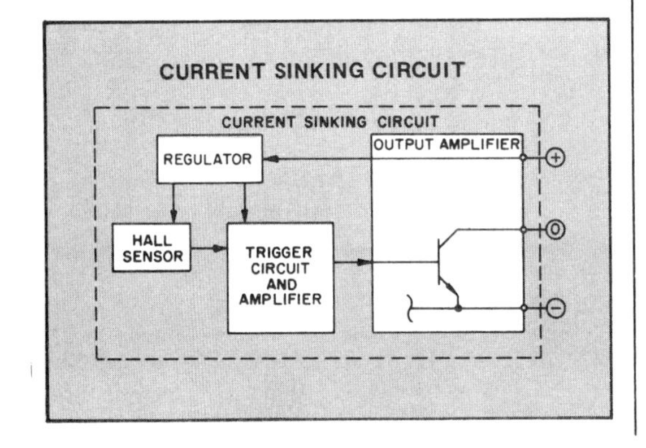

SOURCING

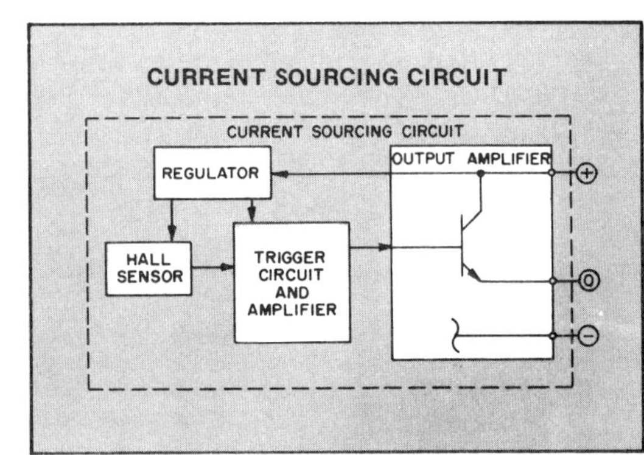

Fig. 5-4. Hall Effect switch (solid state). *(Courtesy of Micro Switch, Freeport, Illinois.)*

ELECTRICAL PARAMETERS

The following parameters are often of importance in the selection or design of the electrical switch to be incorporated into a pressure switch. In the case of purchased pressure switches it is sometimes necessary to compromise on some of these parameters. Unless otherwise noted, the properties apply to both snap action and direct acting switches.

Current

The current rating of a switch is generally the maximum recommended current for continuous duty under normal temperature conditions and for a specified type of load. Electrical loads are resistive, inductive, lamp, or motor, or combinations of these. Current ratings are also given at a rated ac or dc voltage level. Available switches normally considered for use in pressure switches have current ratings ranging from a few milliamperes to approximately 25 amperes.

Voltage

Most of the common electrical switches are good for either ac or dc voltages and have ratings at 120 or 240 volts alternating current (VAC) and 30 volts direct current (VDC) as these are the voltages commonly encountered. Some manufacturers give ratings for various other voltages.

Load

The current rating of the switch must be for the intended load. A current level for resistive load is usually considerably higher than for an inductive or motor load.

Dielectric Strength

In many applications, particularly in aerospace, the switch must withstand high voltage across open contacts without conducting. The usual requirement is to withstand 1050 VAC for one minute. Basically, this is a check on the manufacture of the switch to ensure that it will withstand high voltage transients.

MECHANICAL PARAMETERS

The physical or mechanical characteristics of the basic electrical switch are not often of great importance to the purchaser or user of a pressure switch but may be of the utmost importance to the designer or engineer. Some of these characteristics will be discussed here, and their use, calculation, and so on will be covered in more details in the chapters on pressure switch design.

Actuator

Most of the electrical switches used in pressure switches have metal or plastic plunger actuators as shown in Figure 5-1. Other possibilities much less common are lever, roller, and push button. Toggles, rockers, and similar actuators are seldom considered for pressure switch design.

Housing

In most pressure switches the electrical switch goes inside a housing of the required type; so the housing of the electrical switch is of relatively little importance except in cases of severe environment where a hermetically sealed switch may be desirable. If a stripped pressure switch is to be used, the housing of the electrical switch must be capable of withstanding the expected environment.

Mounting

Most of the electrical switches usually used in pressure switches have two mounting holes as shown in Figure 5-1 and can be mounted into a housing with bolts. Pressure switch housings must be designed to acommodate the chosen electrical switch.

Actuator Forces

There are two forces that the pressure switch designer must be concerned about in relation to the electrical switch actuator. These are the operating force and the release force, illustrated in Figure 5-5. The operating force is the level of applied force that must be reached before the electrical switch will actuate or trip. The release force, smaller than the operating force, is the level the applied force must fall to before the electrical switch will deactivate or untrip. The difference of these forces is called the force differential. A third force, usually of less importance to the pressure switch, is the overtravel force, which is the force required to move the electrical switch actuator to the limit of its overtravel.

Actuator Travel

There are three travel characteristics of the electrical switch actuator that are of prime importance to the designer of pressure switches. The pretravel is the distance traveled from the free position to the operating position (Figure 5-5). Differential travel is the amount of movement from the point where the contacts snap over to the point where they snap back, i.e., the travel between actuation and deactuation. Finally, overtravel is the amount of travel past the operating point. Overtravel and pretravel combine to give the total travel.

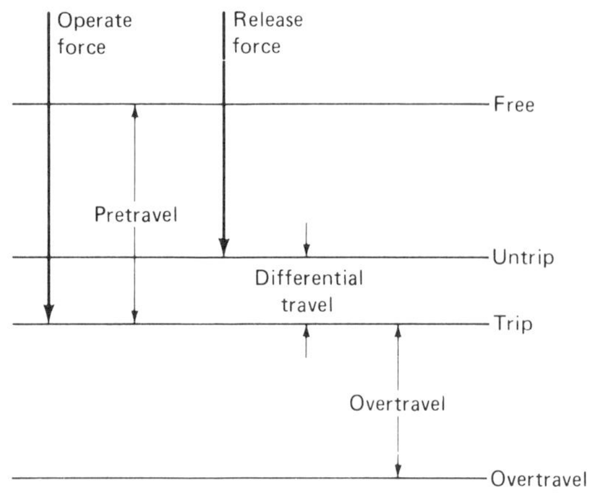

Fig. 5-5. Switch actuator forces.

BASIC SWITCH TECHNICAL DISCUSSION†

ALTERNATION CURRENTS AND DIRECT CURRENTS

The dc rating of a given switch is much less than the ac rating. For example: The Licon* Type 16 Basic Switch (see Figure 5-6) is rated at 10 amperes 250 VAC and only 10 amperes 28 VDC. An arc between contacts is possible when ionization occurs between those contacts. In ac, if the deionization time is less

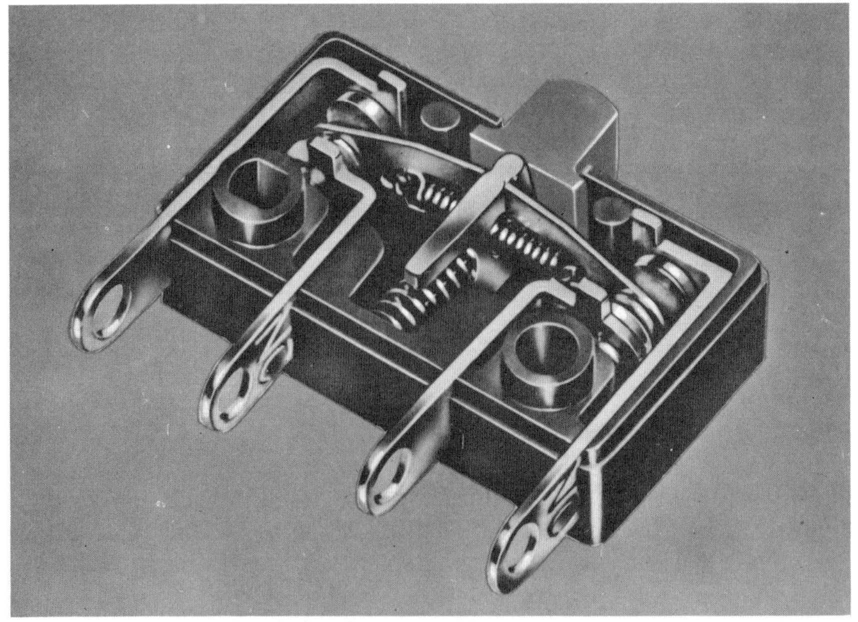

Fig. 5-6. Licon type 16 switch. *(Courtesy of Licon, Div. of Illinois Tool Works, Inc.)*

†The following material has been adapted from publications courtesy of Licon, Division Illinois Tool Works, Inc., Chicago, Illinois. *Licon is a registered trademark.

than the time taken for the voltage to pass through zero and build up, the arc will be quenched. This is usually the case with 60-cycle ac. In 400-cycle ac the interrupting problems may be nearer to dc. Since dc flows as a result of a continuous voltage applied to the circuit and does not periodically pass through zero, it can continuously support an arc. Consequently, switches drop decidedly in amperage ratings going from ac to dc. Double break contacts greatly extend a switch's direct current rating.

CONTACT BOUNCE—CONTACT CHATTER

Whenever two objects collide, there is a force developed that causes them to rebound. The extent of rebound depends on the forces tending to restore them to contact, the relative masses, and the natural frequency of the supporting means.

An MS 25026 switch requires a total transfer time of approximately 10–15 milliseconds. Much of this time is required for the travel from one contact to the other. After initial contact rebounding will occur for approximately 8 milliseconds.

With lamp loads, where the initial current inrush might be 10 or 15 times normal current, a switch exhibiting contact bounce might actually be making the circuit at 15 times normal current for five or six bounces. This means that a switch being tested for 5,000 cycles might actually be closing this high inrush current at six times 5,000 or 30,000 cycles.

CONTACT PRESSURE

Contact pressures in snap action switches vary from less than 10 grams to approximately 100 grams. Coin switches may have as low as 5-gram contact pressure, since the necessary light operating force restricts the design.

High contact pressure reduces resistance. It helps to break through surface film and is particularly important in that it contributes to vibration resistance. With higher initial contact pressures, switches such as Licon* will remain vibration-resistant closer to the trip point.

CONTACT RESISTANCE

Resistance of switch when measured terminal to terminal, and which effectively appears in series with the load, has typical initial values in the range of a few milliohms. These values usually increase during life. The rate of increase is greatly affected by the voltage current, power factor, frequency, and environment in which the electrical load is being switched. Typical military standard per MIL-S-8805 is 25 milliohms.

CONTACT SIZE

Large contacts are desirable for heavy loads because they dissipate the heat from an arc. Silver is a very good contact material because of its excellent conductivity and low electrical resistance. This allows the contact to carry the heat away from the point where the arc is burning and dissipate it over some area. The thickness of a contact has much less effect on its ability to carry loads or to dissipate heat than its surface area has.

DEAD BREAK

Dead break is an open circuit condition in a snap action switch that results from either low contact pressure or actual contact lift-off. It is a particular problem with slow switch actuation.

As the switch plunger is slowly depressed, the switch approaches its trip point, and the contact pressure decreases—this is true of almost every snap action switch. Low contact pressures cause high resistance to current flow. Where low current is present, this high contact resistance may effectively open the circuit.

It is possible, given poor switch design, to have an actual contact lift-off with extremely slow actuation. The moving contact hangs between the stationary contacts and touches neither. The circuit is open, and even a high current will not bridge the gap.

A certain percentage of switches of any design will have dead break. Generally, five design features minimize this condition.

1. Low friction pivot point
2. High contact pressure
3. Contact shape
4. Noncorrosive contacts
5. Inert plastic cases

Licon* has found that the contact shape is important. They are using a flat fixed contact and a radius face moving contact. On the radius face they emboss a dimple that is about 1/32" in diameter and .004" high. This provides a projection of small area that gives high unit contact pressure. Also, by chance alone there is little possibility that a piece of small dirt can intervene. Most dust found in switches is less than .001" thick. With the use of the Licon* dimple there is almost no chance of separation due to dirt. Also, the dimple is so short and close to the volume of the main contact that there is good heat transfer and no derating. This exclusive feature is very important.

Flickering in a switch can usually be eliminated by several interruptions of a 5-ampere current. This procedure will burn off the film or dust that is causing the difficulty.

DOUBLE BREAK

Multiple break, or double break, contacts effectively increase electrical capacity and electrical life of a switch. Arc energy is distributed over more than one location. The heat is, therefore, dissipated at two places, minimizing the possibility of contact weld.

Double break also offers greater interrupting ability. Any contact gap presents resistance to the flow of current in an arc at three places, each of the two contact faces and the contact air gap. Therefore, a double break of .020" contact gap offers resistance at two air gaps of .020" each and at four contact faces. Thus two breaks of .020" each will have much greater interrupting ability than one break of .040".

Multiple contact breaks are often successful, since the voltage drop presented by these multiple breaks may exceed the minimum voltage that will support an arc. At least 18 VDC must be applied across a pair of silver contacts to maintain an arc. Thus, when you present many breaks in the circuit, this minimum arcing voltage is multiplied by the number of breaks in order to get a minimum voltage that would support an arc continuously.

The following table illustrates the increased switch capacity that results from multiple breaks at 120 VDC.

.020-in. gap single break	1/2 ampere
.020-in. gap double break	1½ amperes
.020-in. gap quadruple break	20 amperes

In any multiple break switch the action must be simultaneous. If it is not, the advantage is lost, since the last contact to make or the first contact to break would carry the full load.

FREQUENTLY USED TERMS AND CONFIGURATIONS FOR BASIC SWITCHES

Bounce Rapid rebounding of contact after closing.

Break An opening or interruption of a circuit. Simultaneous interruption of a circuit in two different places is described as double break.

Break-before-make With double throw contacts where the moving contact, in transferring, interrupts one circuit before establishing the other.

Chatter Prolonged undesirable opening and closing of electronic contacts.

Clearance Air space, usually 1/16" minimum, between live metal parts of opposite polarity or to ground.

Cold Flow Change of dimension or distortion caused by sustained application of a force.

Contact Gap The air space between mating contacts when contacts are open.

Corona Discharge of electricity that appears on the surface of the conductor when the potential gradient exceeds a certain value.

Creepage The distance over the surface of an insulator between live metal parts of opposite polarity or to ground. Usually 3/32" minimum.

Detent A catch or holding device.

Dielectric Strength The maximum potential gradient that an insulating material can withstand without rupture.

Dead Break An unreliable contact made near the trip point, at low contact pressure. The circuit is interrupted but the switch does not "snap over."

Double Break A contact arrangement in which the moving switch element bridges across two fixed contacts so that the circuit is broken in two places simultaneously. Sometimes called **shorting bar**.

Double Throw A switch that alternately completes a circuit at each of its two extreme positions. It is both normally open and normally closed.

Snap Action The action of a pair of contacts that open and close quickly enough to immediately extinguish any arc that may form, and which snaps closed with sufficient pressure to firmly establish an electrical circuit.

OTHER ELECTRICAL TYPE SWITCHES†

PHOTOELECTRIC FUNDAMENTALS

The following are guidelines and factors to consider when selecting and applying photoelectric controls.

INTRODUCTION

Most photoelectric controls consist of a light source/photoreceiver combination that provides the input signal to a control base, which amplifies and imposes logic on the signal to transform it into usable electrical output.

There are two main types of controls: self-contained and modular. A self-contained control includes the light source and photoreceiver in the control base, while a modular control uses a light source/photoreceiver combination or reflective scanner separate from the control base. Self-contained retroreflective controls require less wiring and are less susceptible to alignment problems, while modular controls are more flexible in allowing remote positioning of the control base from the input components, and are more easily customized.

SCANNING TECHNIQUES

"Through" Scan. A light source and photoreceiver mounted directly opposite each other so that light from the source shines directly on the receiver provide direct or "through" scanning. Its advantages are:

- Long scanning distance
- Small objects detectable at relatively long distances

Reflective Scan. A light source and photoreceiver mounted on the same side of the object to be detected so that light is reflected to the receiver provide reflective scanning. This technique is used in looking for an identifying mark on the surface of the object or when it is impossible to mount a receiver or retroreflector opposite the light source or reflective scanner. Reflective scans are of two types, diffuse and specular.

1. *Diffuse* (widespread): The reflection emits at various angles from matte surfaces such as kraft paper, wooden pallets, etc. Generally, the light is positioned perpendicular to the reflecting surface, with the photoreceiver at a 45° angle from this surface. Diffuse scanning is recommended when the material to be detected has a slight vertical "flutter."

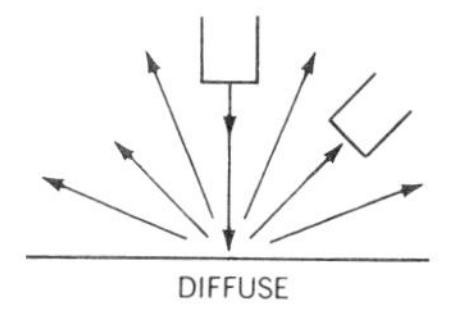
DIFFUSE

2. *Specular:* This type of scanning uses highly reflective surfaces such as mirrors, shiny plastics, or rolled or polished metal plates to reflect to the photoreceiver. The angle at which the light beam strikes the reflective surface must equal the angle of reflection to the photoreceiver. It is not recommended for surfaces having vertical "flutter," and is often used to detect matte or rough-surface materials that appear on smooth surfaces.

†Material in this section has been adapted from publications courtesy of Micro Switch Company, Freeport, Illinois 61032, A Division of Honeywell.

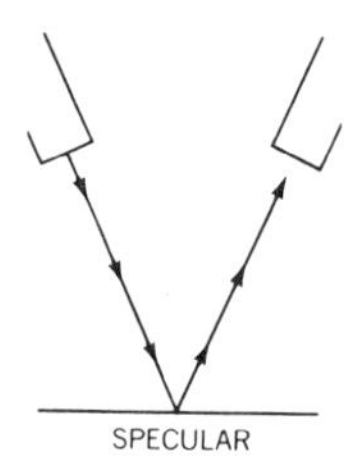

Retroreflective Scan: Retroreflective scanning uses a retroreflective target to return the light beam to the photoreceiver. A retrohead contains both the light source and the photoreceiver, and an object is detected when it passes between the retrohead and the target. The advantages of retroreflective scanning are:

- Single-side wiring
- Noncritical alignment
- Ability to withstand vibration
- Translucent object detection

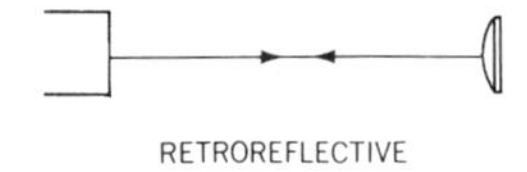

FILTERING

B/G (Blue/Green) Filters. Photoreceivers cannot distinguish colors. They are sensitive only to the quantity of light. They do, however, exhibit maximum sensitivity to the red-light portion of the spectrum. This makes distinguishing red from a white or light-colored background difficult. In these cases, a type BG photocell receiver should be used.

IR Filters. Infrared filtering is also available for several light sources and photoreceivers.

SENSITIVITY ADJUSTMENT

Each photoelectric control has a sensitivity adjustment to determine the light level at which the control will respond.

Conditions that could require the sensitivity to be adjusted to less than fully clockwise include:

- Detection of translucent objects
- High speed response
- High cyclic rate
- Line voltage variation
- High electrical noise atmosphere

OPERATIONAL STABILITY

For most stable operation, the signal ratio (the light seen by the receiver when the object to be detected is in position versus the light seen with no object present) should be 10:1. This ratio pertains to resistance measurements. An opaque object will easily meet this requirement. Translucent material may produce marginal operation. In such cases, using a transition responsive control (which needs only a 2:1 change) and a retroreflective scanner will improve operation.

SCANNING MODE

LOG logic cards and TR logic modules have a mode selector switch to determine whether the output will be energized when the photoreceiver sees light (''light operate'') or when the light is blocked (''dark operate''). Scanning mode may affect scanning distance.

DISABLE/RESET

Each photoelectric control has a terminal to disable the control's response. Closing a normally open (N.O.) connection between this terminal and ground prevents the control from responding to an input signal. This enables the control to be ''gated,'' to perform inspection functions during prescribed time intervals. Making this connection resets the control on all electronic latching relay controls.

ENVIRONMENTAL FACTORS

Environmental factors in control selection and application are given below.

Ambient Light. Use any photoreceiver in fluorescent or moderate incandescent light. In high incandescent light, use a bright light source, and, if possible, shade the photoreceiver or face it away from the ambient light. Where suitable, use a TPC3L narrow vision photoreceiver. For highly variable light conditions, use a transition responsive control or a modulated light source; remove reflective objects from the scanning field. Outdoors, always use an MLS to combat sunlight.

Vibration. Use any light source and photoreceiver where there is little vibration. To compensate for moderate vibration, use a bright light source, and focus it to present a larger image; or use retroreflective scanning. For high vibration, use an LED light source or a self-contained MLS control; shock mount or mount remotely.

Temperature. Temperature may dictate choice of photoreceiver. Use a phototransistor type receiver for temperatures below 32 °F (0 °C) or above 122 °F (60 °C). Any photoreceiver is suitable between 32° and 122°F. For extremely high temperatures up to 302 °F (150 °C), specify a high temperature phototransistor type receiver. It is helpful to cool the receiver with air or water.

Moisture. Use splash-proof units in indirect splash and high humidity; use NEMA 4 controls to withstand high pressure water or detergent steam.

Oil, Paint, Dust. Use a bright light source, and set the control sensitivity well above the operate point. If available with the logic you require, use either a high threshold or a transition responsive control. Direct air across the light source and photoreceiver lenses, and avoid retroreflective scanning. (In extreme cases, consider a proximity control.)

Explosive Atmosphere. Use only explosion-proof light source and photoreceiver.

Line Voltage Variation. If the variance is 10 to 30 volts, use a bright light source, and set the control sensitivity well above the minimum operate point. Use a voltage regulator where variance exceeds 30 volts.

Electrical Noise. Ground all photoreceiver shields and control bases. Use EMI/RFI filtering.

PHOTOELECTRIC DEFINITIONS OF TERMS

Current Sink An output that goes to ground in its ON state.

Dark Operate Mode switch setting that allows control to respond to no light on the photoreceiver.

Diffuse Scan A scanning technique. See discussion of scanning.

Disable To prevent output response to an input signal.

Dwell Duration of one-shot pulse.

Electronic Latching Logic When output turns ON, it will remain on until reset (two types; repeat and nonrepeat; see definitions of repeat cycle and nonrepeat cycle).

Input Signal Duration Length of time light beam is broken or restored.

Interrogate (Enable) To permit output response to input signals during an inspection period; the opposite of disable.

Led Light-emitting diode: solid state light source.

Light Operate Mode switch setting that allows control to respond to light on the photoreceiver.

Maximum Operate Point Sensitivity setting above which control will not respond to the dark signal produced by object to be detected.

Minimum Operate Point (Threshold) Sensitivity setting below which control will not respond to light signal.

Modulated Light High frequency chopped or pulsed light source. Allows rejection of ambient light and increased sensitivity. **MLS** designates modulated light source.

Momentary Start Input signal may be shorter than time cycle (or first time period of dual delay logic).

Nonrepeat Cycle Produces one output cycle for each input cycle.

Off-Delay Logic Adjustable delay (after input signal stops) before relay is de-energized.

On-Delay Logic Adjustable delay (after onset of input signal) before relay is energized.

One-Shot Logic Output energized when input is signaled, and de-energized after dwell time, regardless of input signal duration.

Opaque Does not allow light to pass through.

Reflective Scan A scanning technique. See discussion of scanning.

Repeat Cycle Produces continuous output cycles during each input cycle.

Reset (of Counter) Restores internal count to zero.

Reset (of Latching Logic) Restores latched output to OFF state.

Retroreflective Scan A scanning technique. See discussion of scanning.

Signal Ratio The ratio of dark signal (light blocked) to light signal on a photoreceiver (expressed in ohms for photocell receivers as shown in the installation instructions accompanying each control).

Specular Scan A scanning technique.

Sustained Start Requires input signal to be continuous through time cycle (or first time period of dual delay logic).

Threshold Responsive Type of control that responds to a change in light level (at the photoreceiver) through a preset operate point.

Through Scan A scanning technique. See discussion of scanning.

Time Ratio Ratio of maximum time adjustment (full clockwise) to minimum time adjustment (full counterclockwise).

Transition Responsive Type of control that responds to the rate of change in light level at the photoreceiver.

Translucent Allows light to pass through.

VI Loading Springs

A loading spring is one of the essential components of every pressure switch. Without it an actuation pressure cannot be set. As mentioned in an earlier chapter, however, the loading spring need not always be a separate component; in some cases the pressure sensor may act as the spring. In the case of the bourdon tube switch the tube acts as both sensor and spring. Other switches may use the inherent spring characteristics of bellows or diaphragms. The piston type, however, requires the use of a separate spring to establish a set pressure.

There are three types of springs commonly used in pressure switches and related devices. The helical compression spring wound of circular-cross-section wire is the most common spring element. The negative spring rate characteristic of a Belleville spring makes it useful in some types of pressure switch, and the buckling column spring is finding application in pressure switch design.

In this chapter we shall consider in some detail the design and usage of these types of springs. For further information consult the references at the end of the chapter.

HELICAL COMPRESSION SPRINGS

This type is an open coil helically wound spring that offers resistance to an axially applied compression force (Figure 6-1). It may be wound of wire of any desired cross section, with circular being by far the most common, and may be of straight, tapered or conical, convex, or concave form, with the straight type being most common. In addition, it may be wound with uniform or variable pitch to give a constant or variable rate.

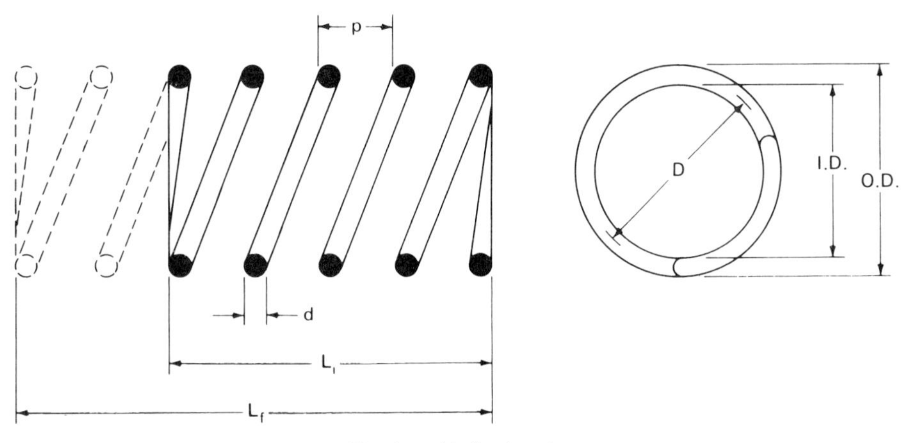

Fig. 6-1. Helical spring.

Because of its resistance to being compressed, a force must be applied to compress or shorten its length. In a pressure switch the applied pressure integrated over the surface area of the pressure sensor provides the force to compress the spring. The deadband is related to the various differences in compressive force necessary to compress the spring some specified amount.

In this section we consider the factors necessary in sizing a spring, and in later chapters we shall relate them to pressure switch parameters.

Basic Spring Equations

The design of a helical compression spring wound of circular cross section wire is based on two fundamental equations,

$$R = \frac{Gd^4}{8D^3N} \tag{1}$$

$$S = \frac{8}{\pi}\frac{F_mDK}{d^3} \tag{2}$$

where symbols have the meanings given in Table 6-1 and as discussed below.

Wire size (d) is the diameter of the spring wire in inches. In most cases one of the standard sizes shown in Table 6-2 should be chosen. Use of odd sizes usually will be costly and can normally be avoided by adjustment of another parameter in equation (1).

Mean coil diameter (D) is often arbitrary within reason and may be adjusted to be compatible with other parameters. In some cases, the spring must work in a fixed bore or over a shaft of fixed diameter and thus must be controlled.

Active coils (N) is the number of coils that can move closer together to effect the compression. The two end coils are not active; hence the total number of coils in the spring is N plus 2.

Torsion modulus (G) is a property of the material from which the spring is made (Table 6-3). It is the ratio of torsional stress to strain.

Spring Rate (R) is the ratio of compressive force to deflection and is a constant for a

Table 6-1. Nomenclature

Symbol		Meaning	Symbol		Meaning
a_1	=	outer radius (in.)	P	=	pressure (psi)
A	=	area of cylinder (in.^{2})	p	=	pitch or coil spacing (in./coil)
D	=	mean coil diameter (in.)	R	=	spring rate (lb/in.)
d	=	wire diameter (in.)	r	=	ratio of outside to inside diameter
E	=	modulus of elasticity (psi)	S	=	stress (psi)
F	=	applied force (lb)	S^1	=	corrected stress (psi)
f	=	running friction (lb)	S_c	=	compressive stress (psi)
f_b	=	breakout friction (lb)	$S_{1,2}$	=	tensile stress (psi)
F_i	=	force on installed spring (lb)	γ	=	dimensionless parameter
F_b	=	force on spring at breakout (lb)	η	=	dimensionless parameter
F_L	=	maximum expected force (lb)	t	=	thickness (in.)
F_m	=	maximum applied force (lb)	X	=	deflection of travel (in.)
F_s	=	force exerted by spring (lb)	Δ	=	differential or increment
F_{sw}	=	force exerted by switch (lb)	δ	=	deflection (in.)
G	=	modulus of elasticity (psi) torsional	δ_{DT}	=	differential travel (in.)
h	=	height (in.)	δ_{OT}	=	overtravel (in.)
K	=	Wahl factor	δ_{PT}	=	pretravel (in.)
L_f	=	free length of spring (in.)	δ_A	=	actuator setting
L_i	=	installed length (in.)	π	=	pi (3.14159 . . .)
L_s	=	solid height (in.)	μ	=	Poisson's ratio
ln	=	natural logarithm	1	=	trip point value
N	=	number of active coils	2	=	untrip point value

given spring when operated within the elastic limit. The spring rate is determined by required change in force and movement:

$$R = \frac{\Delta F}{\Delta X} \quad (3)$$

Compression stress (S) is the maximum stress the spring can withstand while still functioning within the elastic limit of the material.

Wahl factor (K) is a correction figure (Figure 6-2) to the stress to account for additional stress caused by curvature of the wire and shear loading. In many cases where torsional stress is low, it need not be considered; but in case of high stress, rapid cycling, or impact loading it should be used.

Spring load (F_m) is the maximum load the spring will see under anticipated operating conditions. This load directly determines the stress the spring must withstand.

Table 6-2. Preferred Sizes for Spring Wire

1	2	3	4	5	6	7
MUSIC WIRE	HIGH CARBON & ALLOY STEELS	VALVE SPRING QUALITY STEELS	"18-18" CHROME NICKEL AUSTENITIC 300 SERIES	STRAIGHT CHROME MARTENSITIC 400 SERIES	SPRING QUALITY BRASS PHOSPHOR BRONZE BERYLLIUM COP. MONEL & INCONEL	K MONEL & INCONEL X
.004	.032	.092	.004	Same as high carbon & alloy steels, Col. 2	.010	.125
.006	.035	.105	.006		.012	.156
.008	.041	.125	.008		.014	.162
.010	.047	.135	.010		.016	.188
.012	.054	.148	.012		.018	.250
.014	.063	.156	.014		.020	.313
.016	.072	.162	.020		.025	.375
.018	.080	.177	.026		.032	.475
.020	.092	.188	.032		.036	.500
.022	.105	.192	.042		.040	.563
.024	.125	.207	.048		.045	.688
.026	.135	.218	.054		.051	.750
.028	.148	.225	.063		.057	.875
.032	.156	.244	.072		.064	1.000
.042	.162	.250	.080		.072	1.125
.048	.177		.092		.081	1.250
.063	.188		.105		.091	1.375
.072	.192		.120		.102	1.500
.080	.207		.125		.114	1.625
.090	.218		.135		.125	1.750
.107	.225		.148		.128	2.000
.130	.244		.156		.144	
.162	.250		.162		.156	
.177	.263		.177		.162	
	.283		.188		.182	
	.307		.192		.188	
	.313		.207		.250	
	.362		.218			
	.375		.225			
			.250			
			.312			
			.375			

Table 6-3. Properties of Spring Materials. *(Courtesy Lyons' Encyclopedia of Valves.)*

MATERIAL	CONDITION	TORSION MODULUS G (psi)	MAXIMUM UNCORRECTED STRESS (psi)	ELASTIC MODULUS E (psi)
CORROSION RESISTANT ALLOYS				
302 CRES	Spring temper	10,000,000	100,000	28,000,000
302 CRES	Cold drawn	9,500,000	80,000	28,000,000
316 CRES	Spring temper	10,000,000	70,000	28,000,000
316 CRES	Cold drawn	9,500,000	70,000	28,000,000
321 CRES	Spring temper	10,000,000	70,000	28,000,000
410 CRES	Cold drawn	10,000,000	72,000	28,000,000
Carpenter 20	Spring temper	10,000,000	80,000	
17-4 pH	Cold drawn	10,500,000	80,000	29,000,000
17-7 pH	Cold drawn	11,000,000	140,000	29,000,000
Hastelloy C	40% Cold reduced	10,500,000	100,000	
18-8 CRES	Cold drawn	10,000,000	140,000	28,000,000
INCONEL-600	Cold drawn	11,000,000	85,000	31,000,000
INCONEL-625	Spring temper	11,000,000	85,000	31,000,000
K-MONEL-500	Cold drawn	9,200,000	80,000	26,000,000
MONEL-400	Spring temper	9,200,000	65,000	26,000,000
NI-SPAN-C	Spring temper	9,500,000	75,000	27,500,000
NS-355	Spring temper	11,000,000	150,000	
NONFERROUS ALLOYS				
Beryllium copper	Cold drawn	7,300,000	80,000	16,000,000
Beryllium copper	Pretempered	7,300,000	85,000	18,500,000
Phosphor bronze	Hard drawn	6,000,000	70,000	15,000,000
Spring brass	Cold drawn	5,500,000	60,000	15,000,000
LOW ALLOY STEELS				
AISI-1095	Hot rolled	10,500,000	90,000	30,000,000
AISI-5160	Hot rolled	10,500,000	105,000	30,000,000
Chrome vanadium	Oil tempered	11,500,000	110,000	30,000,000
Chrome silicon	Oil tempered	11,500,000	110,000	30,000,000
Music wire	Hard drawn	11,500,000	120,000	30,000,000
Spring wire	Oil tempered	11,500,000	105,000	30,000,000

Table 6-3 (Continued)

MATERIAL	CONDITION	TORSION MODULUS G (psi)	MAXIMUM UNCORRECTED STRESS (psi)	ELASTIC MODULUS E (psi)
LOW ALLOY STEELS				
Valve wire	Hard drawn	11,500,000	105,000	30,000,000
Rocket wire	Hard drawn	11,500,000	150,000	30,000,000
HIGH ALLOY STEELS				
A-286	Spring temper	10,000,000	100,000	
AISI-S1	Cold drawn	11,600,000	90,000	
AISI-H21	Cold drawn	11,000,000	100,000	
AISI-T1	Cold drawn	11,000,000	100,000	
INCONEL X-750	Cold drawn	11,500,000	90,000	31,000,000
INCONEL X-750	Spring temper	11,500,000	120,000	31,000,000
RENE 41	Spring temper	12,000,000	130,000	
S-816	Spring temper	11,600,000	130,000	
250 Maraging steel	Centerless ground	10,000 000	160,000	

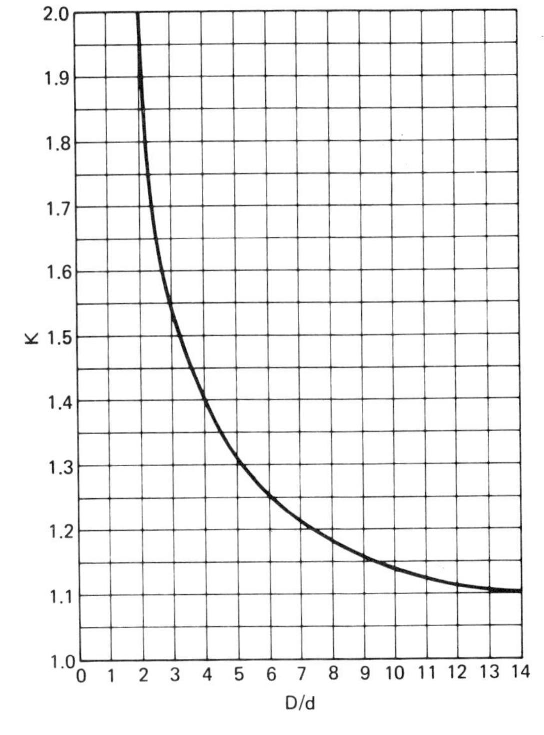

Fig. 6-2. Wahl Factor.

Free length (L_f) is the length of the spring with no compressive load applied.

Installed length (L_i) is the length of the spring when installed in the pressure switch under a specified preload determined by set point requirements.

Spring index (D/d) is the ratio of the mean coil diameter to the wire diameter and is a measure of the stability of the spring. It preferably ranges from 4 to 12. Springs over 12 in index are flimsy and tend to tangle, while those under 4 are difficult to make and are highly stressed.

Length-diameter ratio (L_f/D) is the ratio of free length to mean diameter and is a measure of the tendency to buckle under loading. Generally this ratio should be under 4 to ensure stability under load.

Design Considerations

The following properties, while unrelated to the basic sizing equations discussed above, are of importance in the design or specification of the loading spring for a pressure switch:

Coiling

In most applications the direction of coiling of the spring is arbitrary, with most springs automatically being right-handed. However, in the case of nested springs, they should be of opposite coiling to prevent interlocking.

Ends

In most applications the spring should have squared and ground ends that will be parallel at the installed length and will be perpendicular to the spring axis.

Surface Treatment

The type of treatment of the wire surface, if any, will depend on the type of material, the anticipated environment, and the expected stress level. Steel spring wire should be plated or coated to resist corrosion, stainless steel should be passivated, etc. Where very long fatigue life is required in alloy steels, they may be shot-blasted to increase the surface stress tolerance.

DESIGN PROCEDURE

The following procedure may be used to design or specify a helical compression spring to meet a specific force versus deflection requirement.

Step 1: Write down all given information or requirements pertaining to the spring.

Step 2: Compute the force differential required between two specified forces:

$$\Delta F = F_1 - F_2 \tag{4}$$

Step 3: Determine the distance the spring must deflect under the change in load:

$$\Delta X = X_1 - X_2 \tag{5}$$

Step 4: From Steps 2 and 3 above, compute the spring rate required:

$$R = \frac{\Delta F}{\Delta X} \tag{6}$$

Step 5: From given specifications or arbitrarily, choose a spring material and determine its torsion modulus and maximum stress rating.

Step 6: By careful consideration of allowable geometries and available wire diameters, choose a wire diameter and mean coil diameter and determine the number of active coils from

$$N = \frac{Gd^4}{8D^3R} \tag{7}$$

and the total number of coils, N plus 2.

Step 7: Determine the solid height from

$$L_s = d(N+2) \tag{8}$$

and compare it to the available space for installation. If L_s exceeds the space available, return to Step 5 or 6 and size a new spring. Several tries may be necessary to meet all requirements.

Step 8: Determine the stress on the spring under maximum load conditions from

$$S = \frac{8}{\pi}\frac{F_m D}{d^3} \tag{9}$$

and compare it to the maximum rated stress for the material chosen. If computed stress is high, determine the Wahl factor from Figure 6-2 and compute

$$S^1 = KS \tag{10}$$

If the stress level is too high, return to Step 5 or 6 and size a new spring.

Step 9: From the installed length and force determine the free length of the spring:

$$L_f = L_i + \frac{F_1}{R} \tag{11}$$

Step 10: Determine the ratio L_f/D. If it is greater than 4, consideration should be given to a new spring to reduce the tendency to buckle.

Step 11: Compute the spring index D/d. If it is not in the range of 4 to 12, consider redesigning the spring.

Step 12: If the spring is to work in a cylindrical bore, consider any effects due to a slight increase in mean diameter with a compression. The increase is given approximately by

$$\Delta D = \frac{0.05(p^2 - d^2)}{D} \qquad (12)$$

Step 13: Specify the end configuration, direction of coiling, if relevant, surface treatment, if any, and any other necessary information for manufacture.

EXAMPLE: A spring for a pressure switch application is required to meet the following specifications:

Installed length	1.0 in.
Preload	10.0 lb
Deflection	0.20 in.
Max. load	16.0 lb
Bore diameter	0.75 in.
Material	17-7 PH Cres

Design a suitable spring.

Step 1: Also from Table 6-2:

$$G = 11 \times 10^6, \; S_{max} = 140{,}000 \text{ psi}$$

Step 2: $\Delta F = F_2 - F_1$

$$= 16 - 10$$

$$\Delta F = 6 \text{ lb}$$

Step 3: $X = 0.20$ in. (given information)

Step 4: $\Delta R = \dfrac{\Delta F}{\Delta X}$

$$= \frac{6}{.20}$$

$$= 30 \text{ lb/in.}$$

Step 5: 17-7 PH Cres specified above.

Step 6: Consider that the outside diameter $(D+d)$ of the spring must be less than 0.75 to work in the bore. From Table 6-2 choose $d = 0.072$ and $D = 0.65$. Then:

$$N = \frac{Gd^4}{8D^3R}$$

$$= \frac{(11 \times 10^6)\,(.072)^4}{8(.65)^3\,(30)}$$

$$N = 4.49 \text{ active coils}$$

Step 7: $L_s = d(N+2)$

$$= .072\,(4.49 + 2)$$

$$L_s = 0.47 \text{ in.}$$

Since the installed length can be 1.0 in., the spring will fit in the switch.

Step 8: $S = \dfrac{8}{\pi}\dfrac{F_m D}{d^3}$

$$= \frac{8}{\pi}\frac{(16)\,(.65)}{(.072)^3}$$

$$S = 70{,}954 \text{ psi}$$

Since this is well under the maximum allowable stress of 140,000 psi, the Wahl factor is unimportant, and the spring can be expected to have a very long life under normal operating conditions.

Step 9: $L_f = L_i + \dfrac{F_1}{R}$

$$= 1.0 + \frac{10}{30}$$

$$L_f = 1.33 \text{ in.}$$

Step 10: $\dfrac{L_f}{D} = \dfrac{1.33}{.65}$

$$= 2.05$$

This value is well under 4, so the spring should have no tendency to buckle.

Step 11: $$\frac{D}{d} = \frac{.65}{.072}$$

$$= 9.03$$

Since this value is between 4 and 12, the spring should be stable and easy to manufacture.

Step 12: $$\Delta D = \frac{0.05\,(p^2 - d^2)}{D}$$

To find the pitch:

$$p = \frac{L_f}{N+2}$$

$$= \frac{1.33}{6.49}$$

$$= 0.20$$

Thus: $$\Delta D = \frac{0.05\,(0.20^2 - .072^2)}{.65}$$

$$= .0027$$

Thus the diameter will not increase enough to cause binding in the 0.75 bore.

Step 13: Square and grind ends parallel at 1.0 in. Since 17-7 PH is being used, it should be precipitation-hardened after coiling and then passivated.

BELLEVILLE SPRING

A Belleville spring has several features of possible use in pressure switch design. These features stem from the fact that its load versus deflection characteristics depend upon the geometrical ratio of cone height to thickness, as shown in Figures 6-3 and 6-4. There are three major types of behavior shown. Curve 1, for a small value of h/t (h/t approximately 0.4), shows a linear increase of load with deflection much like that of a helical compression spring. Thus for cases of small deflection and/or heavy loads a Belleville may be used as an alternate to a high-rate helical compression spring.

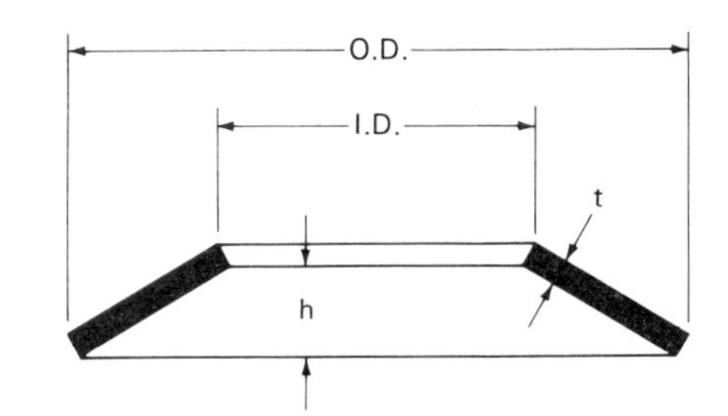

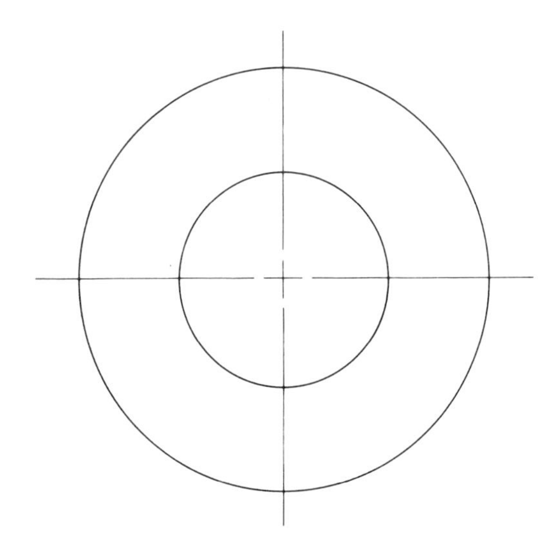

Fig. 6-3. Belleville spring.

Curves 2 and 3 of Figure 6-4 are of the most interest in pressure switch design. For a ratio of h/t in the neighborhood of 1.5, it is seen that the load increases with deflection to a point (1) and then remains approximately constant as the deflection δ increases from about $0.8h$ to about $1.3h$. This is characteristic of a constant load of zero rate spring. As the load is reduced, the spring will snap back over center as shown. Such action is potentially useful in pressure switch actuation if the spring is designed so that the constant load condition occurs at the trip point. For example, such motion could be used to bring direct action contacts together suddenly as the required load is reached.

Curve 3 shows an increase in load with deflection until point A is reached, whereupon

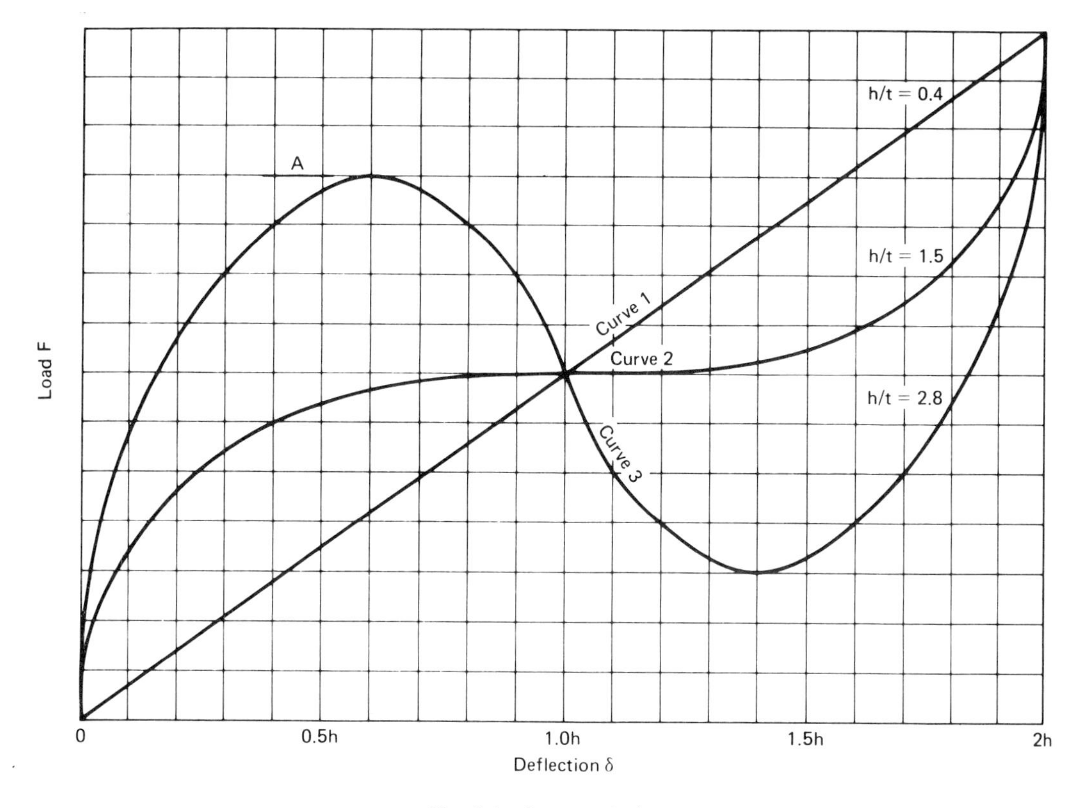

Fig. 6-4. Over centering

the load decreases as deflection increases. This is the action of a negative rate spring. The spring has gone over center and will remain there until the direction of the applied force is reversed. This type of spring action is also useful in pressure switch actuation if a return force is provided, as, for example, by a helical spring.

We shall deal primarily with the zero rate and negative rate springs in this section, as the linear Belleville is analogous to the helical compression spring treated earlier.

Basic Equations

The design of a Belleville spring is fairly complicated compared to a helical compression spring. The following treatment is a summary method suitable for applications where extreme accuracy is not required. For further information, consult the references at the end of the chapter.

Belleville springs may be sized from the following equations:

$$F = C(h-\delta)(h-\delta/2)t + t^3 \tag{13}$$

$$S_c = C[A_1(h-\delta/2) + A_2 t] \tag{14}$$

$$S_1 = C[A_1(h-\delta/2) - A_2 t] \tag{15}$$

$$S_2 = CM[B_1(h-\delta/2) + B_2 t] \tag{16}$$

Where C, A_1, A_2, B_1, B_2, and M are defined by:

$$C = \frac{E\delta}{(1-\mu^2)MA^2} \tag{17}$$

$$A_1 = \frac{6}{\pi \ln r}\left[\frac{r-1}{\ln r} - 1\right] \tag{18}$$

$$A_2 = \frac{6}{\pi \ln r}\left(\frac{r-1}{2}\right) \tag{19}$$

$$B_1 = \frac{r\ln r-(r-1)}{\ln r}\ \frac{r}{(r-1)^2} \tag{20}$$

$$B_2 = \frac{r}{2(r-1)} \tag{21}$$

$$M = \frac{6}{\pi \ln r} \frac{(r-1)^2}{r^2} \tag{22}$$

Equation (13) gives the load as a function of deflection analogous to the spring rate equation (6) for compression springs. Equations (14 through (16) are stress equations for compressive and tensile stress. The stress may change from compressive (equation 14) to tensile (equation 15) if the deflection δ is large and the term $h - \delta/2$ becomes negative. The compressive stress is numerically the larger and may be used above as the stress criterion unless the tensile stress is used for some other reason.

The constants in equations (17) through (22) can be calculated in terms of material and geometrical parameters; or when extreme accuracy is not required, they may be determined from Figures 6-5 and 6-6.

Figure 6-7 is a series of plots of load versus deflection for various ratios of height to thickness. It is seen that for ratios of h/t greater than approximately 2.8 the curve peaks and the load drops for further deflection. This is the overcentering or snap action discussed earlier. Ratios under 0.4 will give a fairly linear response, which is useful for cases where heavy loads and small deflection are required.

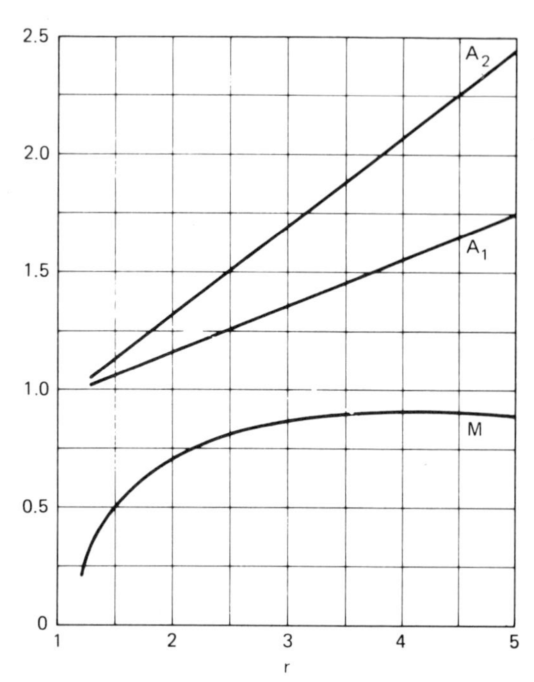

Fig. 6-5. Belleville spring constants.

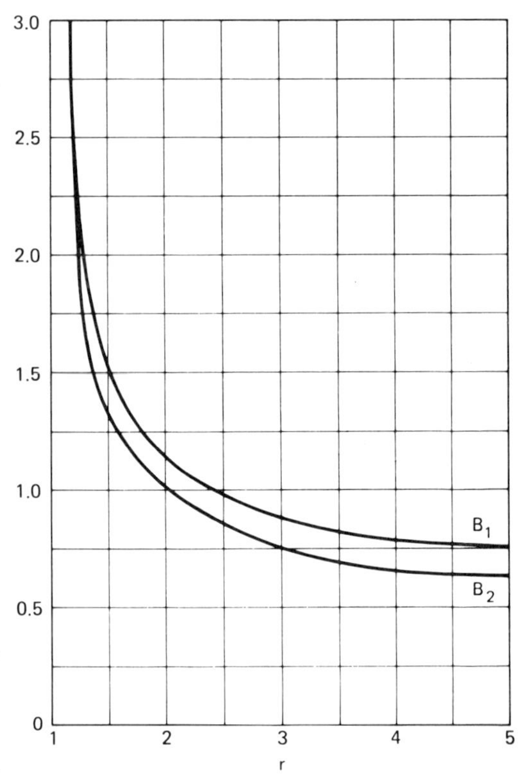

Fig. 6-6. Belleville spring constants.

Design Procedure

The design of a Belleville spring often involves a fair amount of trial and error, and careful design and consideration of essential requirements is needed. The following procedure may be used to design a zero rate Belleville spring.

Step 1: Write down all requirements the spring must meet together with properties of a suitable spring material.

Step 2: From requirements of Step 1 or arbitrarily, choose values of r, E, μ, and h/t (near 1.5).

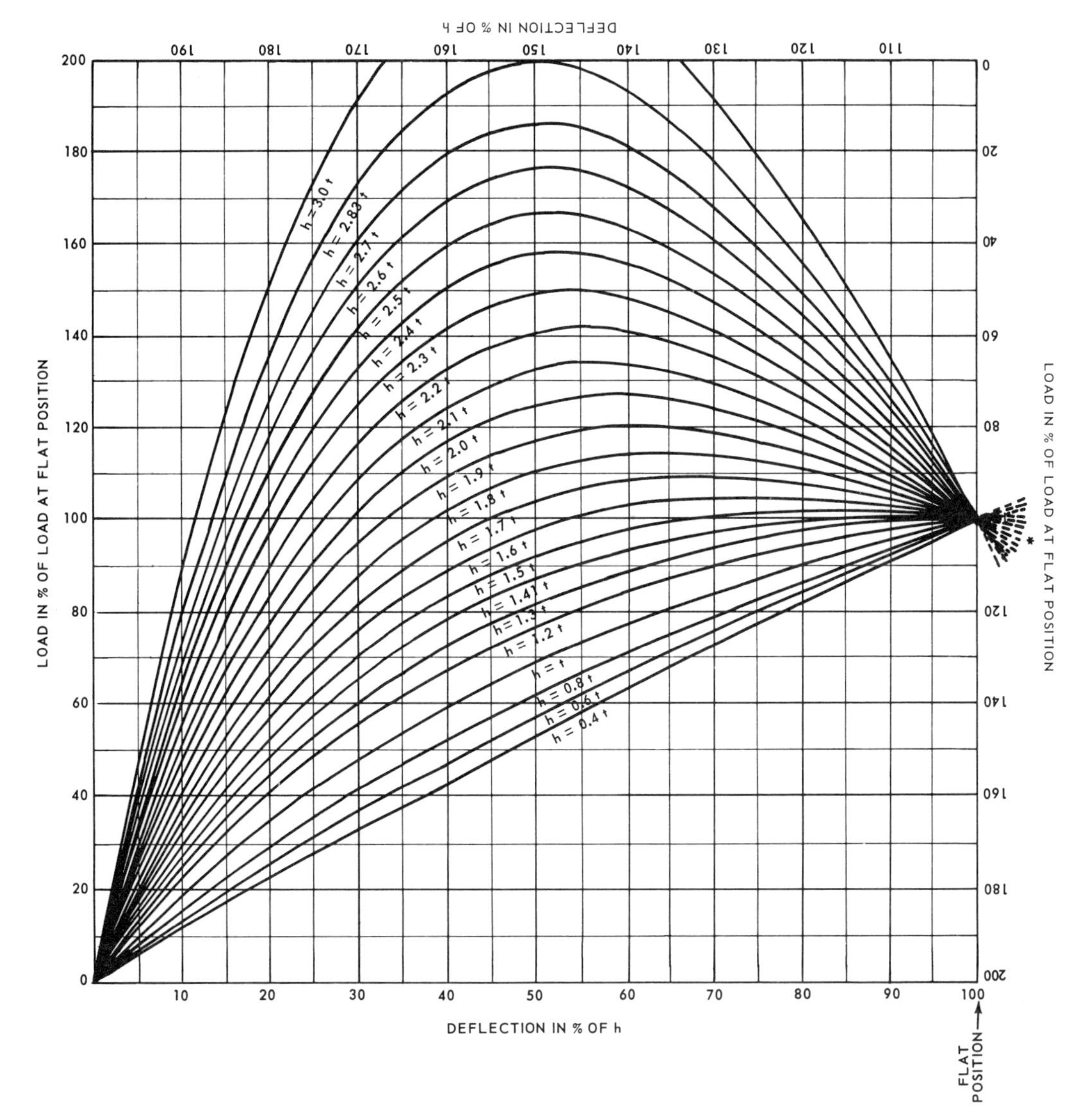

*Dotted lines indicate that the curves are symmetrical beyond the flat position.

Fig. 6-7. Load deflection characteristics.

Step 3: Compute M from equation (22), or read it graphically from Figure 6-5.

Step 4: Using $h = 1.5t$ (or a value chosen in Step 2) and $\delta = h$ in equation (13), set F equal to the required load at the flat point ($\delta = h$), and solve for the thickness t. If the value obtained is unfeasible for any reason, return to Step 2 and repeat the procedure until a satisfactory value is found. Try to keep r near 2.0 if possible.

Step 5: Compute A_1, A_2, and C as a function of h from equations (17) through (19), or read it graphically from Figures 6-5 and 6-6. If the stress S_2 is required, compute B_1 and B_2.

Step 6: Using the last value of t obtained, determine h.

Step 7: Compute the compressive stress from equation (14), and compare it to the allowable values in Table 6-3 for the chosen material.

If necessary, return to Step 2 and size a new spring. If the problem requires, compute stresses S_1 and S_2.

Step 8: From the last value of r obtained and any requirement of Step 1, determine appropriate outside and inside diameters. Using the last value of t obtained, compute h. This completes the spring sizing.

For a negative rate spring, use the above procedure substituting $h = 3t$ (approximately) for $h = 1.5t$. This problem will be considered in the next chapter.

EXAMPLE: A pressure switch using a Belleville spring has a load of 45 lb at the actuation point. The outside diameter is 1.0 in. and the material is 302 cres. Design a suitable spring of zero rate.

Step 1:

$$F = 45 \text{ lb}$$

$$D_1 = 1.0 \text{ in.}$$

$$\mu = 0.30$$

$$E = 28 \times 10^6 \text{ psi}$$

$$S_c = 250{,}000 \text{ psi max.}$$

Step 2: Let $r = 2$, $h = 1.5t$.

Step 3:

$$M = \frac{6}{\pi \ln r}\,\frac{(r-1)^2}{r^2}$$

$$= \frac{6}{\pi \ln 2}\,\frac{(2-1)^2}{2^2}$$

$$M = 0.689$$

Step 4:

$$F = \frac{Eht^3}{(1-\mu^2)MA^2}$$

$$= \frac{1.5Et^4}{(1-\mu^2)MA^2}$$

$$45 = \frac{1.5(28 \times 10^6)t^4}{(1-.30^2)\,(.689)\,(.5)^2}$$

$$t = 0.020 \text{ in.}$$

Step 5: a)

$$A_1 = \frac{6}{\pi \ln r}\left[\frac{r-1}{\ln r} - 1\right]$$

$$= \frac{6}{\pi \ln 2}\left[\frac{2-1}{\ln 2} - 1\right]$$

$$A_1 = 1.22$$

b)

$$A_2 = \frac{6}{\pi \ln r}\left(\frac{r-1}{2}\right)$$

$$= \frac{6}{\pi \ln 2}\left(\frac{2-1}{2}\right)$$

$$A_2 = 1.38$$

c)

$$B_1 = \frac{r\ln r - (r-1)}{\ln r}\,\frac{r}{(r-1)^2}$$

$$= \frac{2\ln 2 - (2-1)}{\ln 2}\,\frac{2}{(2-1)^2}$$

$$B_1 = 1.11$$

d)

$$B_2 = \frac{r}{2(r-1)} \quad \text{(optional)}$$

$$= \frac{2}{2(2-1)}$$

$$B_2 = 1.0$$

e)

$$C = \frac{Eh}{(1-\mu^2)MA^2}$$

$$= \frac{(28 \times 10^6)h}{(1-.30^2)(.689)(.5)^2}$$

$$C = (17.86 \times 10^7)\,h$$

Step 6:

$$h = 1.5\,t$$

$$= 1.5\,(.020)$$

$$h = 0.30 \text{ in.}$$

Step 7: $S_c = C[A_1(h - h/2) + A_2 t]$

$$= (17.86 \times 10^7)(.030)[1.22(.030/2) + 1.38(.020)]$$

$$S_c = 245{,}932 \text{ psi}$$

Examination of constants B_1 and B_2 and equation (16) shows that this will be the greatest stress and well under S_c max.

Step 8: O.D. = 1.0 h = 0.030
I.D. = 0.5 t = 0.020

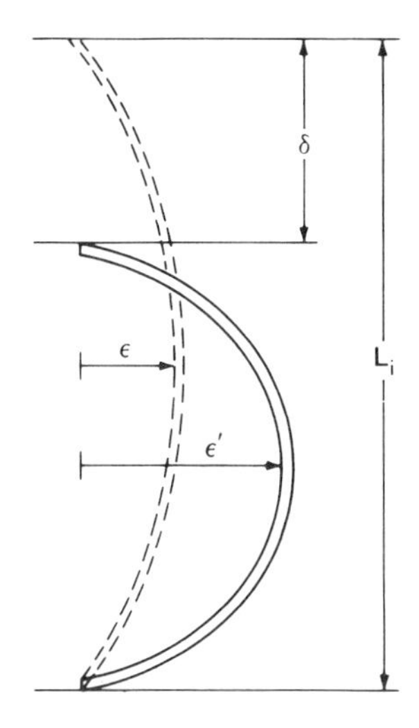

Fig. 6-8. Buckling column spring.

BUCKLING COLUMN SPRING

A third type of loading spring that is of some use in pressure switches and related devices is the buckling column spring. Two variations of this spring are of interest. The buckling column spring shown in Figure 6-8 is a flat metal spring loaded longitudinally. Increasing force causes the spring to deflect laterally as shown. The buckling coil spring shown in Figure 6-9 is a variation of the helical compression spring that is intentionally made to buckle. The spring also may serve as an electrical conductor. By causing the spring to buckle and make contact at a predetermined force level, the device may act as a pressure switch.

The strong point of these springs seems to be the potential for a great savings in size and weight, which is of extreme interest to the aerospace industry. Much work, however, remains to be done with this concept. Their principal drawback is the difficulty of predicting the actual buckling point of the spring. The following procedure, adopted from Reference 1, gives the theoretical load necessary to buckle a helical spring. However, actual results depend strongly on how the ends of the spring are fixed.

The critical load necessary to cause buckling is given by

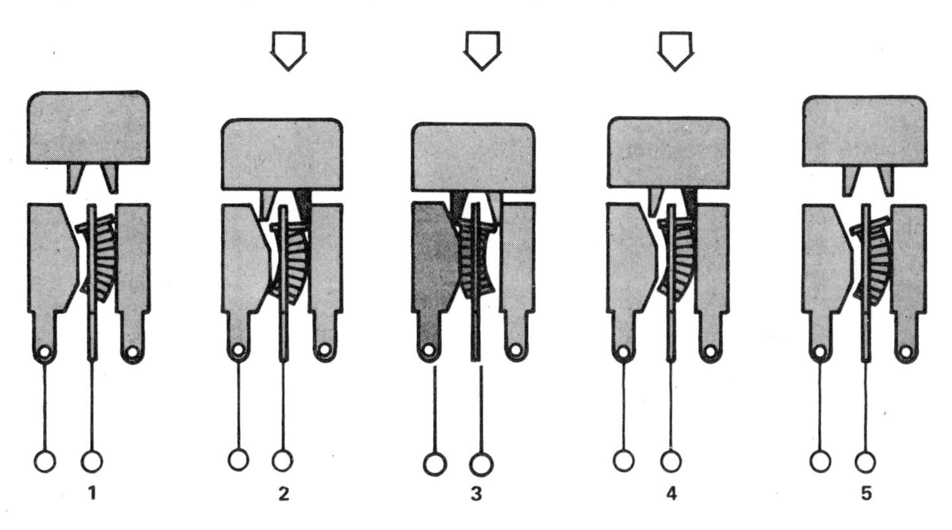

Fig. 6-9. Buckling coil spring. *(Courtesy of Illuminated Products, Inc., Anaheim, California.)*

$$F_{cr} = C_B \alpha_0$$

where

$$\alpha_0 = \frac{Gd^4L_f}{8D^3N}$$

and under the assumption that $E = 2.6\ G$; so that:

$$C_B = 0.812 \left(1 \pm \sqrt{1 - 6.87 \frac{D}{L_f}}\right)^2$$

From the rate equation we have then:

$$\delta_{cr} = C_B L_f$$

as the critical deflection. These equations lead to case 2 of Figure 6-10. Hence, to determine buckling parameters we may, as an alternate approach, find F_{cr} from Figure 6-10 and compute

$$F_{cr} = R\delta_{cr}$$

This case corresponds to one end fixed and one end guided and also to the case of both ends hinged. In the case of a spring with both ends fixed (case 1), the value of L_f/D in equation (11) must be doubled; and in case of one end fixed and the other end free, the ratio L_f/D must be halved.

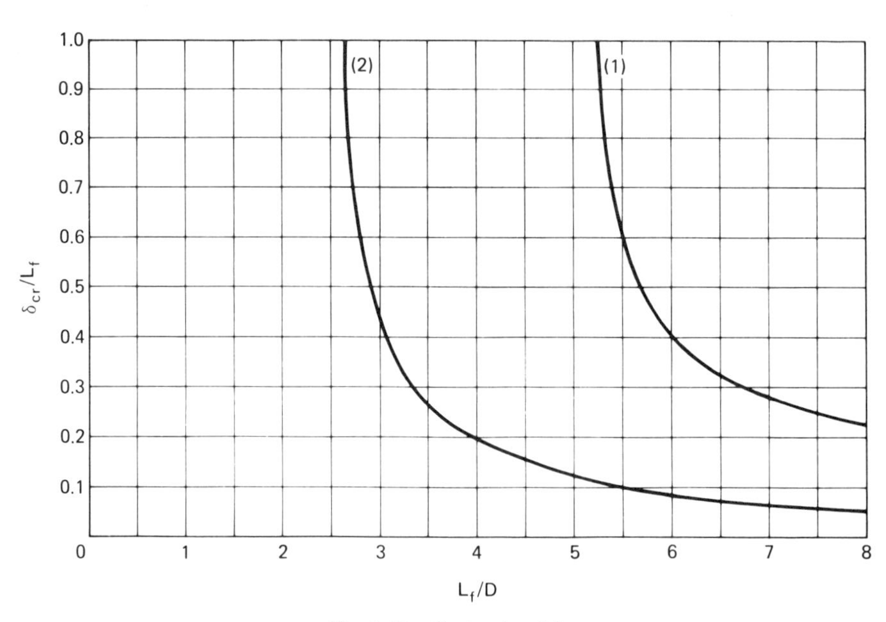

Fig. 6-10. Spring buckling.

REFERENCES

1. Wahl, A. M., *Mechanical Springs*, 2nd Ed., McGraw-Hill Book Co., New York, 1963.
2. Spring Manufacturers Institute, *Standards for Compression, Extension, Torsion, and Flat Springs*, Bristol, Connecticut, 1967.
3. Associated Spring Corporation, *Design Handbook —Springs and Custom Metal Parts*, Bristol, Connecticut, 1967.
4. Rockwell International, *Mechanical Spring Design Guide*, Clawson, Michigan, 1973.
5. U.S. Government Printing Office, *Military Standard 29A*, Washington, D.C.
6. Associated Spring Corporation, *Solving Spring Design Problems with Belleville Spring Washers*, Bristol, Connecticut, 1966.
7. Lyons, J. L. and Askland, C. L., *Lyons' Encylopedia of Valves*, Van Nostrand Reinhold Co., New York, 1975.

VII Piston Actuated Pressure Switches

In this chapter we consider some design aspects of piston actuated pressure switches. We are primarily concerned with piston actuators used in conjunction with helical compression loading springs. A few other types are briefly surveyed at the end of the chapter.

Often a pressure switch is required to have a specified differential (deadband) between trip and untrip points, although sometimes only a trip point is specified. We shall consider the first type in this chapter, as its design presents the greater challenge, and because the procedures outlined will also apply to the case where only one point is specified.

As pointed out in earlier chapters, this type of pressure switch is normally used only for sensing fairly high gage pressures.

In the approach to be presented here we introduce two dimensionless parameters η and γ, related to the compression force on the installed spring with no pressure on the piston, and to the breakout friction, respectively. This enables us to determine a pressure less than the untrip point at which the piston will break out and move up to the trip point, thereby eliminating the breakout friction from the calculation of spring rate.

DESIGN FACTORS

The set point(s) and differential of the pressure switch will depend primarily on three factors —the parameters of the snap action electrical switch to be used, seal friction, and parameters of the loading spring.

Electrical Switch

Normally the electric switch is chosen to satisfy required electrical ratings, anticipated environmental factors, and similar criteria. Once a suitable switch has been chosen, the following parameters should be determined:

Operating force
Release force
Pretravel
Differential travel
Overtravel

Usually they can be obtained from manufacturers' literature. These values are usually needed in sizing the load spring to give the required differential.

In some cases, direct acting contacts may be desired. In this instance movement of the contacts will be determined by properties of the loading spring. Usually in this case a negative rate spring, such as a Belleville spring, is recommended.

Seal Friction

In any real device with moving parts there will be friction forces acting to retard the movement. In a piston type pressure switch the piston seal must be slightly compressed in order to effect a seal between piston and cylinder. Hence, there will be seal friction as the piston moves in the bore. Actually, there are two cases of friction to consider, breakout friction and running friction.

Breakout friction acts to prevent the movement from starting and is usually considerably larger than the running friction, which exists after movement has started. However, our approach to the problem will be such that under stated conditions the breakout friction can be eliminated from the spring calculations. This is done by allowing the piston to break out and begin moving before the untrip pressure is reached. Thus, only running friction (assumed constant) acts and need be considered when sizing the spring. Conditions for this approach will be given in a later paragraph.

Seal friction will depend on several factors, the most important of which is the surface finish of the cylinder wall. Generally, this surface should have at least a 16-microinch finish. For low pressure operation it should be even smoother, if possible. Other factors to consider are the type of seal, squeeze of the seal, frequency of cycling, and any unusual environmental condition such as temperature extremes.

Most seals used in pressure switches are either standard "O"-rings, Teflon rings, or similar materials. These seals are normally placed in standard size grooves in the piston which will allow proper clearance and seal squeeze in the cylinder. Manufacturers' literature should be consulted for seal and groove dimensions. As a rough rule of thumb, a dynamic seal should be squeezed approximately 15% of its cross-sectional diameter for use under average conditions. For very high pressures, more squeeze is needed.

To estimate the amount of friction to be expected for Teflon seals the curves in Figure 7-1 may be used. They present average values taken from several manufacturers' catalogs and are intended only as general guidelines. The curves are based on 16-microinch surface finish, room temperature, and manufacturers' recommended installation.

Figures 7-2 through 7-4 give some friction values for some typical elastomeric O-rings. Sixteen-microinch surfaces, room temperature, and standard groove sizes are assumed. Again for more exact values, seal manufacturers' literature should be consulted.

In cases of extreme environmental conditions or other unusual situations it is usually necessary to determine the friction experimentally.

Since the friction varies with seal diameter, the first step in the pressure switch design (after defining the problem) is to choose a diameter and thus determine breakout and running friction. Then to eliminate breakout friction from the spring sizing compute

$$\gamma = \frac{f_b}{F_2}$$

and require $\gamma < 1$. If this condition is not met, a new seal and/or diameter must be chosen or f_b reduced in some manner, such as by lubrication.

In summary, accurate knowledge of seal friction is essential to the design of this type of pressure switch.

Loading Spring

The central problem in the design of most piston type pressure switches is the proper sizing of a compression spring. This spring will determine the trip and untrip points and hence the differential.

Helical Compression Spring

The following procedure will size a compression spring that begins compressing sufficiently before the untrip point that breakout friction may be neglected. Refer to Chapter 6 for more information on spring design and for properties of spring materials.

Step 1: Determine the forces on the piston at the trip and untrip points due to pressure from

$$F = P \cdot A$$

and compute the difference

$$\Delta F = F_1 - F_2$$

This difference must be (and nearly always will be) much larger than the difference

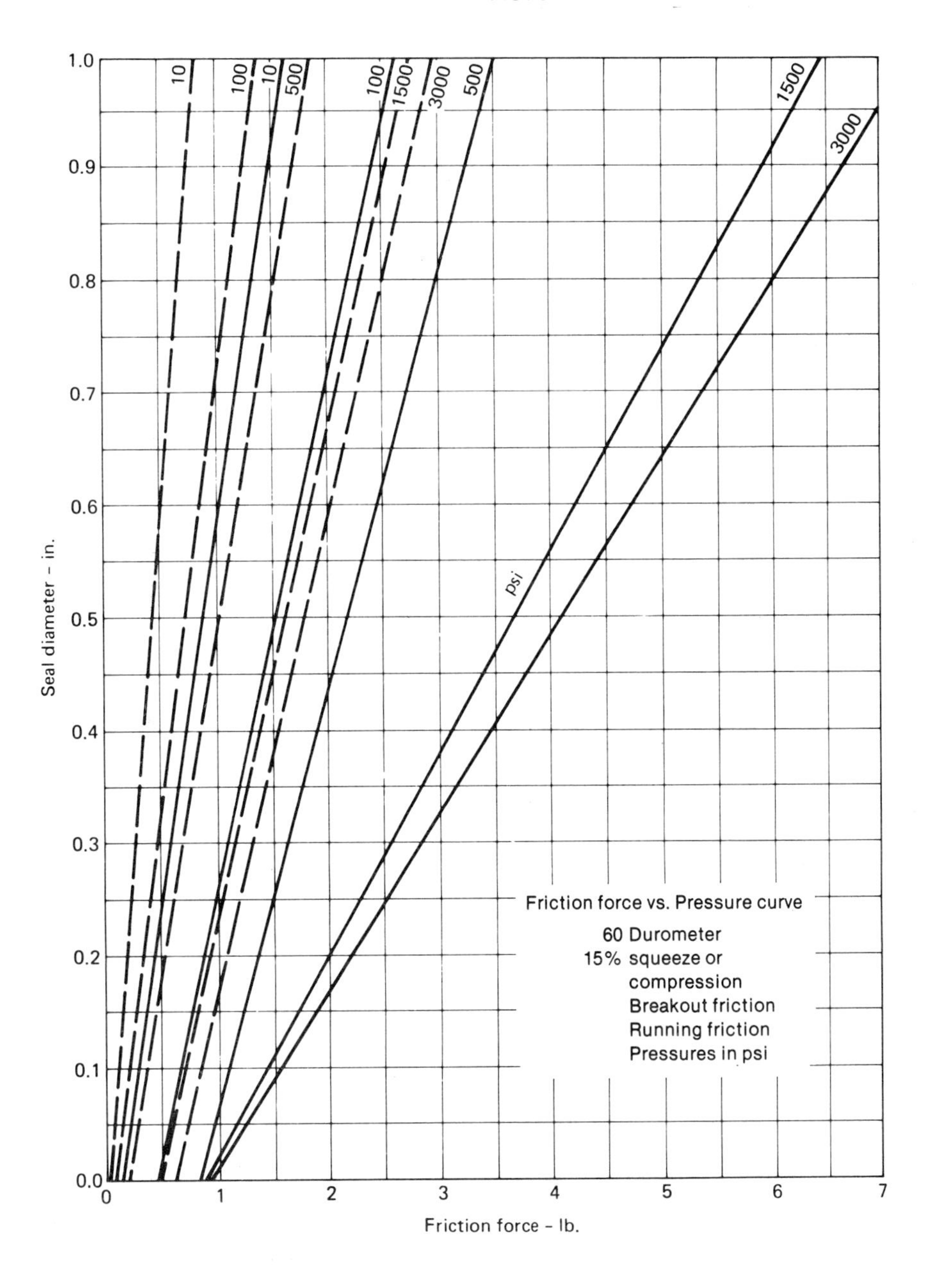

Fig. 7-1. Friction force vs. pressure curve.

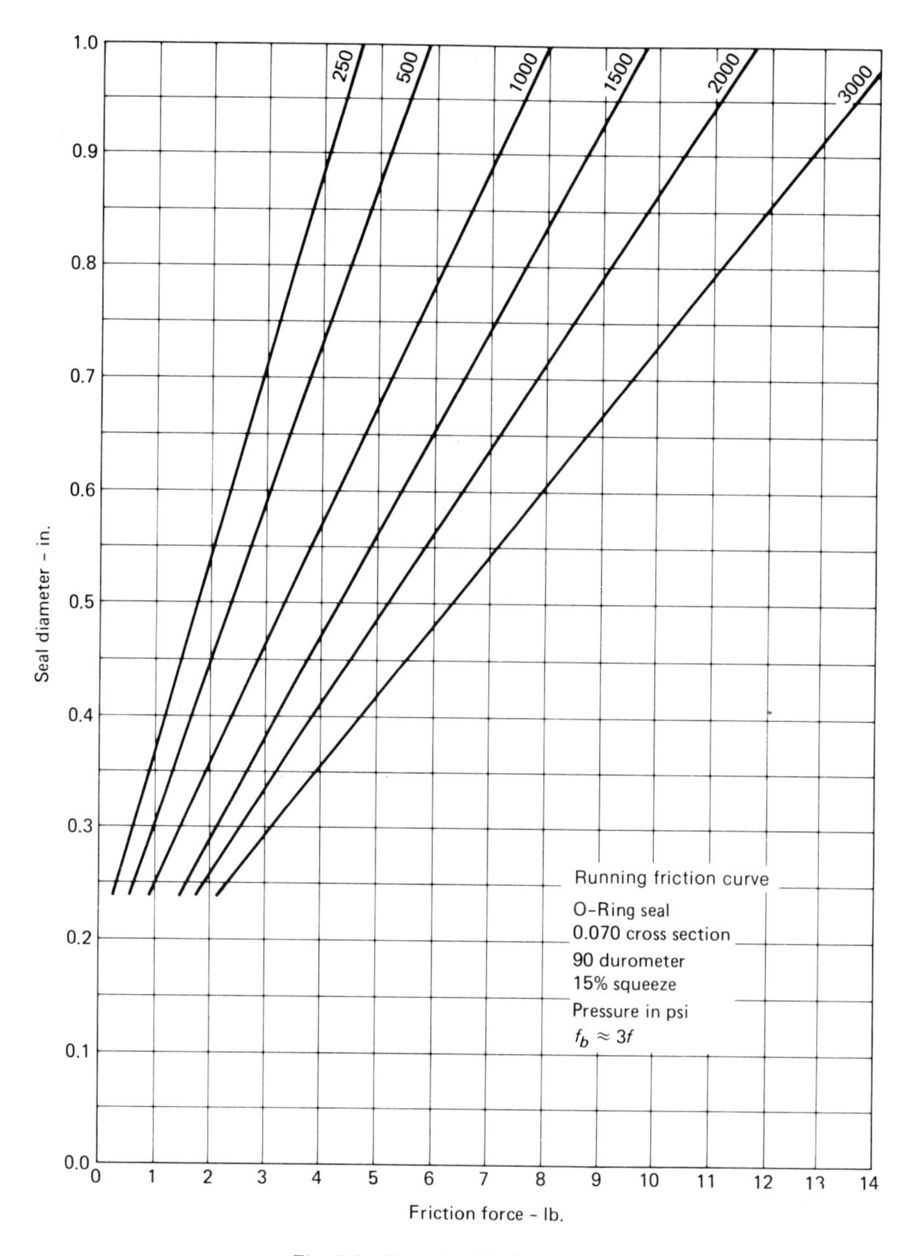

Fig. 7-2. Running friction curve.

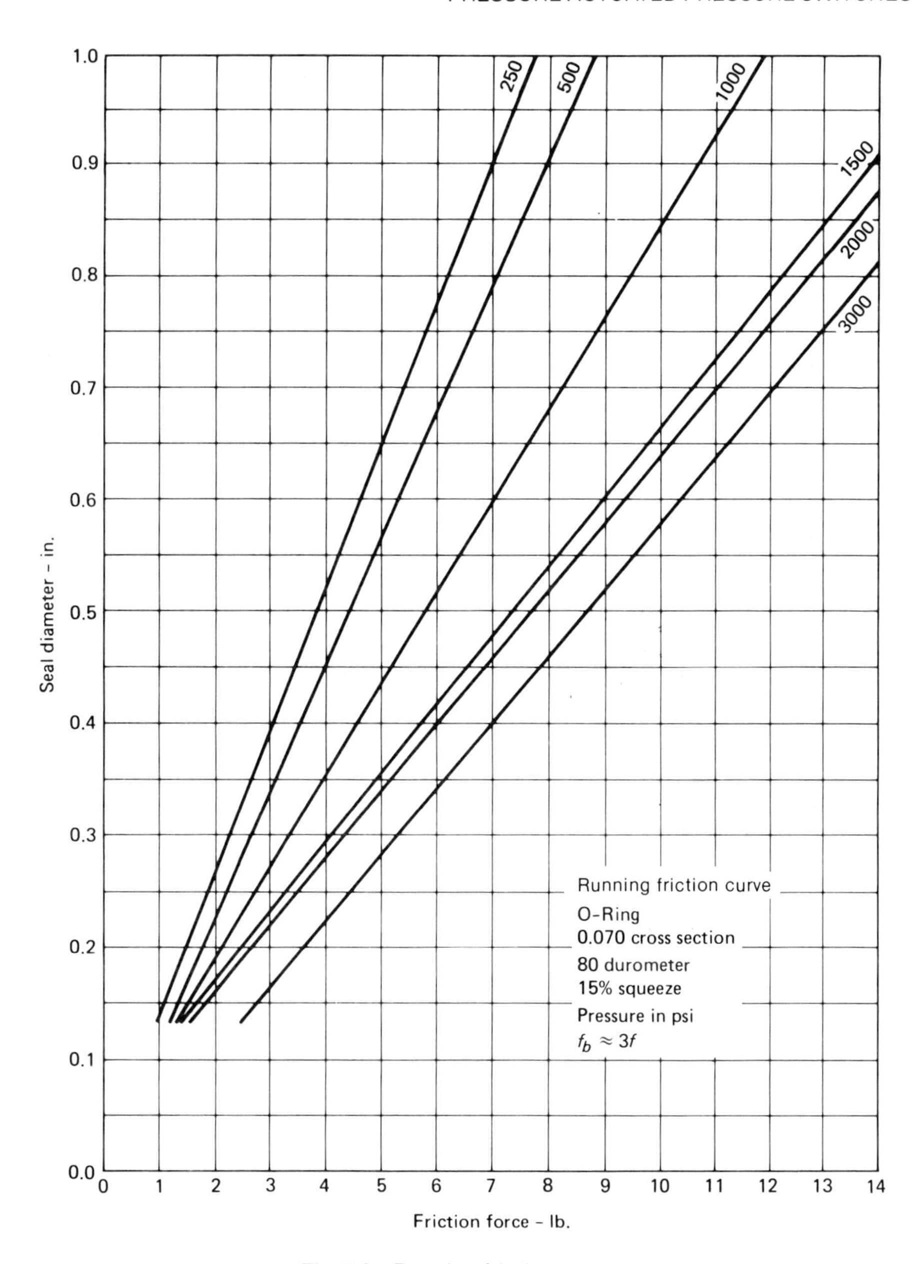

Fig. 7-3. Running friction curve.

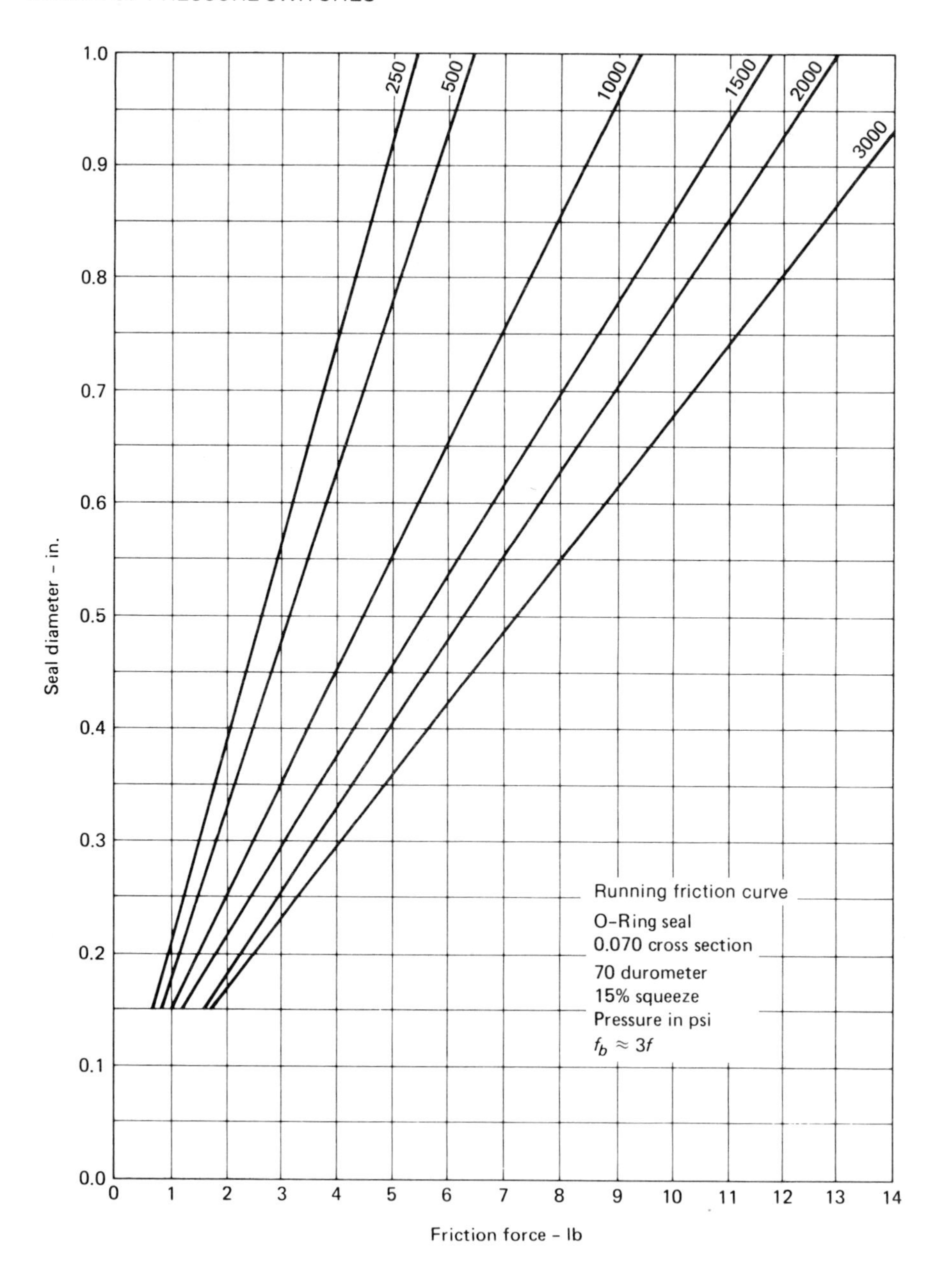

Fig. 7-4. Running friction curve.

between the operating force and release force of the snap action electrical switch chosen. If necessary, increase the diameter and hence piston area to ensure this.

Step 2: From the preceding sections the seal friction and electrical switch force differential were determined. Thus, referring to Figure 7-5, the force differential of the spring is

$$\Delta F_s = \Delta F - 2f - \Delta F_{sw}$$

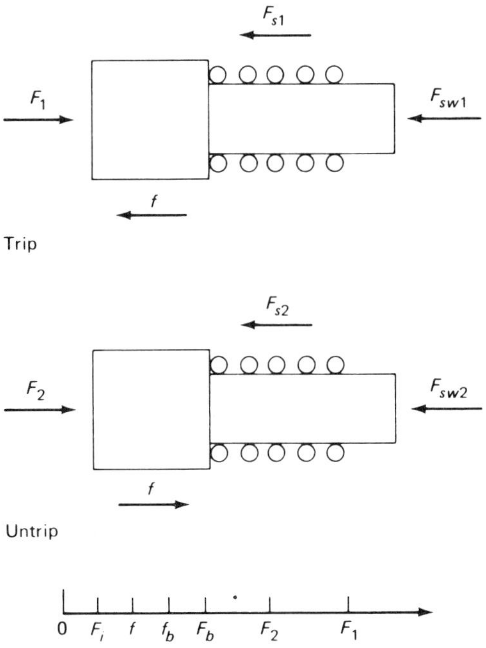

Fig. 7-5. Piston switch forces.

Step 3: Determine the distance the piston must travel between trip and untrip points. For a snap action switch this is the differential travel δ_{DT}. For direct acting contacts the travel is fairly arbitrary consistent with electrical ratings, geometric considerations, etc. As a general guide allow .010 to .030 travel, if possible.

Step 4: Determine the required spring rate from

$$R = \frac{\Delta F_s}{\delta_{DT}}$$

$$= \frac{\Delta F = 2f = \Delta F_{sw}}{\delta_{DT}}$$

Step 5: The physical parameters of the spring may be found from

$$R = \frac{Gd^4}{8D^3N}$$

Step 6: Determine the maximum stress on the spring from

$$S = \frac{2.55 F_L D K}{d^3}$$

The Wahl factor K may be found from Figure 6-2, in Chapter 6. If the computed stress exceeds the maximum allowable for the chosen spring material, either choose a stronger material or change the coil and wire diameters. Several trials may be needed to meet all requirements.

Step 7: From envelope restrictions or arbitrarily, choose an installation length L_i.

Step 8: Choose an installation force F_i and a parameter η such that $F_i = \eta F_2$ and $\eta + \gamma < 1$ so that the force at breakout will be

$$F_b = (\eta + \gamma) F_2$$

Step 9: Determine the solid height of the spring from

$$L_s = (N + 2)d$$

The solid height must satisfy

$$L_s < L_i - \frac{F_1 - \eta F_2}{R}$$

to ensure sufficient movement to reach the trip point.

Step 10: Determine the free length of the spring from

$$L_f = L_i + \frac{\eta F_2}{R}$$

and compute the ratio L_f/D. If this ratio is greater than approximately 4, the spring may be subject to buckling and should be resized.

Step 11: If the spring is installed in a fixed-diameter bore, allow for increase in the mean coil diameter when the spring compresses. The increase is given approximately by

$$\Delta D = \frac{0.05(p^2 - d^2)}{D}$$

This completes the design of the loading spring for the piston actuator.

Other Considerations

It is generally considered poor practice to let a spring compress solid under load. To prevent this it is best to incorporate a mechanical stop into the design. Also, the use of a stop will give overtravel protection to the electrical switch. The amount of piston travel past the upper set point should not exceed the overtravel rating of the electrical switch or let the spring compress solid. The maximum force on the spring is then

$$F_L = F_1 + R\delta_{OT}$$

which should be used in the calculation of spring stress. The piston movement should stop after moving a maximum distance

$$\delta = \delta_{OT} + \frac{F_1 - F_b}{R}$$

Since it is quite difficult in practice to produce all parts to nominal dimensions, it is usually necessary to allow for tolerance buildup. It is generally easiest to provide for any needed adjustment of spring setting by allowing for adjustment of the installed length L_i and thereby the installed load.

Also, it is usually necessary to be able to adjust the amount of pretravel δ_A of the actuator. Thus, the trip point can be set experimentally very near the nominal. The setting of the actuator (Figure 7-6) should be approximately

$$\delta_A = \frac{F_1 - F_b}{R} - \delta_{PT}$$

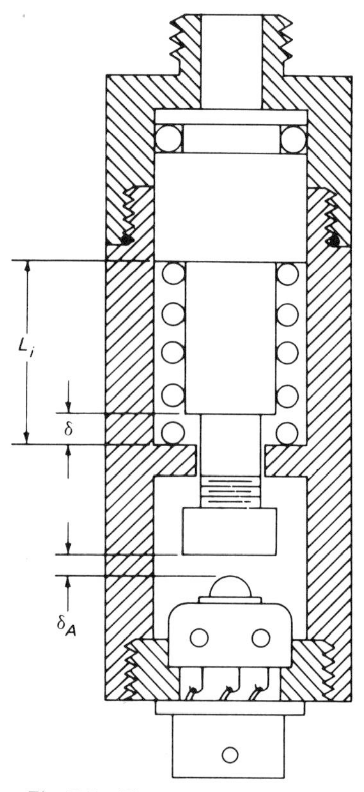

Fig. 7-6. Piston actuated switch.

EXAMPLE: A pressure switch must meet the following requirements:

Trip	142 psi
Untrip	125 psi
Current	3 amperes
Cylinder dia.	0.5 in.

Design a suitable switch.

Possible Solution

a) A suitable electrical switch has the following parameters:

Operating force	0.33 lb
Release force	0.06 lb
Differential travel	0.004 in.

b) For a Teflon seal the breakout and running friction are, from Figure 7-1:

$$f_b = 1.5 \text{ lb}$$

$$f = 0.75 \text{ lb}$$

$$\gamma = \frac{f_b}{F_1}$$

$$= \frac{1.5}{24.5}$$

$$= .061$$

Thus, the forces at the trip and untrip points are:

$$F_1 = P \bullet A$$

$$= 142\,\pi\,\frac{(0.5)^2}{4}$$

$$F_1 = 27.88 \text{ lb}$$

$$F_2 = 125\,\pi\,\frac{(.5)^2}{4}$$

$$F_2 = 24.54 \text{ lb}$$

c) The required spring rate is:

$$R = \frac{(F_1 - F_2) - 2f - \Delta F_{sw}}{\delta_{DT}}$$

$$= \frac{(27.88 - 24.54) - 2(.75) - (.33 - .06)}{.004}$$

$$= 390 \text{ lb/in.}$$

d) Suitable spring parameters satisfying

$$R = \frac{Gd^4}{8D^3N}$$

are:

$$G = 11{,}500{,}000 \text{ psi}$$

$$D = 0.48 \text{ in.}$$

$$d = 0.080 \text{ in.}$$

$$N = 1.31 \text{ active coils}$$

e) The stress, assuming a maximum load of 160 psi (31 lb), is:

$$S = \frac{2.55 F_L DK}{d^3}$$

$$= \frac{2.55(31)(.48)(1.2)}{(.08)^3}$$

$$S = 93{,}000 \text{ psi}$$

Thus a high stress material such as 17-7 PH Cres is needed.

f) The solid height is:

$$L_s = (N+2)\,\text{d}$$

$$= (1.3 + 2\,(.080)$$

$$L_s = 0.26 \text{ in.}$$

g) Choose an installed length of 0.40 in. and choose $\eta = 0.5$. Then:

$$L_s < L_i - \frac{F_1 - \eta F_2}{R}$$

$$.26 < .40 - \frac{27.9 - 12.25}{390}$$

$$.26 < .36$$

Thus the spring will not compress solid before the trip point is reached.

h) The free length is:

$$L_f = L_i + \frac{\eta F_2}{R}$$

$$= .40 + \frac{(.5)(24.5)}{390}$$

$$L_f = .432 \text{ in.}$$

i) Diametral increase is:

$$\Delta D = \frac{0.05(p^2 - d^2)}{D}$$

$$= \frac{0.05(.114^2 - .08^2)}{.48}$$

$$\Delta D = 0.007 \text{ in.}$$

NEGATIVE RATE SPRING—ELEMENTARY ANALYSIS

In this section we shall consider the design of a Belleville spring actuated pressure switch utilizing direct acting contacts. The extension to pressure switches using snap action electrical switches may be accomplished by either matching spring parameters to the electrical switch parameters or by providing for movement of the electrical switch.

The case to be considered here is that of a fixed trip point and a free unspecified untrip point such as with a manual reset switch. The design of a pressure switch with specified differential is quite difficult, and usually a great deal of trial and error is needed.

This approach takes seal friction into account. However, in practice the amount of friction is often very small and could be neglected compared to the pressure load.

Step 1: List all specified parameters, material properties and any other known information.

Step 2: From Step 1 or arbitrarily choose a seal diameter and from the specified trip point compute the force from

$$F_1 = P_1 \cdot A$$

Step 3: For the chosen seal determine the breakout friction and running friction from manufacturers' data, Figures 7-1 through 7-4, or experiment. Compute

$$\gamma = \frac{f_b}{F_1}$$

and require $\gamma << 1$. If necessary, change seals until the condition is met.

Step 4: Compute the force on the Belleville spring at the trip point from

$$F_{b1} = F_1 - f$$

Step 5: Using specified or appropriately chosen values of r, E, and μ compute

$$M = \frac{6}{\pi \ln r} \frac{(r-1)^2}{r^2}$$

Step 6: From Figure 6-7 for a negative rate spring with $h = 2.83t$, the load at $\delta = h$ is:

$$F_b = \frac{F_{b1}}{2}$$

Step 7: From Step 6 data determine the washer thickness t from

$$F_b = \frac{2.83\, Et^4}{(1-\mu^2)MA^2}$$

Step 8: The movement δ of the piston sensor must be such that firm electrical contact is made. To ensure this, the travel must be somewhat less than $2h$, the full travel of the Belleville spring. Let

$$\delta = 2Eh$$

where E is near but not equal to 1.0.

Step 9: Determine the stress constants A_1, A_2, and C from Figure 6-5 or from

$$A_1 = \frac{6}{\pi \ln r}\left[\frac{r-1}{\ln r} - 1\right]$$

$$A_2 = \frac{6}{\pi \ln r}\left(\frac{r-1}{2}\right)$$

$$C = \frac{E\delta}{(1-\mu^2)MA^2}$$

Step 10: Compute the stress in compression S_c from

$$S_c = C[A_1(h-\delta/2)+A_2t]$$

and compare it to the allowed value for the chosen spring material. If the stress is too large, return to Step 1 and size a new spring.

Step 11: Compute the spring parameters h, O.D., and I.D. This completes the Belleville spring calculations.

A sample of this is given in the following example.

EXAMPLE: A pressure switch is required to trip at a pressure of 60 psi. A Belleville spring is used with direct acting contacts. The seal is 1.0-in.-diameter Teflon. Design a suitable switch.

Possible Solution:

Step 1:

$$P = 60 \text{ psi}$$

$$\text{O.D.} = 1.0 \text{ in.}$$

$$E = 30 \times 10^6 \text{ (steel)}$$

$$\mu = .30$$

$$r = 2$$

$$f_b = 2.1 \text{ lb}$$

$$f = 1.2 \text{ lb}$$

$$S_c = 250{,}000 \text{ psi (steel)}$$

Step 2:

$$F_1 = P_1 A$$

$$= 60 \frac{\pi(1.0)^2}{4}$$

$$= 47.1 \text{ lb}$$

Step 3:

$$\delta = \frac{f_b}{F_1}$$

$$= \frac{2.1}{47.1}$$

$$\delta = .04$$

Step 4:

$$F_{b1} = F_1 - f$$

$$= 47.1 - 1.2$$

$$F_{b1} = 45.9 \text{ lb}$$

Step 5:

$$M = \frac{6}{\pi \ln r}\frac{(r-1)^2}{r^2}$$

$$= \frac{6}{\pi \ln^2}\frac{(2-1)^2}{2^2}$$

$$M = .689$$

Step 6:

$$F_b = \frac{F_{b1}}{2}$$

$$= \frac{45.9}{2}$$

$$F_b = 23 \text{ lb}$$

Step 7:

$$F_b = \frac{Eht^3}{(1-\mu^2)MA^2}$$

$$= \frac{2.83Et}{(1-\mu^2)MA^2}$$

$$= \frac{2.83(30\times 10^6)t^4}{(1-.3^2)(.689)(.5)^2}$$

Solving for t:

$$t = .014 \text{ in.}$$

Step 8: Choose $E = .90$. Then:

$$\delta = 2Eh$$

$$= 2(2.83)Et$$

$$= 2(2.83)(.90)(.014)$$

$$\delta = .071$$

Step 9:

$$A_1 = \frac{6}{\pi \ln r}\left[\frac{r-1}{\ln r} - 1\right]$$

$$= \frac{6}{\pi \ln 2}\left[\frac{2-1}{\ln 2} - 1\right]$$

$$A_1 = 1.22$$

$$A_2 = \frac{6}{\pi \ln r}\left(\frac{r-1}{2}\right)$$

$$= \frac{6}{\pi \ln 2}\left(\frac{2-1}{2}\right)$$

$$A_2 = 1.38$$

$$C = \frac{E\delta}{(1-\mu^2)MA^2}$$

$$= \frac{30 \times 10^6(.071)}{(1-.3^2)(.689)(.5)^2}$$

$$C = 1.365 \times 10^7$$

Step 10:

$$S_c = C[A_1(h - \delta/2) + A_2 t]$$

$$= 1.365 \times 10^7 \left[1.22\left(.040 - \frac{.071}{2}\right) + 1.38(.014)\right]$$

$$= 338{,}656 \text{ psi}$$

$$S_c = 596{,}778 \text{ psi at } \delta = h$$

As can be seen, this stress S_c is much too high. The parameters selected will have to be changed until $S_c < 250{,}000$.

VIII Bellows Actuated Pressure Switches

The design of bellows actuated pressure switches is in many ways much more difficult than the design of piston actuated switches as discussed in the previous chapter. Because of the difficulty of predicting the actual spring rate of a bellows, it is much more difficult to control a deadband or set point.

A second drawback is the need to purchase the bellows, as it is generally unfeasible to fabricate one's own. Since available bellows are usually restricted to a given catalog series, the optimum bellows may not be available. Of course, one may resort to ordering a special bellows from a manufacturer, but usually at a substantially higher cost.

For the purposes of this chapter we shall assume—perhaps optimistically—that a bellows meeting our specifications may be obtained. We look first at some of the important factors in the selection of a bellows.

Selection Parameters

The proper bellows for a given pressure switch depends on several factors such as the pressurized media, operating and proof pressures, temperature, and other factors discussed in Chapter 3. These parameters will affect such things as the choice of material, spring rate, geometry, etc.

Materials

The material of which the bellows is to be made will be dependent on many factors. It must first of all be chemically compatible with the pressurized fluid. For corrosive fluids the common choices are stainless steels, nickel and its alloys, and so on. Whenever there is any doubt about compatibility, tests should be run. Be aware also that the destructive effect of the fluid on the bellows may in some instances be dramatically accelerated by increased pressures, temperatures, and flexing of the bellows. The corrosion charts presented in the appendix may be useful as guidelines but should be used with caution.

A second consideration in the choice of bellows material is the temperature to which it will be exposed. For very high temperature operation consideration must be given to the possibility of "heat treating" the bellows, particularly softening or annealing. Normally, however, this will not be a problem with pressure switches. Special care should also be exercised if the switch is expected to see temperature shock, i.e., sudden and large changes in temperature, as this may increase the chance of structural failure. For extreme temperature conditions consider first the use of stainless steel.

As noted in Chapter 2, bellows are normally restricted to fairly low pressures. If the required set point of a pressure switch is much above 2,000 psi, the bellows should be dropped in favor of a piston or bourdon tube type. Of course, special designs can be made to function at much higher pressures, but the cost of such designs is usually prohibitive, and we shall not consider them here. It should also be determined whether the application of the pressure will be of impulse type or will be fairly slow. If the former, a high-impact-strength material such as steel should be chosen. Most of the bellows vendors' catalogs will list the allowable pressures for their bellows. Often these are fairly conservative ratings.

A fourth consideration dependent on the nature of the bellows material is the fatigue life or the expected cycle life of the bellows. In the case of safety switches and others that are seldom if ever actuated, this is of little importance. On the other hand, for process control switches that must actuate often, the fatigue life may be of great economic consequence. The higher the stress the shorter the life expectancy; so attempt to be conservative in determining allowable stress. Where stress is high, consider bellows made of stainless steel and similar alloys. At low stress levels cheaper materials such as copper alloys may be sufficient. Examples of bellows stress calculations will be given in a later section.

MATHEMATICAL DESIGN OF AN ELECTRO-DEPOSITED METAL BELLOWS

HOW TO ESTABLISH A BELLOWS DESIGN

When a bellows is to be selected by catalog or other data from any particular company, you should determine all of your requirements for that bellows as follows:

1. Kind of flexing required of the bellows: Specify extension, or compression, or or bending, or swiveling, or parallel-ends off-set, or torque, or speed of rotation. Provide a drawing or sketch showing related fittings and extremes of flexing where possible. This is very important to the manufacturer to enable him to work out a reliable design.
2. Specify the amount of compression or extension or flexing in fractions of an inch or in degrees, or by dimensions on the flexing diagram (maximums).
3. Specify pressure difference between inside and outside of the bellows, maximum instantaneous, and whether higher pressure is applied inside or outside the bellows.
4. Specify whether rigid stops will limit the extension or compression of the bellows to its rated stroke, or if the bellows will be required to withstand pressure unrestrained. This is very important, since a restrained bellows will give a much better performance.
5. The spring rate in pounds per inch, or conversely the amount of force available to flex the bellows the desired amount, should be specified.
6. Specify required useful life of the bellows expressed as the number of flexing cycles, and define the flexing cycle.
7. Extremes of temperature, both low and high, should be stated.
8. Corrosive conditions that apply should be described.
9. The method to be used to join the bellows to end fittings, such as soldering, brazing, welding, cementing, should be specified.
10. Specify vibration or shock to be experienced by the bellows.
11. Specify types and lengths of ends.

INTRODUCTION TO ELECTRO-DEPOSITED METAL BELLOWS

There are five major types of metal bellows: rolled, hydroformed, welded, chemically deposited and electro-deposited. Electro-deposited bellows are manufactured by forming a mandrel to the shape of the inside of the bellows, depositing the proper thickness of spring-quality metal onto it, trimming the ends, and dissolving the mandrel.

In this section the mathematical design of

a bellows will be restricted to the electrodeposited nickel type. If other types are involved, such as previously mentioned, it is advised that you contact the respective manufacturing groups for their data. For the electrodeposited nickel bellows, the following design criteria may be observed.

Symbols used are defined in Table 8-1.

Table 8-1. Nomenclature

O = bellows outside diameter, in.
I = bellows inside diameter, in.
t = bellows wall thickness (average), in.
N = number of active convolutions in the bellows
E = Young's modulus of elasticity for the bellows metal. Use 23,350,000 for our electrodeposited nickel.
Y = Young's modulus for a cylindrical bellows
S = maximum permissible stroke for the bellows, in.
s = maximum permissible stroke per convolution, in.
n = length of one convolution of the bellows, in.
L = length of the active convolutions portion of the bellows. This is the overall length less ends, less 0, 1/2, or 1 convolution, depending on what convolution faces are immobilized by end fittings joints, in.
P = pressure applied to the bellows, or pressure rating of the bellows, lb/sq in. differential
π = 3.1416
A = angle subtended by a bellows bent in a natural circular arc, degrees
r = spring rate of one convolution, lb/in.
d = specific weight of the bellows metal, lb/cubic in., or 0.321 for nickel
f = deflection of a bellows with application of force, in.
R = spring rate of the whole bellows, lb/in.
w = weight per unit length of the bellows, lb
W = total force acting on the bellows, lb
z = vibration frequency in cycles/sec
a = acceleration in gravities. This equals in./sec/sec divided by 386.
g = the acceleration of Earth's gravity, 386 in./sec/sec

RATING BELLOWS FOR COMBINED STROKE AND PRESSURE

Where the working pressure exceeds 40% of the nominal pressure rating of the bellows, select the permissible stroke (axial) from the pressure/usable stroke chart. For example, assume a bellows rated at 100 psi (from the pressure formula) is to work at 80 psi in service. Enter the chart with 0.8 (for 80%) on the pressure scale and read out 0.67 on the usable stroke scale. Multiply this value by the rated stroke (from the stroke formula), and get the usable stroke for the bellows at 80% working pressure.

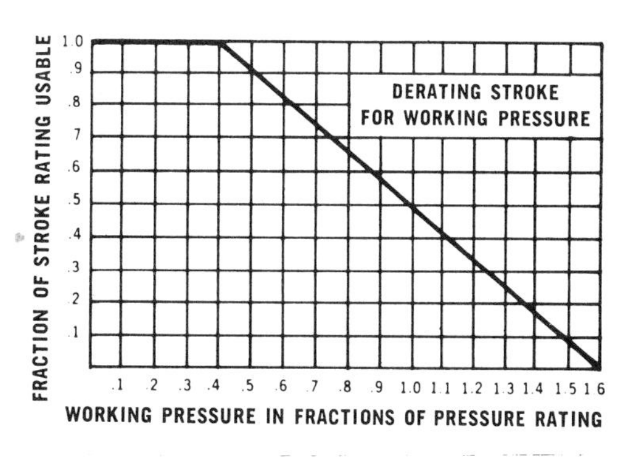

Derated stroke for working pressure graph.

LIFE EXPECTANCY

The life expectancy of a metal bellows is expressed in stroke cycles and not in time or speed of repetition of the cycles.

The data in the life expectancy chart, based on careful and extensive life test data on bellows, are conservative.

LIFE EXPECTANCY

Minimum Life Expectancy In Cycles	LIFE FACTOR, as a fraction of the bellows stroke at 100,000 cycles life expectancy:	
	In Compression	In Off-set Rotation
1,000	1.50	1.70
10,000	1.25	1.40
100,000	1.00	1.00
1,000,000	.84	.82
10,000,000	.78	.74
100,000,000	.75	.73
Infinity	.72	.72

EXAMPLE: Suppose a given bellows design requires a minimum life expectancy of 1,000,000 cycles at a compression stroke of 0.313″. The table shows a life factor of 0.84 for this case. This means that the permissible stroke is 0.84 times the formula value. The formula value, 0.313, divided by 0.84 = 0.372″. Entering this value in the stroke formula shows that a bellows 19% longer would be required.

DERATING BELLOWS STROKE FOR PRESSURE AND LIFE

Obtain the fraction of stroke usable from the graph for derating stroke for working pressure (page 81). Assume this fraction is 0.65. Next, extract the life factor from the life expectancy table for the required life. Assume this is 1.25. The bellows stroke rating would be (0.65 × 1.25) times the formula value of the stroke.

SPRING RATE

$$R = \frac{4.3E(O+I)t^3}{(O-I-t)^3N} \text{ lb/in}$$

This formula gives values for bellows with convolutions having parallel side walls. For bellows with "V" grooves the rate is one-third greater.

This formula gives a straight line compression versus force characteristic and represents the spring resistance due to the bending of the convolution walls.

PRESSURE RATING

$$P = \frac{1.25 \times 10^6 t^2}{(O-I-t)^2} \text{ psi}$$

The above formula gives nominal pressure rating. Proof pressure is 1.75 times the above. Burst pressure is 2.50 times the above.

STROKE RATING

$$S = \frac{.0010(O-I-t)^2N}{t} \text{ in. compression, for 100,000 cycles life expectancy}$$

The rating in extension is 70% of the above. The equation applies to all types of convolutions and is based on bending stress limits, except V type, which can be used only in compression.

EFFECTIVE AREA

$$\text{Effective area} = 0.785\frac{(O+I)^2}{4} \text{ sq in.}$$

This formula is not theoretically accurate but gives results close to actual bellows values.

CRITICAL BUCKLING PRESSURE

With increasing pressure applied inside a bellows whose ends are fixed, a critical pressure, P_c, will be reached at which the bellows will suddenly bow sideways. Below this P_c the bellows will not buckle; above it the bellows will buckle outward without control and damage itself at a few percent more pressure than P_c. The critical pressure is given by the following formula:

$$P_c = \frac{6nr}{L^2}$$

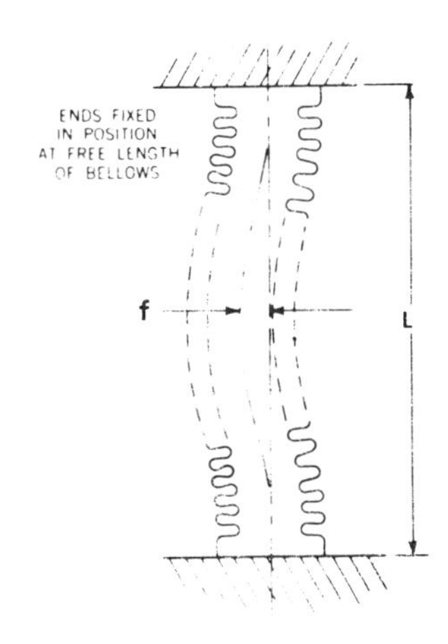

ALLOWABLE CIRCULAR ARC BENDING

This calculation assumes a natural circular arc bend.

$$A = 71.6\, Ns/O \text{ degrees}$$

The value given by the formula may exceed the angle attainable by the bellows, unless the stroke per convolution, s, is limited to the value at which bellows convolutions touch.

The formula gives the value for 100,000 cycles life expectancy. For any other value use the life factor from the table, and multiply it by the formula value of the angle.

OFF-SET BENDING WITH ENDS PARALLEL

$$\text{Off-set} = 0.52\, N^2 n\, s/O \text{ in.}$$

The above formula is for bellows with bend angles smaller than 30°. Note that in this arrangement the middle third of the bellows convolutions are nearly straight and unstressed while the end thirds get sharp bends.

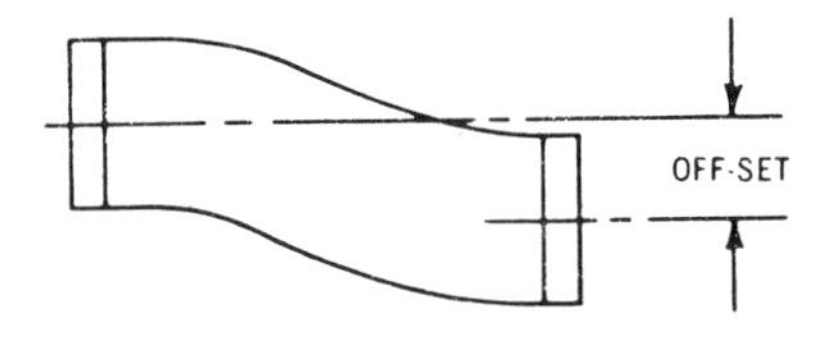

Since the number of convolutions, N, varies with the length of the bellows, the allowable off-set varies with the square of the active length.

This type usage is encountered in flexible shaft couplings.

The formula value is for 100,000 revolutions. For any other value, multiply the formula value by the life factor from the off-set rotation column of the life expectancy table.

CANTILEVER BENDING

$$f = \frac{5.25\, WL^3}{D^2 nr} \text{ in.}$$

where W = deflecting force at the outer end, lb, and L = distance of reflecting force from fixed base of bellows, in.

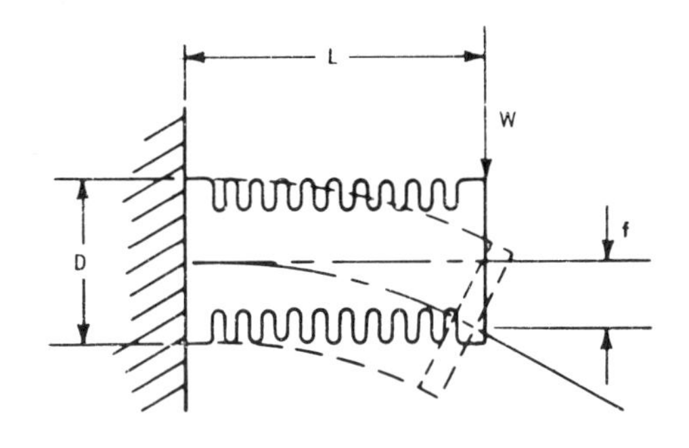

WEIGHT PER UNIT LENGTH

$$w = \frac{\pi td}{2n} [(O - t)^2 - (I + t)^2 + n(.75 \times O + I)] \text{ lb}$$

YOUNG'S MODULUS, Y, FOR CERTAIN BELLOWS FORMULAS

Certain acceleration formulas that follow require a modulus, Y.

$$Y = \frac{4nr}{\pi O^2} \text{ lb/sq in./in.}$$

DEFLECTION UNDER STEADY ACCELERATION

Note: These formulas are based on the assumption the bellows contains no filling except gas. If the bellows is liquid-filled the weight per unit length, w, must be increased by $.3026 \times (O + I) \times$ (Sp. Wt.) where Sp. Wt. is the specific weight of the fluid in lb per cubic in.

Cantilever

$$f = \frac{2.0\, wL^4 a}{YO^4} \text{ in.}$$

where a = acceleration, in gravities.

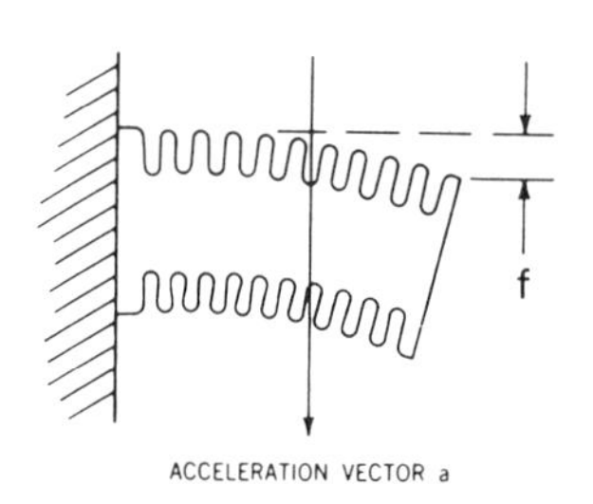

One End Fixed, One Freely Supported

$$f = \frac{.09\, wL^4 a}{YO^4} \text{ in.}$$

where a = acceleration, in gravities.

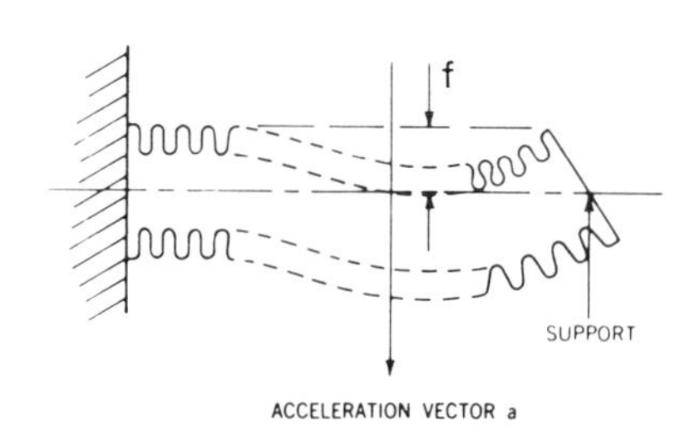

Both Ends Freely Supported

$$f = \frac{.224\, wL^4 a}{YO^4} \text{ in.}$$

where a = acceleration, in gravities.

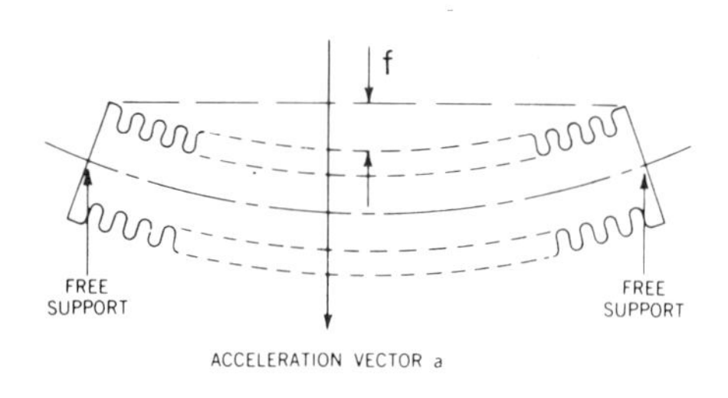

Both Ends Fixed

$$f = \frac{.0448\, wL^4 a}{YO^4} \text{ in.}$$

where a = acceleration, in gravities.

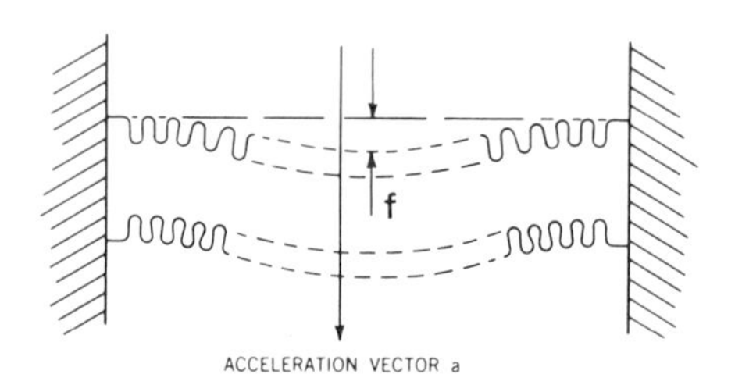

RESONANT VIBRATION FREQUENCIES

At Right Angles to the Axis

$$\text{Frequency} = z = \frac{I}{6.28} \sqrt{\frac{g}{f}} \text{ cycles/sec}$$

where g = 386 in./sec/sec, the accleration of gravity, and f = deflection of the middle of the bellows due to acceleration from the preceding formula.

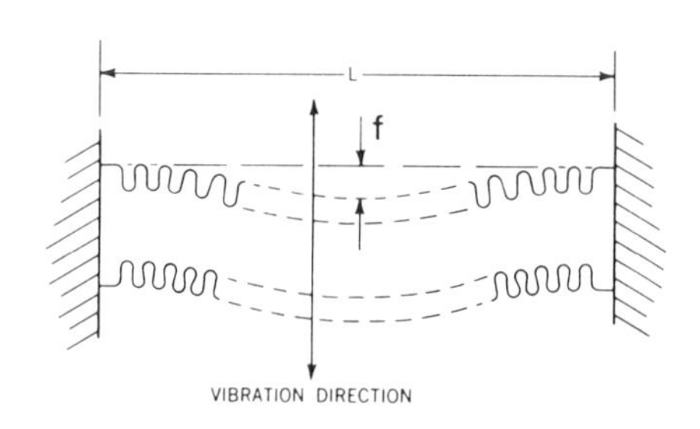

In Line With the Axis

$$\text{Frequency} = z = \frac{I}{6.28}\sqrt{\frac{18nrg}{wL^2}}\ \text{cycles/sec}$$

This frequency equation is based on the approximation that the middle third of the bellows vibrates against the two end thirds. The above formula gives the fundamental mode. Harmonics of this will occur at higher frequencies in the order 2-3-4, etc.

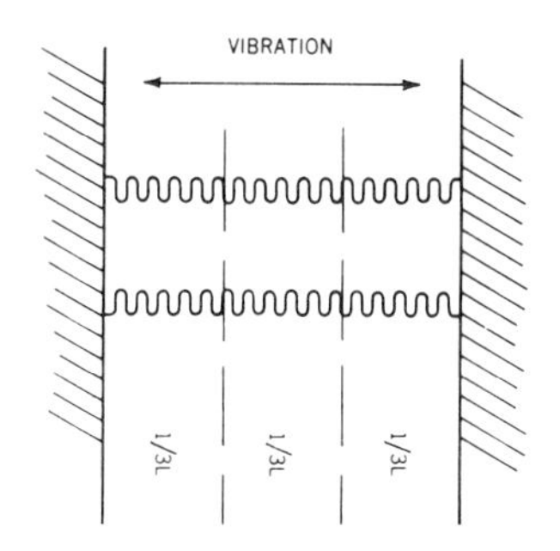

TYPICAL APPLICATIONS OF BELLOWS

Hydraulic Multiplier and/or Remote Transmission

Two bellows connected by a long capillary and liquid-filled can transmit motion or force either way—and at the same time multiply or divide.

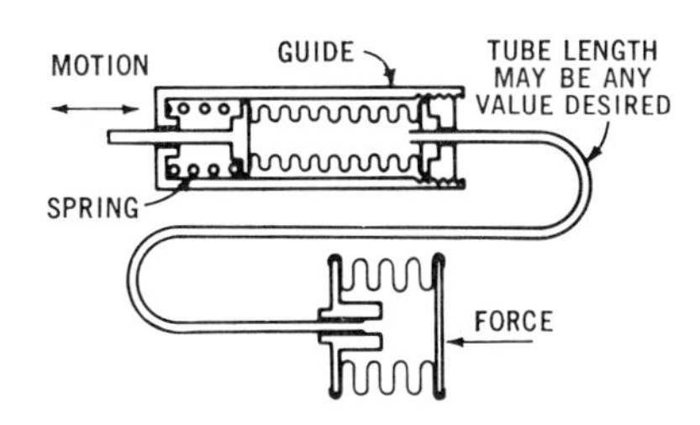

Metallic Seals for Motion into a Hermetically Sealed Housing

Laboratory vacuum systems, manipulators, crystal-growing apparatus:

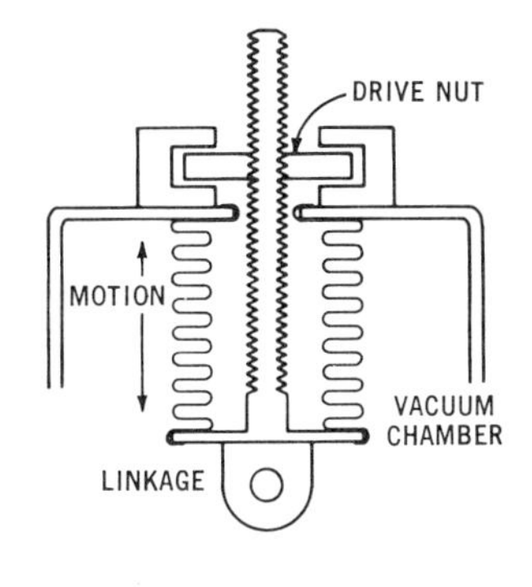

Electrical circuit breakers, push button switches:

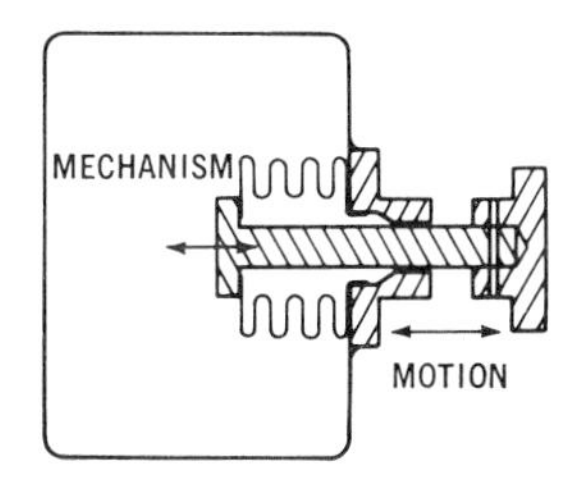

To transmit motion out of the high pressure zone of a differential pressure transducer:

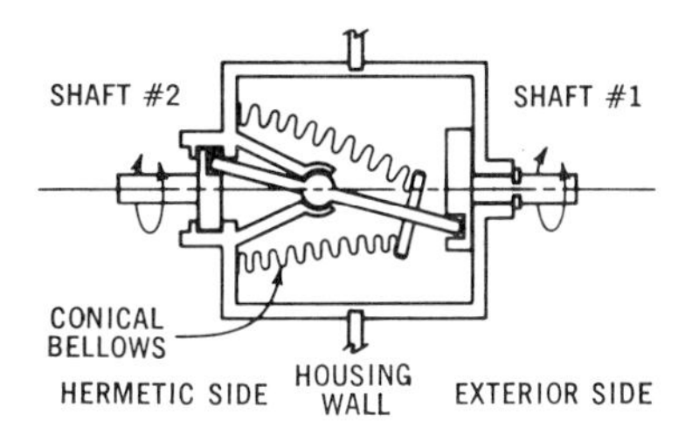

Pressure-Responsive Devices

Pressure switches, pressure gages, pressure actuators, pressure-electric transducers:

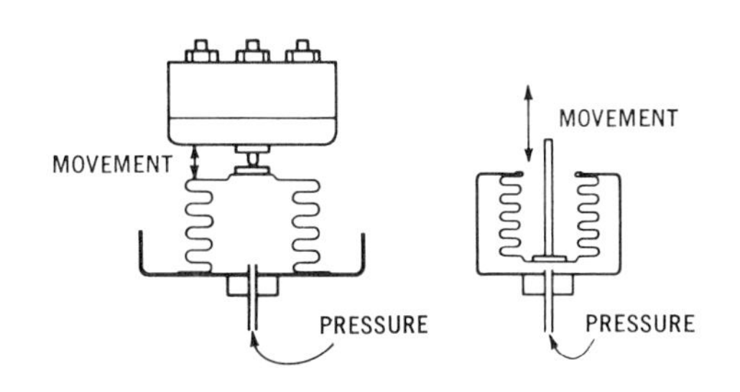

Full range aneroid, 0 to 15 psia:

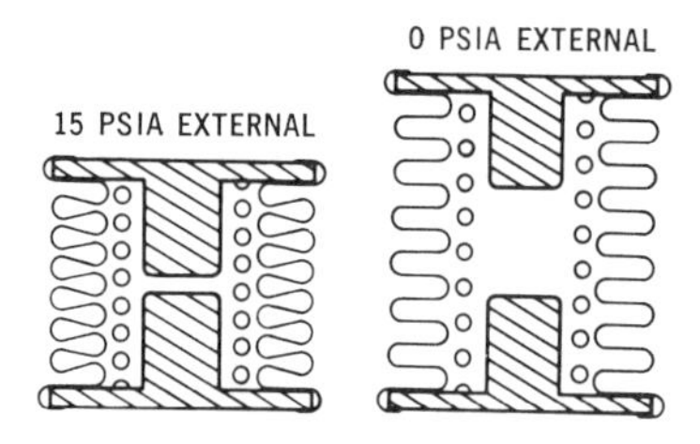

Short range aneroid, which works only from 0 psia to a small fraction of one atmosphere:

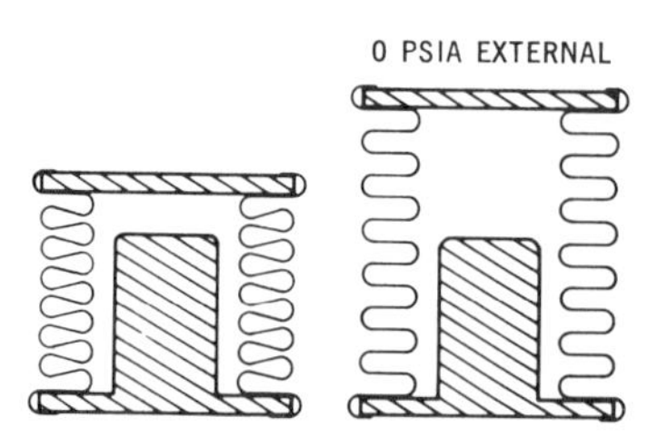

Some Other Applications

- A bellows will isolate vibration in piping between machinery and delicate apparatus.
- Thermostatic action is obtained by a bellows, capillary, and bulb liquid-filled and sealed.
- Thermal expansion in metal piping can be absorbed by a short bellows inserted in it.
- A gold-plated miniature bellows makes the most efficient spring in microwave electronic assemblies.
- Miniaturization of missile, satellite, or experimental equipment can be done with electro-deposited nickel bellows to .040″ O.D.

Typical bellow assembly. *Courtesy of Servometer Corp., Cedar Grove, New Jersey.)*

CHARTS AND NOMOGRAMS FOR DESIGNING WELDED DIAPHRAGM METAL BELLOWS†

These nomograms and coefficient curves were prepared from theoretical analysis of bellows technology, reinforced empirically by highly controlled test data. Their use over a period of time shows that values obtained for each design parameter assure proper

†The information in this segment has been checked by consultants, by engineers, and by a decade of experience in the application of metal bellows. To the best knowledge it is correct and reliable. However, no offer of guarantee of results using this information is given because there is no control over such use, and author and consultants are not responsible for any infringement of any patent or trademark by any user of this information. *(Portions of this material were reproduced through courtesy of Servometer Corp.)*

for each design parameter assure proper bellows performance for all functional requirements. The basic types are the flat plate and the nesting ripple configurations. They have the following inherent characteristics:

	FLAT PLATE	NESTING RIPPLE
RESISTANCE TO PRESSURE	GOOD	VERY GOOD
LONG STROKE CAPABILITY	FAIR	EXCELLENT
LINEARITY OF FORCE OUTPUT WITH PRESSURE	EXCELLENT	GOOD
LINEARITY OF STROKE WITH PRESSURE	EXCELLENT FOR SHORT STROKE	GOOD

These characteristics show why the flat plate is used, for example, in pressure-sensing applications, while the nesting ripple is used in volume compensation applications where a long stroke in a short capsule-free length is generally required.

SPRING RATE DETERMINATION

When the span, *S*, material thickness, *h*, and inside diameter, *ID*, are known, the spring rate, *K*, of a bellows of known number of convolutions and diaphragm configuration is easily found. First, find spring rate per convolution, *k*, using the spring rate nomogram (Figure 8-1). The values found are based on a modulus of elasticity of 29×10^6 psi. For other modulus values, the *k* value must be corrected. It varies directly with the modulus.

Then, determine *K*, the spring rate in lb/in., for the given diaphragm configuration as follows:

Flat Plate Spring Rate: Divide the spring rate value *k* obtained from the nomogram by the number of convolutions.

Nesting Ripple Spring Rate: Using Figure 8-2, find the spring rate coefficient, C_k, and divide by the number of convolutions. Note: For materials not heat-treated or hardened, deduct 10% from the *k* value for spring rate obtained from the nomogram.

PITCH DETERMINATION

Figures 8-3 and 8-4 determine the bellows convolution stroke and pitch for the two configurations. The factors mainly responsible for pitch limitation are permanent set and flipping, along with special performance characteristic demands such as spring rate linearity. The maximum parameters define the pitch that can be kept without occurrence of flipping or springback. It is necessary,

NOMENCLATURE

OD	= Outside diameter
ID	= Inside diameter
Span	= Depth of convolution measured from *OD* to *ID*, = $(OD - ID)/2$. The ratio of span to *OD* should be less than 1/3.
P	= Pitch, the distance from the center of one weld bead to the center of the adjacent bead
NP	= Nested pitch
h	= Diaphragm thickness
N	= Number of convolutions; a convolution is two diaphragms welded together at *ID*.
K	= Spring rate, the ratio of force to stroke expressed in lb/in.
Mean diameter	= $(OD + ID)/2$
EA	= Effective area, that surface on which pressure acts to produce thrust

$$EA = \pi \left(\frac{OD + ID}{4}\right)^2$$

A/K	= Effective area/spring rate = stroke per psi

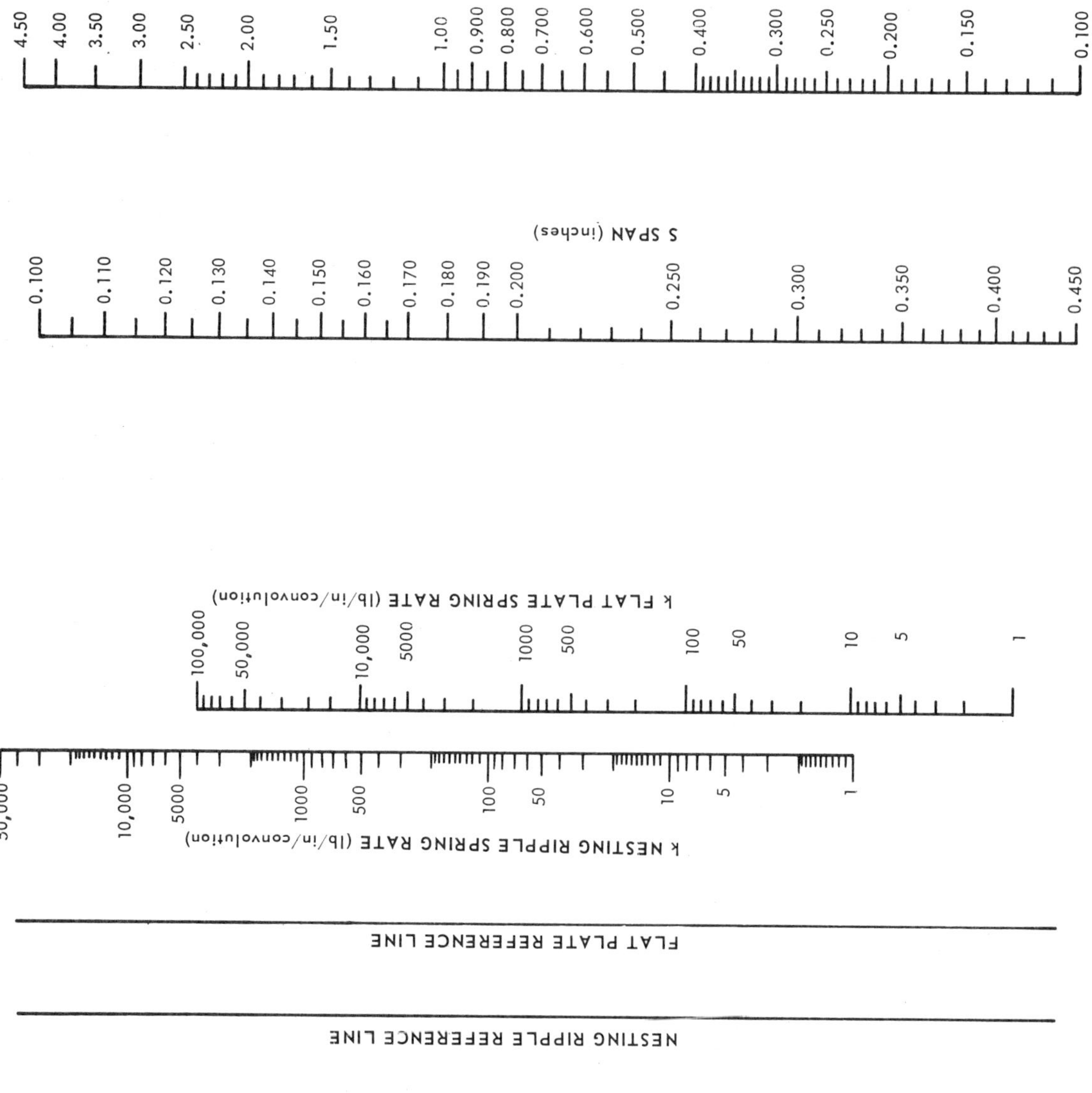

Fig. 8-1 Spring rate nomogram. (Based on a modulus of 29×10^6 psi.)

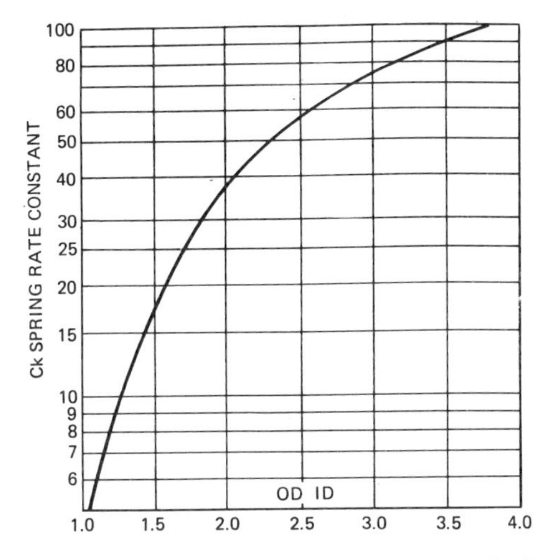

Fig. 8-2. Ripple bellows spring rate constant.

however, to calculate the diaphragm stress to ensure that it does not exceed material yield strength when the diaphragm is deflected through the required stroke.

Flat Plate Pitch: With the values of span and stroke in compression known, coordinates are drawn on the pitch for flat plate curve (Figure 8-3). If the point of intersection on the lines drawn vertically from the span and horizontally from the stroke in compression values falls in the area below the maximum curve, a pitch may be calculated for the bellows. If the point falls above the maximum curve, a new span or stroke must be selected.

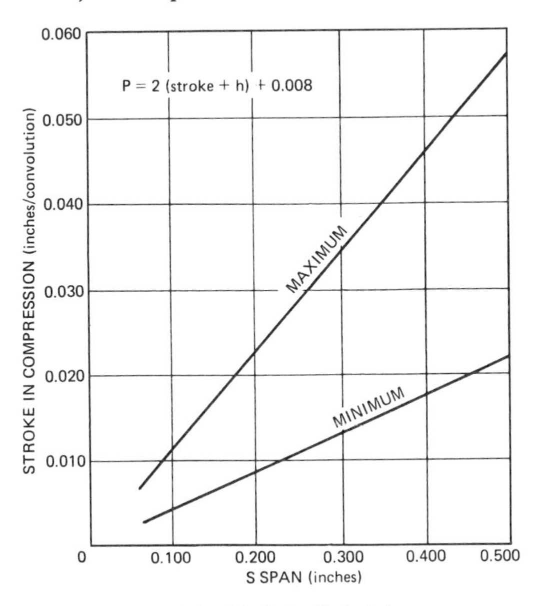

Fig. 8-3. Pitch for flat plate.

The pitch, P, is calculated by substituting the stroke per convolution value, found in Figure 8-3, in the formula:

$$P = 2\,(\text{Stroke} + h) + 0.008$$

Pitch for Nesting Ripple: Procedure for finding the stroke of a nesting ripple configuration is the same as for the flat plate.

To find the pitch, substitute the value for the stroke, found in Figure 8-4, in the formula:

$$P = \text{Stroke} + NP$$

where $NP = (2h)(1.60)$.

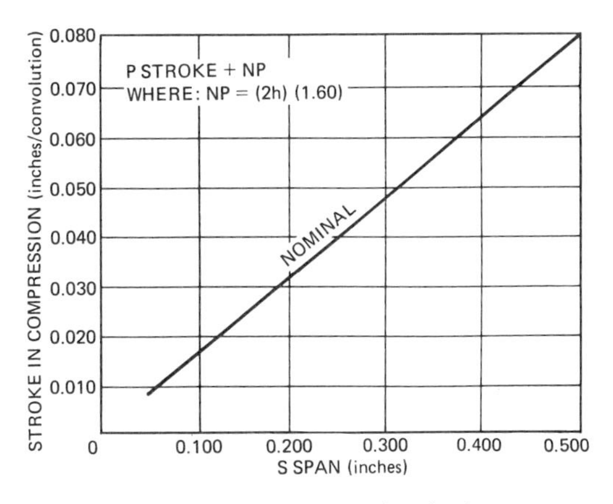

Fig. 8-4. Pitch for nesting ripple.

STRESS LEVEL AND CYCLE LIFE RANGE DETERMINATION

When all other parameters have been determined, the life expectancy should be examined before the final specifications are set. Cycle life is a function of pressure and deflection stress levels.

The nomogram in Figure 8-5 is used to find uncorrected pressure stress and uncorrected deflection stress for both contours—the flat plate and nesting ripple. Align h, material thickness, with ID, inside diameter, intersecting both pressure stress and deflection stress reference lines. Then, align the pressure stress reference line intersection with ΔP, differential pressure, intersecting the uncorrected pressure stress in psi. Next, align the deflection stress reference line intersection

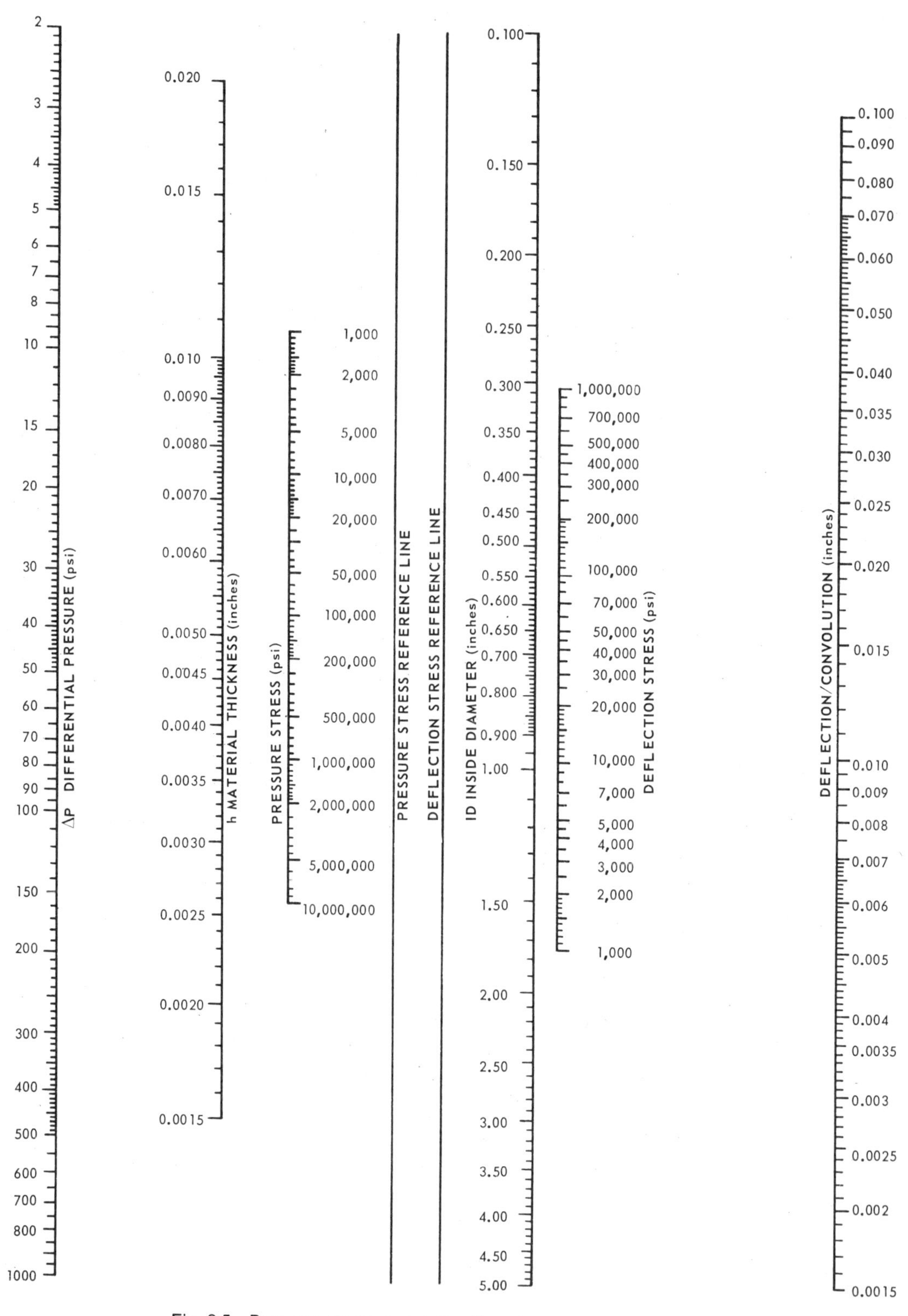

Fig. 8-5. Pressure stress and deflection stress nomogram.

with deflection/convolution, intersecting the uncorrected deflection stress in psi.

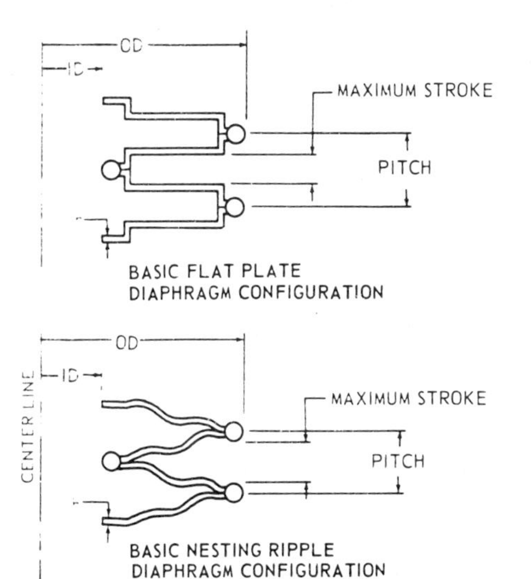

The two stress factors now are multiplied by a coefficient stress factor, found in Figure 8-6 for the flat plate configuration and Figure 8-7 for nesting ripple configuration, to obtain the actual pressure stress and deflection stress for the bellows *ID* and *OD*. Using the appropriate curve, project vertically from the *OD/ID* value, intersecting the C_pOD, C_pID and $C_\delta OD$, $C_\delta ID$ curves. From these intersections, project horizontally to the right for

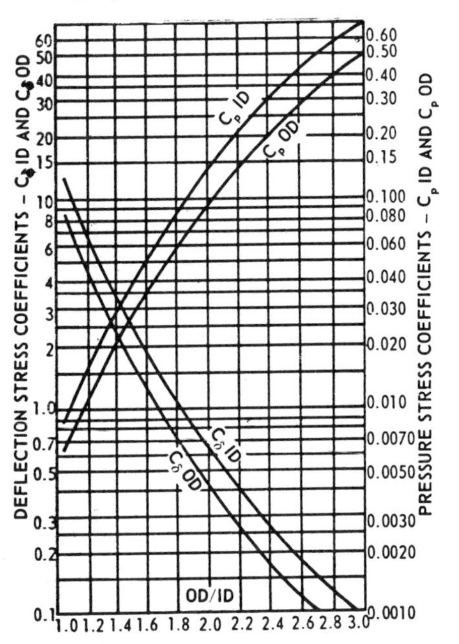

Fig. 8-6. Flat plate deflection and pressure stress coefficients.

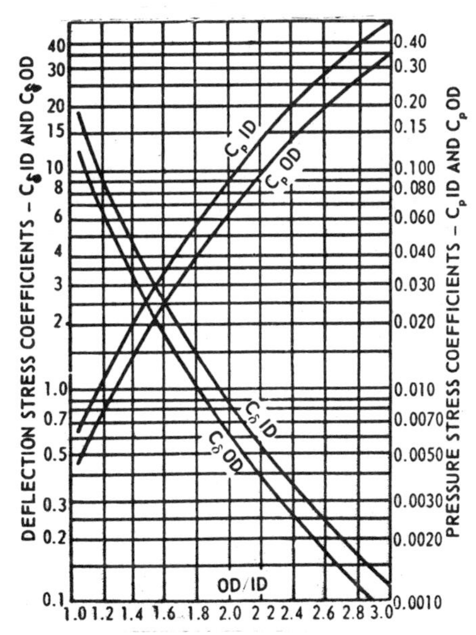

Fig. 8-7. Nesting ripple deflection and pressure stress coefficients.

the pressure stress coefficients and horizontally to the left for the deflection stress coefficients. These stresses must be qualified as to whether they represent compression stresses (−) or tension stresses (+), depending on the deflection or pressure conditions. This is determined from the table below.

STRESS CONDITION	JOINT STRESS (+) OR (−)	
	ID JOINT	OD JOINT
COMPRESSION STROKE	−	−
EXTENSION STROKE	+	+
MAXIMUM DIFFERENTIAL PRESSURE (external)	+	−
MAXIMUM DIFFERENTIAL PRESSURE (internal)	−	+

To calculate absolute stresses, from which the bellows cycle life range is determined, multiply all joint stresses that are in tension (+) by 2 (stress concentration factor). This factor (2) also is used when deflection stress and pressure stress are combined and the algebraic addition indicates a tension stress (+). When the pressure and deflection stresses have been assigned plus or minus values, the total stress per convolution is found by an algebraic addition of the stresses.

When the maximum and minimum stress levels (stress level cycling range) for the *OD* and *ID* of the joint of the bellows design are established, the maximum stressed joint (where failure may be expected to occur) must be determined to find the expected cycle life range. Next, using the minimum and maximum stresses found in the algebraic addition of stress, find the mean stress and the alternating stress as follows: Mean stress = [(Max. stress) + (Min. stress)]/2. Alternating stress = [(Max. stress) − (Min. stress)]/2.

Using Figure 8-8 or 8-9, the expected cycle life range for a particular bellows design can be found. The two materials are called out on the curves. On either curve, project upward from the calculated value of mean stress and horizontally from the calculated value of alternating stress. The intersection represents the expected cycle life.

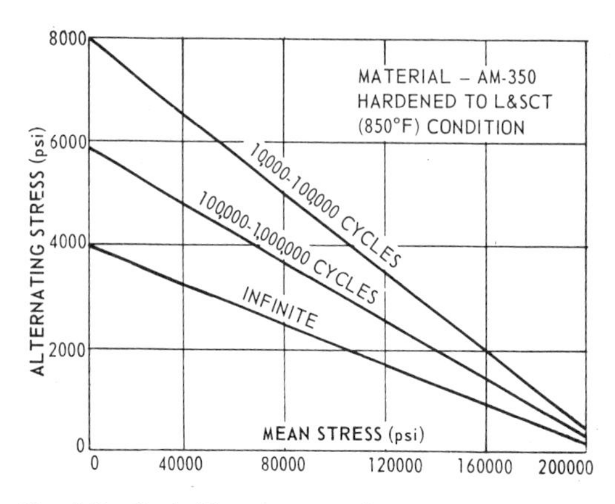

Fig. 8-8. Cycle life range as a function of alternating stress versus mean stress.

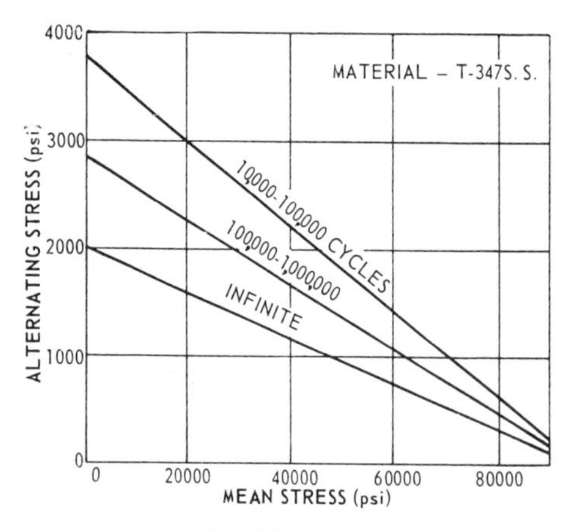

Fig. 8-9.

EXAMPLE CALCULATION: The example below illustrates a typical volume-compensating application. In this particular type of application, the bellows must compensate for fluid changes resulting from temperature variations.

Given Design Values: Case diameter = 1.25 in.; maximum bellows $OD = 1.00$ in.; required volume change (ΔV) = 0.20 cubic in.); maximum length (including fittings) = 0.680 in.; maximum pressure change (ΔP) = 20 psi; required life = 5,000 cycles.

Solution: Assume a nesting ripple configuration and AM-350 HT material. Also, assume a span of 0.200 in. and material thickness of 0.0025 in. Because larger spans and thicker materials are easier to weld, good design practice dictates starting here and working toward smaller spans and lighter gages if necessary.

With span and OD known, the ID can be found from $S = (OD - ID)/2$: $ID = OD - 2S = 1.00 - 2(0.2) = 0.60$ in.

Next, calculating the effective area:

$$EA = \pi \left(\frac{OD + ID}{4}\right)^2 = \pi \left(\frac{1.00 + 0.60}{4}\right)^2$$

$$= 0.50 \text{ sq in.}$$

The stroke required $= \Delta V/EA = 0.20/0.50 = 0.40$ in. Now, using Figure 8-4, find the stroke in compression per convolution by projecting upward from span = 0.2 to the curve, thence to the left to read 0.032 in. The number of convolutions now can be found:

$$N = \frac{\text{Stroke}}{\text{Stroke/Convolution}} = \frac{0.400}{0.032} = 13$$

The maximum length is found from:

$$N[(\text{Stroke/Convolution}) + P_N/\text{Convolution})]$$

where:

$$\begin{aligned} NP/\text{Convolution} &= 2h(1.60) \\ &= 2(0.0025)(1.60) \\ &= 0.008 \end{aligned}$$

Thus, maximum length = (13) (0.32 + 0.008)

= 0.520 in. This value is within the 0.680 in. length available and provides a safety factor. Good design calls for a 5% factor to allow for variables such as EA and nested length. To find the spring rate K, use Figure 8-1, aligning $h = 0.0025$ with span = 0.20, intersecting the proper reference line.

Align this point with $ID = 0.60$ and read $k = 8$ lb/in./convolution. Using Figure 8-2, project upward from $OD/ID = 1.0/0.6 = 1.67$ to the curve, thence to the left to read $C_k = 25$. Then, $K = k(C_k)/N = (8)(25)/13 = 15.4$ lb/in.

With K known, the ΔP over the full stroke can be found:

$$\begin{aligned}\Delta P &= (k)(\text{Stroke})/EA \\ &= (15.4)(0.40)/0.50 \\ &= 12.3 \text{ psi}\end{aligned}$$

This is within the maximum allowable of 20 psi. Next, evaluating life expectancy, the bellows is in maximum stress at full stroke = 0.400 in. (= 0.031 in./convolution) and 12.3-psi external pressure. The bellows is in minimum stress when both pressure and deflection are zero.

Using Figure 8-5, align $h = 0.0025$ with $ID = 0.600$, intersecting the reference lines. Align the pressure stress reference intersection with $\Delta P = 12.5$, and read the uncorrected pressure stress = 140,000. Align the deflection stress reference intersection with deflection/convolution = 0.032, and read the uncorrected deflection stress = 56,000. Using Figure 8-7, project upward from $OD/ID = 1.67$ and read the stress coefficients: $C_pID = 0.043$: $C_pOD = 0.029$; $C_\delta ID = 2.1$; $C_\delta OD = 1.5$. Next, calculating the actual pressure stresses:

$$ID_{max} = (+140{,}000)(0.043) = +6000 \text{ psi}$$

$$OD_{max} = (-140{,}000)(0.029) = -4000 \text{ psi}$$

and the actual deflection stresses:

$$ID_{max} = (-56{,}000)(2.1) = -118{,}000 \text{ psi}$$

$$OD_{max} = (-56{,}000)(1.5) = -84{,}000 \text{ psi}$$

Summing algebraically for the maximum stress:

$$\begin{aligned}ID_{max} &= -118{,}000 + (+6{,}000) \\ &= -112{,}000 \text{ psi}\end{aligned}$$

$$OD_{max} = -84{,}000 + (-4{,}000) = -88{,}000 \text{ psi}$$

All minimum stresses are zero. The maximum joint stress as calculated occurs on the ID joint (−112,000 psi). This value is well below the yield point of the material chosen.

Next, calculating:

$$\text{Mean stress} = (-112{,}000 + 0)/2 = -56{,}000 \text{ psi}$$

Using Figure 8-8, project upward from mean stress = 56,000 and horizontally from alternating stress = 56,000. The intersection falls in the 10,000 to 100,000 cycle area, so the design easily meets the 5,000 life cycle requirement.

Because the selected pitch falls under the maximum pitch curve shown in Figure 8-4, flipping or springback of the diaphragm will not occur.

REFERENCES

1. Howell, Glen W. and Weathers, Terry M. Aerospace Fluid Component Designers Handbook, Revision C, November 1968. TRW Systems Group, 1 Space Park, Redondo Beach, CA.
2. Men of Engineering; Servometer Corporation Catalog, Servometer Corporation, 501 Little Falls Rd., Cedar Grove, N.J. 07009.
3. Robertshaw Controls Co., 333 N. Euclid Ave., Anaheim, CA 92803.
4. Bellows Design Manual; Babcock & Wilcox Co., Belfab Unit, 305 Pentrass Blvd., Daytona Beach, FL 32015
5. Matheny, J., *Machine Design,* **34,** No. 1, 1962. Penton Publishing Co., 1111 Chester Ave., Cleveland, OH 44114.
6. *Design News,* A Cahners Publication, 221 Columbus Ave., Boston, MA 02116.

Reproduced by permission of *Design News.* Written by Stephen J. Zierak, Manager of Engineering, and Peter Judersleben, Project Manager, Metal Bellows Corporation, Sharon, Massachusetts.

IX Bourdon Tube Actuated Pressure Switches

The bourdon spring is the heart of a majority of our pressure and temperature measuring instruments. The tube, oval or elliptical in cross section, is coiled, sealed at one end, with pressure admitted to the other. Increasing pressure forces the two walls to bulge in the direction of their original circular cross section, straightening the coil. We define the reaction movement of the free end of the bourdon spring as tip travel. This tip travel distance is exactly proportional to the amount of pressure, or vacuum, applied. The linear motion of the tip is translated to uniform angular motion by a pointer over a dial by means of a mechanical or electrical linkage.

Tip travel is affected by the length of the bourdon spring, its radius, the cross-sectional shape of the tube, wall thickness, and elastic characteristic of the alloy used. The ultimate goal in the bourdon spring is to adjust these variables to realize uniform tip travel for a series of pressure ratings while holding constant those variables that affect the choice of case, movement, linkage, and mounting parts. However, the apparent simplicity and straightforward operation of a finished spring belie the complexity encountered in a precise mathematical analysis of the complicated forces and physical factors in parameter definitions.[1,2].

BOURDON TUBING SELECTION AND DESIGN, A SIMPLIFIED APPROACH

Here is an approach to the design of flat oval bourdon springs based on the following empirically developed relationship:

$$\text{Tip travel} = K\frac{PR^2L\theta}{St^3E}$$

where K is a constant, P is the applied pressure or vacuum, θ is the free angular length of the spring, R is the radius of curvature of the free spring, L is the outside diameter (O.D.) of the long axis, S is the outside diameter (O.D.) of the short axis, t is the wall thickness of the tube, and E is the modulus of the elasticity of the tube material.

The formula was developed for flat oval tubes (Figure 9-1) with sides as nearly parallel as possible, and a general ratio of O.D. major axis to O.D. minor axis of approximately 3.7 to 4.3 to 1. This shape is favored for most applications based on stability, uniformity, reproducibility, accuracy, yield, useful range, fatigue resistance, ease of manufacture, and most reliable overall pressure characteristics.

By J. L. Myer, formerly Superintendent Melting and Refining Dept., U.S. Mint, and E. B. Bitzer, Chief Metallurgist, Handy & Harman Tube Co.

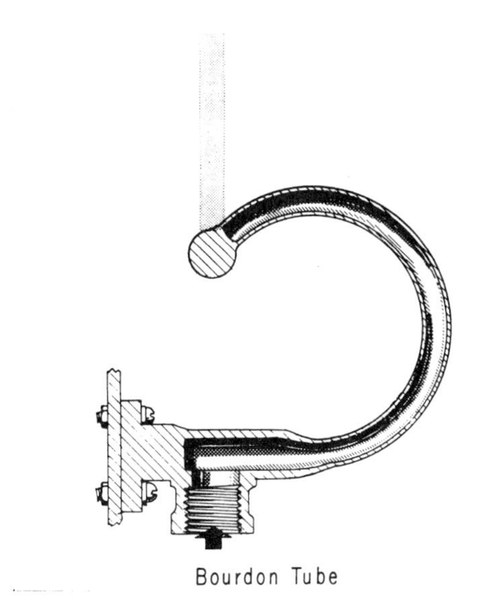

Fig. 9-1. Typical bourdon spring and its cross section. *(Courtesy of Delaval Turbine, Inc., Barksdale Controls Division, Los Angeles, California.)*

ANGULAR LENGTH, θ

The relationship between angular length and tip travel was determined from the proportionality shown in Figure 9-2. The origin of the curve at 0° tip travel and 35° angular length indicated the effective length is less than that measured for the free length. Apparently the soldered or welded joints of the spring at each end not only immobilize the angular portion of the spring that actually lies within the socket and tip, but also affect the spring immediately adjacent to the joint. The deviations of the curve from straight lines are thought to have resulted from irregularities of joining the tip as the experimental angular lengths were repeatedly cut shorter. For thin-wall tubing at low pressures, the mere weight of the tip and solder joint seems to cause a sluggishness of response on long springs.

Noted on the curves also are the maximum pressure values beyond which the spring would not be reliable for accurate pressure readings. This limit, of course, represents the practical end of the useful elastic working range of the metal in the spring and most especially at the shoulders. If the uncoiling of a bourdon spring is envisioned as two curved plates attempting to unroll, and prevented from doing so by the shoulder sections of the tubing that join them together, the longer springs will be seen to have greater tip travel, while they create far greater stressing of the shoulder sections. Therefore, long springs give the advantage of greater tip travel, but at the expense of a decrease in working pressure and fatigue life.

RADIUS OF SPRING CURVATURE, R

The way that tip travel is affected by radius of curvature is shown in Figure 9-3. Replotted on logarithmic coordinates, tip travel appears to vary approximately as the square of the radius of curvature including a minor adjustment, which, again, is thought to relate to the rigidifying effect of the mountings.

LONG AXIS, L

A linear relationship for tip travel versus the O.D. of the long axis is suggested in Figure 9-4. If a compensating factor is subtracted from the long axis O.D. equivalent to approximately two times the short axis O.D., the curves pass through the origin or zero point. Considering the curving nature of the shoulder areas, it seems reasonable that they would detract from the useful flat working surfaces somewhat in the manner implied.

The graph also indicates maximum working pressures according to the long axis values. These pressures are seen to become large for narrow springs, and small for wide springs. As was discovered with the angular length relationship, the features that allow large tip travel again do so at the expense of lower maximum pressures, so that one is obtained at the expense of the other.

SHORT AXIS, S

Figure 9-5 shows an inverse but linear relationship for tip travel and the short axis. Springs with large values of the short axis take on a "box girder" type of structure

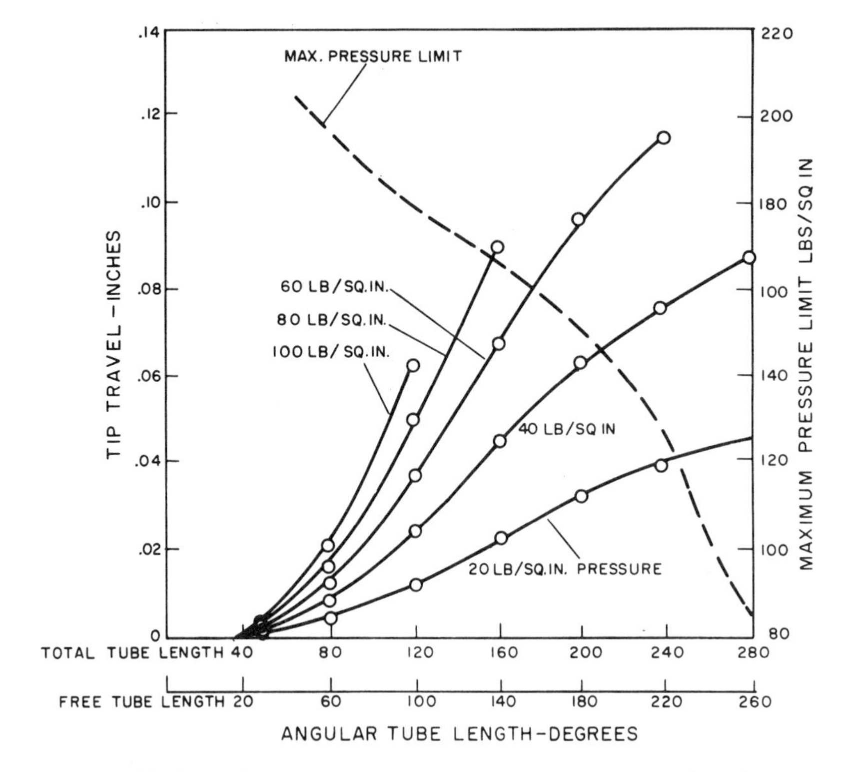

Fig. 9-2. The relationship between tip travel and angular length.

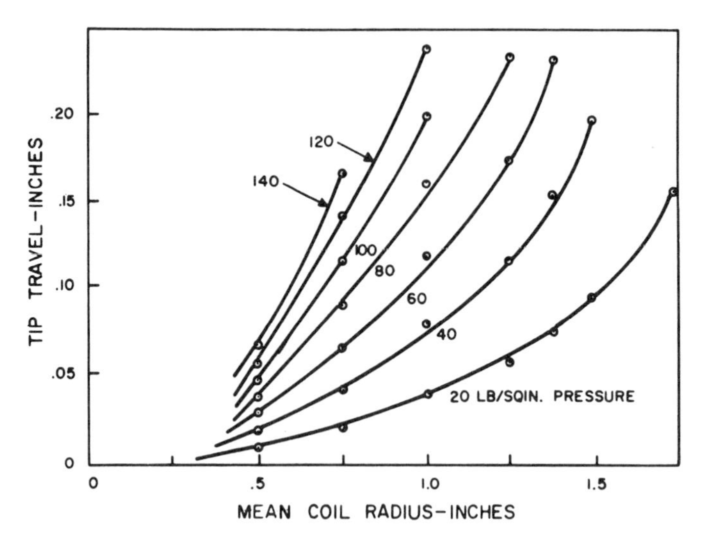

Fig. 9-3. The relationship between tip travel and coil radius.

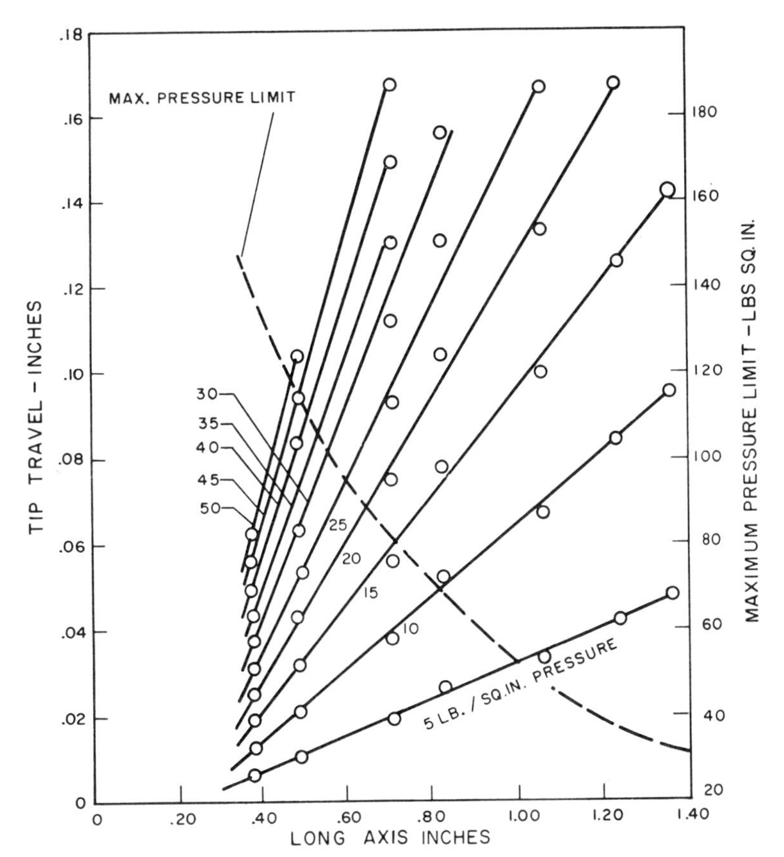

Fig. 9-4. The relationship between tip travel and long axis.

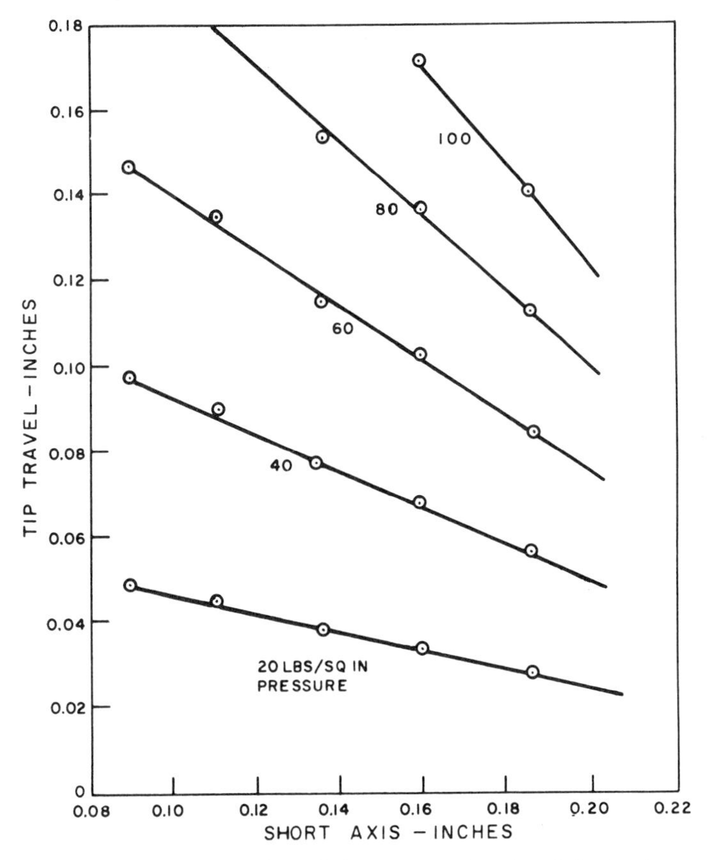

Fig. 9-5. The relationship between tip travel and short axis.

with considerable rigidity, and resist the tip travel motion, holding it to quite low values as shown.

The lower limit of the short axis dimension is, of course, dictated by twice the wall thickness measurement. The practical upper limit of the short axis dimension is fixed by the major axis of the tube, in which event it would have a fully circular cross section, a design actually used for very high pressure springs.

WALL THICKNESS, *t*

The relationship between tip travel and wall thickness was developed, holding the O.D. of both the major and minor axes constant, so that any addition to wall thickness extended the walls inwardly, and decreased both I.D. measurements simultaneously. Theoretically, there may be no real justification for doing this, but from a practical standpoint, there are excellent reasons. First, holding the O.D. of all tubing in process for such gages to a standard size greatly simplifies all of the drawings, tooling, and gaging problems involved. Second, by maintaining uniform outside sizes for all pressures in the range under discussion, standardized slotted sockets and female tips can be used, along with uniform soldering fixtures and coiling setups. Furthermore, high pressure tubes (which have heavy walls) need very little inside pressure volume to exert the necessary force to actuate the springs properly. On the other hand, low pressure (thin-wall) tubing gives a maximum of pressure volume inside, where it is most needed for such bourdon springs.

Tip travel was measured for increasing pressure for different wall thicknesses and plotted as shown in Figure 9-6. A common or standard tip travel within the safe working pressure for all tubes was selected, and the corresponding working pressure replotted against wall thickness on logarithmic coordinates. Figure 9-7 shows a linear relationship with a slope of approximately 3, indicating that tip travel varies with the inverse cube of the wall thickness.

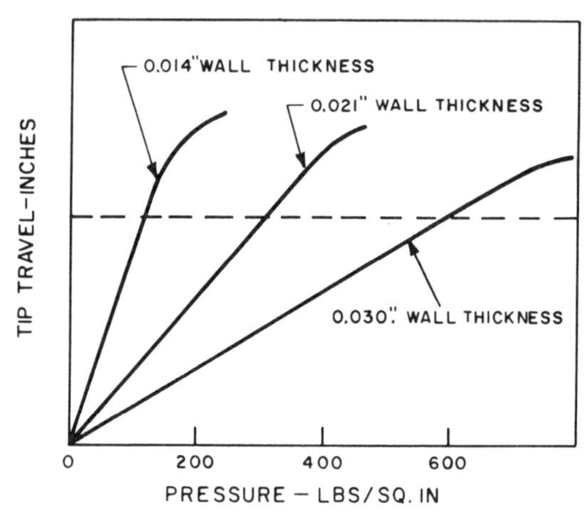

Fig. 9-6. The relationship between tip travel and pressure for various wall thicknesses.

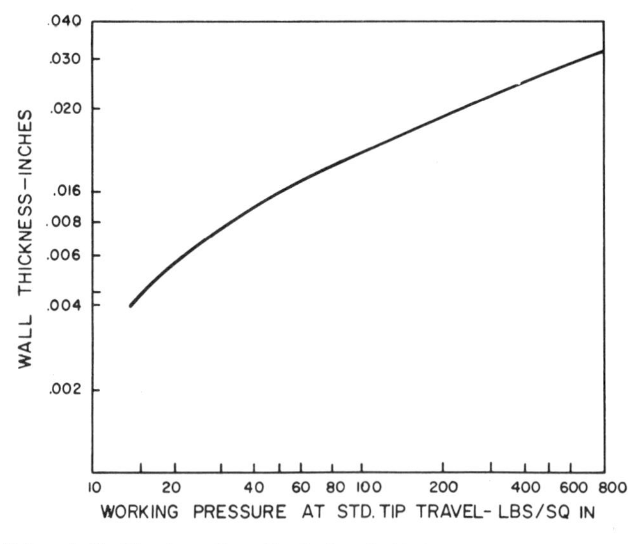

Fig. 9-7. Figure 9-6 replotted to show the effect of wall thickness on a selected working pressure.

MODULUS OF ELASTICITY, *E*

Metals subject to stress behave elastically in the initial range of loading. The modulus of elasticity is a measure of deflection for a given load in the elastic range. For example, steel with a modulus of 30,000,000 psi stretches 0.00033 in./in. under 10,000 psi load, while beryllium copper with a modulus of 19,000,000 psi stretches 0.00053 in./in.

The modulus of elasticity stays essentially constant for each material and is not greatly affected by cold work or heat treatment. It is subject to slight change with abnormal operating temperatures, and this is reflected in the functioning of bourdon springs in extremes of heat or cold. Under such operating conditions a suitable compensator is linked into the indicating system, or a constant-modulus alloy such as Ni-Span-C is used.

The tip travel for any given pressure varies inversely with the modulus of elasticity of the material from which the spring is made. This gives the designer considerable latitude, as an adequate choice of materials can cover wide ranges of operating characteristics that would be difficult to obtain in other ways. Judicious alloy selection can assist in holding finished bourdon springs to a common coiling and cross-section size for optimum results in the standardized manufacturing and instrument set-up.

DISCUSSION OF VARIABLES

The exponential powers by which the variables influence the tip travel are the key to spotting where the greatest care must be exercised in making bourdon springs of predictable and uniform response. Accordingly, wall thickness and radius of spring curvature stand out as obvious controlling variables. The radius of curvature is especially difficult to control because of springback which occurs when the coiled tube is released from the coiler. There is no royal road to success in this respect; dogged efforts are required to keep tubing mechanical properties, dimensions, and coiler tension adjustments all under extremely close control. Wall thickness control is equally difficult. Seamless tubing can have appreciable eccentricity, with wall thickness varying up to ±10% in any cross section. Redrawing improves this somewhat, but often not sufficiently for precise control. Through selection, tubing with ±5% or less variation is available. Control of average wall thickness by weight appears to offer even better precision and follows logically from the cubic relationship shown for the wall thickness. The latter is a measure of the spring mass that is effectively deflected by application of pressure.

Adjustments of long axis and short axis dimensions appear to be of equal influence upon inspection of the formula. However, considering the slopes of the curves for each and recognizing that the short axis is usually one-fourth the length of the long axis, a few thousandths of an inch out of tolerance turns out to be much worse when it occurs on the short axis than on the long axis. The long axis and short axis variations are not compensating, a fact not often appreciated. If the short axis is made undersize in the shaping or coiling operation, the long axis of necessity will be correspondingly oversize. The undersize short axis, and the oversize long axis as well, each increase the tip travel of the spring for a given pressure.

DESIGN OF A STANDARDIZED BOURDON SPRING SERIES

The suggested approach is to hold constant all of the nominal variables in the tip travel equation except the wall thickness. Under these conditions, the wall thickness versus pressure relationship can be established and used as the only variable requiring selection and adjustment.

The type of gage assembly into which the series is to fit will largely determine the angular length and to some extent the radius of curvature. The radius should be kept as large as the case dimensions will permit in keeping with good practice regarding alignment and transfer of the linear tip travel motion to the movement mechanism.

The modulus of elasticity can be accommodated by judicious alloy selection as previously noted. Here a practical compromise among strength, corrosion resistance, and cost must often be made. Strength of the alloy selected is typically enhanced by either strain hardening (cold work), precipitation hardening, or Martensitic transformation (quenching and tempering of steel). A convenient ratio of long to short axis dimensions must be established, preferably staying as close as possible to a 4 to 1 ratio.

Establish the wall thickness-pressure relationship by making prototypes with different wall thickness while keeping the long and short axis outside dimensions constant. Carefully mount the springs and obtain the pressure-tip travel characteristics as in Figure 9-6. From these curves, determine the safe working pressure as mentioned above, and note the tip travel for that pressure.

With the safe working pressure and the corresponding tip travel now known for each of the springs of different metal thicknesses, fix a uniform optimum value for tip travel that is within safe limits for all the spring types, and standardize it for the series. Turn now to Figure 9-7 and enter the data just obtained under the relationship of wall thickness to safe working pressure (at constant tip travel for the full pressure to be put on the dial reading as working pressure). These four points may then be used to project the complete line relating all pressure values to their corresponding correct wall thickness for that particular series. In this manner, it becomes a simple matter to select the necessary wall thickness or, specifically, weight per foot for any and all pressures desired in that series, holding all the other dimensional factors at constant values, or within acceptable tolerances.

SUMMARY

An empirically developed formula provides a simplified approach to the design of flat oval bourdon springs. Wall thickness is selected as the variable to achieve a series of pressure ratings while holding constant other variables, thus permitting standardization and interchangeability of external hardware. Wall thickness is raised to the third power, in the formula developed, magnifying the importance of this dimension. The cubic function suggests that mass is the controlling variable, and specification of average wall by weight is indicated to realize improved predictability and uniformity of spring response.

REFERENCES

1. Jennings, F. B., Theories on Bourdon Tubes, *Transactions of the ASME,* January 1956.
2. Mason, H. L., Sensitivity and Life Data on Bourdon Tubes, *Transactions of the ASME,* January 1956.

Diaphragms for Pressure Switches

A. NONMETALLIC DIAPHRAGMS

A diaphragm is a thin, dividing partition. It is used to separate two areas, one that is moving and one that is fixed, to keep any exchange of fluids or gases from occurring between the two chambers or areas.

Diaphragms are separated into three classes: (1) diaphragms that are used as a dividing membrane where no differential pressure exists between the chambers; (2) static diaphragms, which are used as a divider between two fluids and which have virtually no motion; (3) dynamic diaphragms, which are used for sealing devices between fixed and moving members, and which normally transmit a pressure.

The diaphragms used as dynamic sealing membranes and pressure transmitters make up the group used most often, and this chapter will discuss them further. This type of diaphragm works the same way as a sliding-contact packing, the difference being there is no leakage (and there is no breakaway friction, they require no lubricant, etc.). There are equivalent diaphragms for all the different sliding-contact packings.

Diaphragms can also be divided into two other groups: rolling and flat. The rolling diaphragm has a 180° convolution so that during the translation thc diaphragm rolls off the wall of the piston onto another piston wall. Flat diaphragms have no convolutions or have convolutions of less than 180°.

FLAT AND CONVOLUTED DIAPHRAGMS

When a diaphragm is called flat, this does not mean that it actually is flat. Fluid pressure still forms convolutions during the working cycle. When you have relative movement of the diaphragm assembly, flat diaphragms are distorted from a single plane. This distortion needs an elongation of the cord length of the fabric.

Usually flat diaphragms are made of a flexible material, but designs should avoid elongation of the diaphragm during its working stroke. You can avoid this limitation by using one or more convolutions. This will cause the diaphragm to flex and not to stretch. Diaphragms are generally made from fabric permeated with an elastomer. Most fabric and elastomer combinations withstand flexing indefinitely, but stretching could cause fatigue failures. See Figure 10-10 for flat diaphragm pressure switch.

Design Considerations

The design for a flat diaphragm should begin with thought to several precautions (Figure 10-1). The following questions need to be evaluated:

- How much stroke will be required?
- What are the pressures involved?

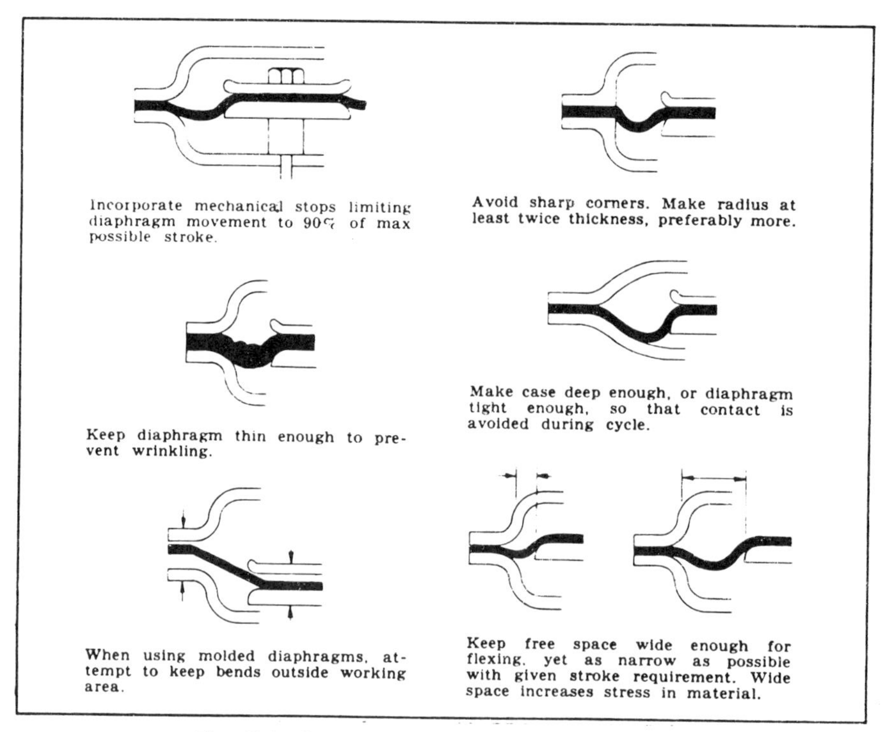

Fig. 10-1. Design precautions for flat diaphragms.

- How important is effective area shift?
- What is the fluid's nature that will contact the diaphragm?
- Will a limiting factor be caused by available space?
- What is the required life expectancy?

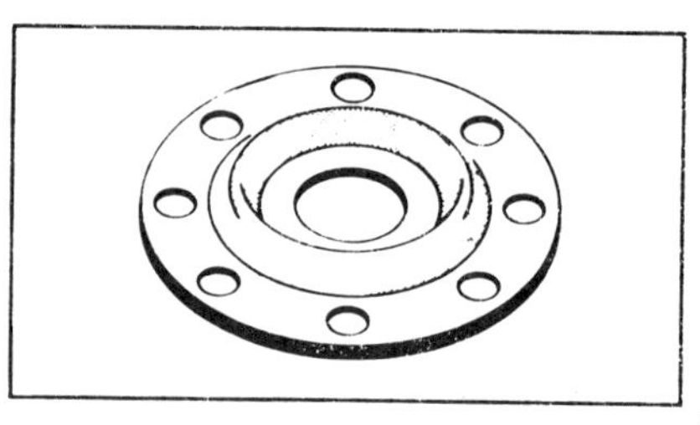

Fig. 10-2. Convoluted flat diaphragm. Convolutions increase allowable stroke per given diameter but increase expense.

Stroke

Most of the flat diaphragms that are heavy duty and need a long life have some sort of convolutions molded in. These let the diaphragm flex instead of stretching the fabric.

Usually, the travel available on convoluted diaphragms (Figure 10-2) is approximately 1.5 times the convolution height. Travel on dished diaphragms (Figure 10-3) is approximately twice the height of the dish. The effective area of a convoluted or dished diaphragm becomes more constant during operation as the convolution approaches a pure semicircle.

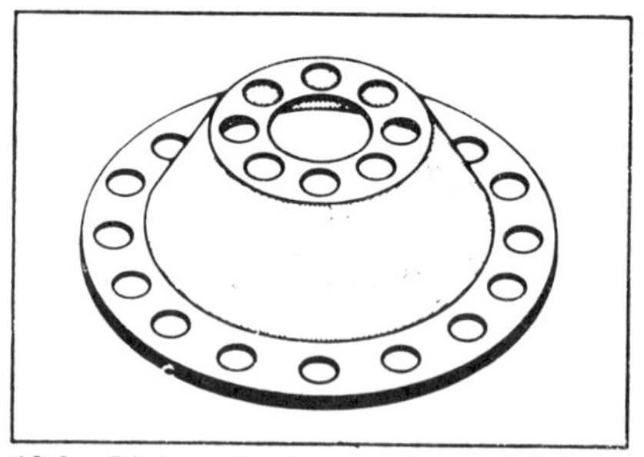

Fig. 10-3. Dished diaphragm. Deep dish increases working stroke. Design of application is more difficult because of danger of scuffing.

Practical limits are placed on how large the dish or convolution can be. If you have higher convolutions than the convolution width, you increase molding difficulty. Generally dished diaphragms are proportioned so 1/4″ of height is allowed for each inch of diameter.

Sometimes the small light-duty diaphragms for pumps or actuators are made perfectly flat. When this is done, stretching of the diaphragm material during operation cannot be avoided. Maximum available stroke is usually considered to be 7 to 9% of the diaphragm diameter. However, this figure depends on the flexibility of the material.

Pressure

The strength and resilience of the material and the design configuration used decide the amount of pressure a diaphragm can safely endure. Tensile strength of a diaphragm depends more on the fabric used in the membrane than it does on the elastomer.

The resilience of the diaphragm also affects the tensile strength. A material with a low modulus of elasticity, or a resilient material, will elongate more with a given load. The elongation causes a reduction in the cross-sectional area and thus reduces the load-bearing ability.

The burst strength, or pressure-bearing ability, of a diaphragm is affected by the amount of curvature in it. Usually those with more curvature are able to take higher pressures for the same material thickness. A full semicircular convolution with a 180° curve is the optimum shape. The partial convolution, which is pressure-formed in flat diaphragms, is generally about 60°. For a partial convolution the stress increase in the fabric is 1.42 times that for a 90° convolution, and twice as much as for a 60° convolution, compared to a 180° convolution. Later in this chapter we shall give the equations for calculating fabric stress in a 180° convolution.

A thicker membrane is usually required for flat diaphragms than for rolling diaphragms, since the stress levels are higher for a given pressure. The thicker membranes are not as sensitive to small pressure changes, and hysteresis losses are greater.

Area Shift

The flat diaphragm's effective pressure area is more at the two ends of its stroke than it is at the middle. Some devices rely on proportionate displacement so that a flat diaphragm cannot be used.

If the cross-sectional shape of the unsupported area of the diaphragm has a true semicircular cross section at the bottom of the convolution, the area of any fabric-and-elastomer diaphragm is held constant, since the circular shape is the natural configuration for a pressure-filled elastic membrane.

The convolution that is formed in flat diaphragms by pressure reaction is only part of a semicircle. Therefore, the geometry of the cross section of the convolution will shift from a part semicircular configuration in its mean stroke position to a substantially straight-line configuration at either the bottom or the top of the stroke. These variations cause substantial changes in the effective working area.

Design of Assemblies

A complete design involves the specification of associated parts beside selecting a diaphragm type, thickness, diameter, and material.

The center of the diaphragam is usually held with plates, and the outer portion is attached to the housing at the edges. Attached to the center plates and guided by the housing is a push rod, which transfers motion to the diaphragm, or is moved by pressure from behind the diaphragm.

Center plates may be attached by cementing, but these cemented plates cannot accept high stresses. The cementing eliminates the center hole and makes sealing easier. Bolted center plates are used where diaphragm replacement might be required.

Springs are generally used in conjunction with diaphragms to give a return or to give a reference force for pressure regulation. These springs generally ride on the center disc.

Flange pressures on the clamped portion of the diaphragm have to be sufficient to hold the diaphragm and to cause a seal on the edges. In this respect you can compare diaphragms to a gasket.

There is a correct "squeeze" or precompression for each diaphragm application, which will give adequate sealing without undue distortion of shape. This value should be determined, even if by trial and error, and should be adhered to on all production units.

ROLLING DIAPHRAGMS

Rolling diaphragms are used as if they were long-travel bellows, or radially sealed hydraulic pistions with zero leakage.

Rolling diaphragms can be used where:

- Very low hysteresis loss is required.
- High breakout friction is offensive, or moving friction should be kept low.
- The spring gradient of metal bellows is not needed, or the required stroke is greater than that obtainable by metal bellows.
- A slight leak is objectionable in cylinders or actuators.
- Response to slight pressure variations is needed.
- The contained fluid or gas carries abrasive particles in a cylinder or an actuator.

Operation

The rolling diaphragm has a full 180° convolution. When the pressure increases in the loading chamber (Figure 10-4), the piston will move down, causing the diaphragm to roll off of the piston side wall and onto the cylinder wall with a frictionless action. Almost the entire pressure load is supported by the piston head, and just a small amount of the pressure is felt by the diaphragm. Contained pressure holds the rolling diaphragm against the walls of the cylinder and piston.

This pressure causes a force against the convolution of $F = P \times C$ for each inch of circumferential length, where F = pressure force, lb; P = pressure, psi; C = convolution width, in. This pressure force, F, is supported equally by the fabric members in both walls of the rolling diaphragm. So, the tensile force, F_t, acting in either wall is, for each unit of circumferential length, in lb/in.:

$$F_t = \frac{PC}{2}$$

Since fabric stress and tensile stress, S_t, are identical, the previous equation can be expressed by:

$$S_t = \frac{PC}{2}$$

Rolling diaphragms can be used in many

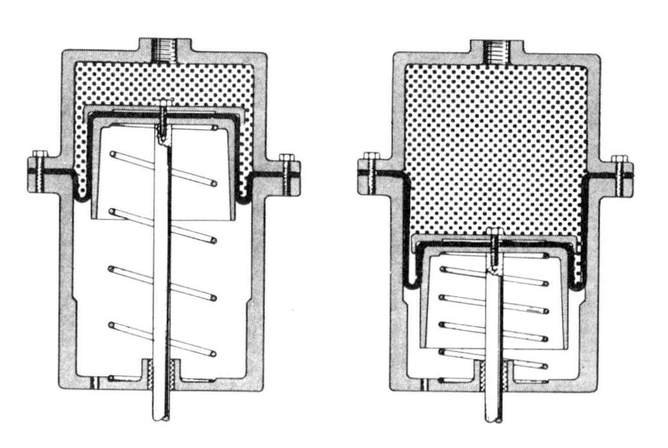

Fig. 10-4. Rolling diaphragm. As the piston moves down, the diaphragm rolls off the piston side wall onto the cylinder side wall. Diaphragms are usually made in relatively thin thicknesses, 0.0010 to 0.035 in., from fabric overlay. Fibers are oriented to allow free circumferential elongation but no axial elongation.

medium- and high-pressure applications, owing to the narrow convolution widths, and the resulting low stress values in the fabric.

Pressure Reversal

There is one limitation to rolling diaphragms. They cannot be used in an application where the normally low-pressure side exceeds, at any time, the pressure on the normally high-pressure side by more than 1/2 psi.

Pressure reversal can cause side wall distortion, overstressing, and scuffing action (see Figure 10-5). Usually, any pressure reversal causing mutliple pleats will not be corrected by application of pressure on the high-pressure side. Early failure will be the result of operation with pleats. Special rolling diaphragms are available that will flip-flop under pressure reversals without causing multiple pleats.

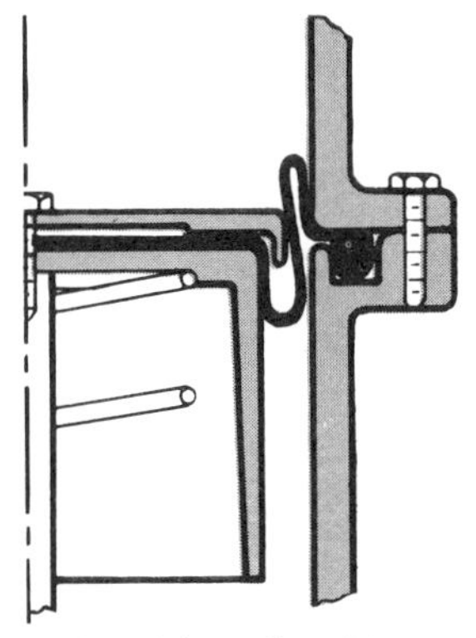

Fig. 10-5. Pressure reversal may cause rolling diaphragm to wrinkle, and scuff against the side wall.

Standard Sizes

Table 10-1 gives the rolling diaphragm sizes. Basically there are two types of rolling diaphragms: (1) those molded in the shape of a top hat (Figure 10-6); and (2) those with

Fig. 10-6. Most rolling diaphragm applications use "top hat" type diaphragms, where convolution is made by hand inversion during assembly.

Table 10-1.

TYPICAL SIZES

for CLASSES 1, 3, 4, 5, and

CLASSES 3C, 4C, 5C (PRE-CONVOLUTED DESIGN)

Cylinder Bore D_C	Piston Diameter D_P	Convolution C	Effective Pressure Area A_E	TYPICAL HEIGHTS **H** for CLASSES 1, 3, 4					MAXIMUM Height K for Classes 3C, 4C, 5C
.37	.25		.08	.31					
.44	.31		.11	.31					
.50	.38		.15	.38					
.56	.44	1"/16	.20	.44					
.62	.50		.25		.50				.100
.69	.56		.31		.56				
.75	.62		.37		.62				
.81	.69		.44			.69			
.87	.75		.52			.75			
.94	.81		.60		.62	.81			
1.00	.81		.64	.44	.62	.81	1.00		
1.06	.87		.74	.44	.62	.87	1.06		
1.12	.94		.83	.44	.69	.94	1.12		
1.19	1.00		.94	.44	.50	.69	1.00	1.19	
1.25	1.06		1.05	.44	.50	.75	1.00	1.25	
1.31	1.12		1.17	.44	.56	.81	1.06	1.31	
1.37	1.19		1.29	.44	.56	.87	1.12	1.37	
1.44	1.25		1.42	.44	.62	.94	1.19	1.44	
1.50	1.31		1.55	.44	.62	.94	1.25	1.50	
1.56	1.37	3"/32	1.69	.44	.69	1.00	1.31	1.56	.150
1.62	1.44		1.84	.44	.69	1.00	1.37	1.62	
1.68	1.50		1.99	.44	.75	1.06	1.44	1.68	
1.75	1.56		2.15	.44	.75	1.06	1.44	1.75	
1.87	1.69		2.49	.44	.81	1.12	1.50	1.87	
2.00	1.81		2.85	.44	.81	1.25	1.62	2.00	
2.12	1.94		3.24	.62	.87	1.31	1.75	2.12	
2.25	2.06		3.65	.62	.94	1.37	1.81	2.25	
2.37	2.19		4.08	.62	1.00	1.44	1.87	2.37	
2.50	2.31		4.54	.62	1.06	1.50	2.00	2.50	
2.62	2.31		4.79		1.06	1.56	2.12	2.62	
2.75	2.44		5.28		1.12	1.62	2.25	2.75	
2.87	2.56		5.80		1.12	1.69	2.31	2.87	
3.00	2.69		6.35		1.19	1.75	2.37	3.00	
3.12	2.81		6.92		1.25	1.87	2.50	3.12	
3.25	2.94	5"/32	7.51		1.31	1.94	2.62	3.25	
3.37	3.06		8.13	1.00	1.37	2.00	2.75	3.37	.250
3.50	3.19		8.78	1.00	1.44	2.12	2.81	3.50	
3.62	3.31		9.45	1.00	1.50	2.19	2.87	3.62	
3.75	3.44		10.1	1.00	1.50	2.25	3.00	3.75	
3.87	3.56		10.9	1.00	1.56	2.37	3.12	3.87	
4.00	3.69		11.6	1.00	1.62	2.44	3.25	4.00	
4.12	3.62		11.8	1.00	1.62	2.50	3.31	4.12	
4.25	3.75		12.6	1.00	1.69	2.62	3.37	4.25	
4.37	3.87		13.4	1.00	1.69	2.69	3.50	4.37	
4.50	4.00		14.2	1.00	1.81	2.75	3.62	4.50	
4.62	4.12		15.0	1.00	1.81	2.81	3.69	4.62	
4.75	4.25		15.9	1.00	1.87	2.87	3.75	4.75	
4.87	4.37		16.8	1.00	1.87	2.94	3.87	4.87	
5.00	4.50		17.7	1.00	2.00	3.00	4.00	5.00	
5.12	4.62		18.7	1.12	2.00	3.12	4.12	5.12	
5.25	4.75		19.6	1.12	2.12	3.19	4.25	5.25	
5.37	4.87		20.6	1.12	2.12	3.31	4.37	5.37	
5.50	5.00	1"/4	21.6	1.12	2.25	3.37	4.44	5.50	.375
5.62	5.12		22.7	1.25	2.25	3.44	4.50	5.62	
5.75	5.25		23.8	1.25	2.37	3.50	4.56	5.75	
5.87	5.37		24.9	1.25	2.37	3.56	4.75	5.87	
6.00	5.50		26.0	1.25	2.50	3.62	4.81	6.00	
6.25	5.75		28.2	1.50	2.75	4.00	5.00	6.25	
6.50	6.00		30.6	1.50	3.00	4.25	5.25	6.50	
6.75	6.25		33.2	1.75	3.25	4.50	5.50	6.75	
7.00	6.50		35.8	1.75	3.50	4.75	5.75		
7.25	6.75		38.5	2.00	3.75	5.00	6.00		
7.50	7.00		41.3	2.00	4.00	5.25	6.25		
7.75	7.25		44.2	2.25	4.25	5.50	6.50		
8.00	7.50		47.2	2.25	4.50	5.75	6.75		

molded-in convolution (Figure 10-7), which usually have a limited stroke.

Both of these types are available in four separate rim configurations, which are grouped in classes. Table 10-2 defines classes 1, 3, 4, and 5 for the top hat type. The classes for the molded-convolution type are the same, but a "C" is added after the class number.

Design of Assemblies

The design of a rolling diaphragm application involves many of the same principles used in the design of flat diaphragm assemblies, with some exceptions.

Table 10-3 gives recommended flange designs, cylinder diameters, piston diameters, and so on. These are for all types and classes of rolling diaphragms except that diaphragms with molded convolutions are only available in one maximum height for each size range. Table 10-4 shows their stroke.

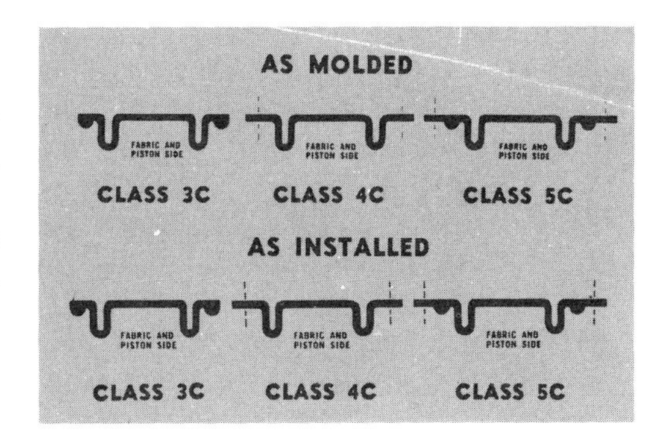

Fig. 10-7. Class-C rolling diaphragms have molded convolutions. They usually have less stroke than top hat types.

Table 10-2.

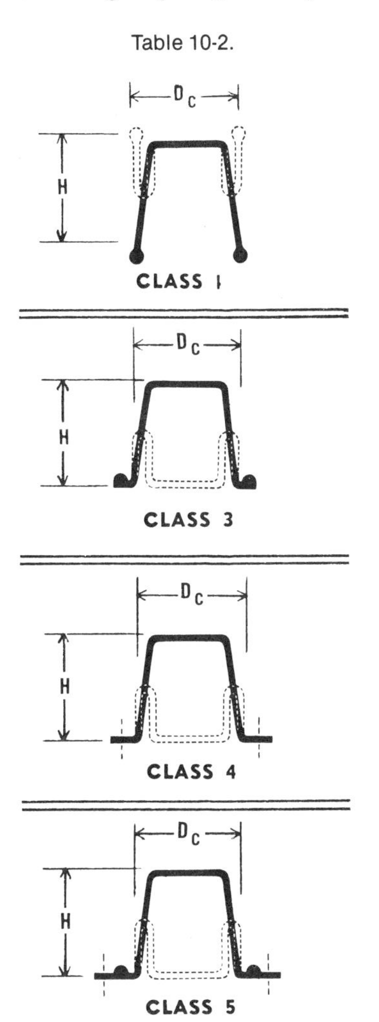

Plain Bead: Used where minimum housing O.D. is required. Square groove should be used, with volume of groove equal to volume of bead. Conventional bolted-flange construction can be completely eliminated, and the diaphragm held by flange retainer plate. Or, it can be used with threaded bevel ring.

Beaded Rim: Used when a soft gasket action is required to provide tight pressure sealing against roughly machined or warped flange surfaces. Has a bead around entire periphery of a rim having narrow radial width. Volume of the retaining groove should be equal to the volume bead.

Plain Rim: Most commonly used. Applied in applications that have flat mating surfaces between the cylinder and the bonnet. Rim of the diaphragm also serves as a gasket to prevent leakage. The metal flange faces should be flat or serrated (concentric V-grooves spaced 1/32 in. apart with a depth of 0.006 in.) to prevent pull-out of rim under high pressure. Flange loading pressure should usually not exceed 1,000 psi.

Extended Beaded Rim: Used when the working pressures are in excess of 150 psi and when flange clamping surfaces are warped or rough.

TYPICAL DIMENSIONS

Table 10-3. Rolling Diaphragm Dimensions

(The associated figures are only for dimensional reference. Actually when mating flanges are in position shown, the beads of the Rolling Diaphragm completely fill the bead grooves of the flange.)

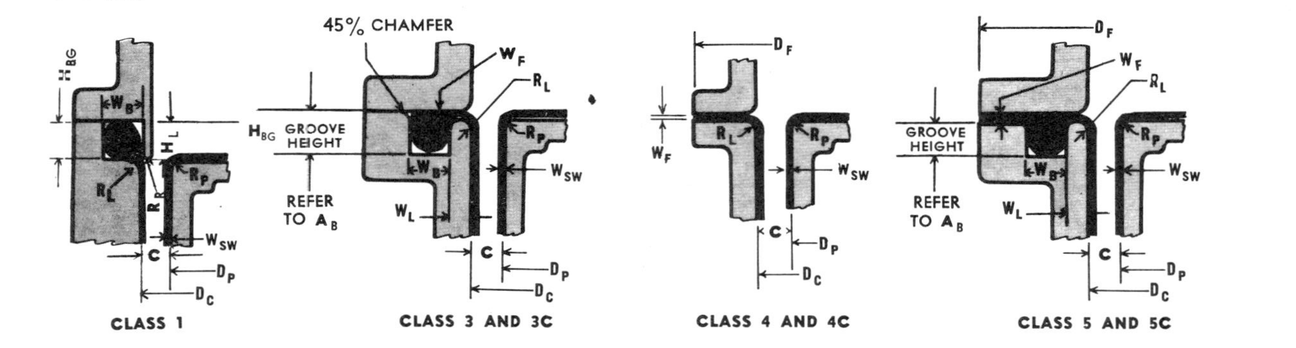

DIMENSIONS FOR ROLLING DIAPHRAGM DESIGNS

Cylinder Bore D_C	Piston Diameter D_P	Convo-lution Width C	Maximum Up-Stroke S_A	Maximum Down-Stroke S_B	Classes 1, 3, 4, 5 Min. Piston Skirt Length L_P	Classes 1, 3, 4, 5 Min. Cylinder bore length L_C	Classes 3C, 4C, 5C Min. Piston Skirt Length L_P	Classes 3C, 4C, 5C Min. Cylinder bore length L_C	Max. Standard Outside Flange Diameter Classes 4, 4C D_F	Max. Standard Outside Flange Diameter Classes 5, 5C D	Blend Radius R_L & R_P	Class I Bead Groove Dimensions R_R	Class I Bead Groove Dimensions H_L	Class I Bead Groove Dimensions H_{BG}	Clamp Lip Width W_L
.37 to .99	$D_C - \frac{1}{8}$	$\frac{1}{16}$	$H - \frac{9}{32}$	$H - \frac{9}{32}$	$\frac{H + S_A}{2}$	$\frac{H + S_B}{2}$	$\frac{5}{32}$	$\frac{5}{32}$	$D_C + .75$	$D_C + 1.31$	$\frac{1}{32}$	—	—	—	.062
1.00 to 2.50	$D_C - \frac{3}{16}$	$\frac{3}{32}$	$H - \frac{13}{32}$	$H - \frac{13}{32}$	$\frac{H + S_A}{2}$	$\frac{H + S_B}{2}$	$\frac{7}{32}$	$\frac{7}{32}$	$D_C + 1.0$	$D_C + 1.50$	$\frac{1}{16}$	.025	.103	.100	.125
2.51 to 4.00	$D_C - \frac{5}{16}$	$\frac{5}{32}$	$H - \frac{5}{8}$	$H - \frac{5}{8}$	$\frac{H + S_A}{2}$	$\frac{H + S_B}{2}$	$\frac{11}{32}$	$\frac{11}{32}$	$D_C + 1.5$	$D_C + 2.00$	$\frac{3}{32}$	.032	.131	.125	.187
4.01 to 8.00	$D_C - \frac{1}{2}$	$\frac{1}{4}$	$H - \frac{29}{32}$	$H - \frac{29}{32}$	$\frac{H + S_A}{2}$	$\frac{H + S_B}{2}$	$\frac{1}{2}$	$\frac{1}{2}$	$D_C + 2.0$	$D_C + 2.75$	$\frac{1}{8}$	.045	.203	.200	.250

Cylinder Bore D_C	Sidewall Thickness W_{SW}	Flange Thickness W_F	Bead Width Classes 1, W_{BG}	Bead Width Classes 3, 3C, 5, 5C W_{BG}	Bead Groove Area Classes 1 H_{BG}	Bead Groove Area Classes 3, 3C, 5, 5C H_{BG}
.37 to .99	.015 ± .002	.020	— —	.109 ± 003	— —	.082 ± 002
1.00 to 2.50	.017 ± .003	.020	.125 ± 002	.141 ± 003	.096 ± 002	.118 ± 002
2.51 to 4.00	.024 ± .004	.030	.156 ± 002	.219 ± 003	.122 ± 002	.178 ± 002
[illegible]	[illegible]	[illegible]	[illegible]	[illegible]	[illegible]	[illegible]

Table 10-4. Stroke versus Height Chart (Use only for Classes 3C, 4C, 5C)

CYLINDER BORE D_C	CONVOLUTION C	MAXIMUM STROKE S_A OR S_B	MAXIMUM HEIGHT K
.37— .99	1/16	.075	.100
1.00—2.50	3/32	.085	.150
2.51—4.00	5/32	.150	.250
4.01—8.00	1/4	.250	.375

Flex Life: While a rolling diaphragm is working, it is subjected to a change in circumference (see Table 10-5). The circumference increases when it rolls from the smaller-diameter piston to the larger-diameter cylinder wall.

The number of expected total strokes before failure is dependent on the mean operating level of pressure and on the percent of total stroke used, assuming that temperature or environmental conditions do not degrade the fabric or elastomer.

When the application needs high flex life, the fabric should be prepared in such a way as to provide free circumferential elongation. If freedom in circumferential elongation is not enough, serious stress buildups will result. Axial fabric cords should be free from elongation under maximum load.

Shaft Torque: If the possibility exists that the piston shaft may be subjected to torque during assembly, a rotating slip joint should be used to prevent turning action from wrinkling the rolling diaphragm (see Figure 10-8). Twist can also be applied to the piston by some compression springs, so that when compressed they will rotate through some angular displacement. If this twisting effect is more than 2°, one end of the spring should be supported on a ball bearing thrust plate.

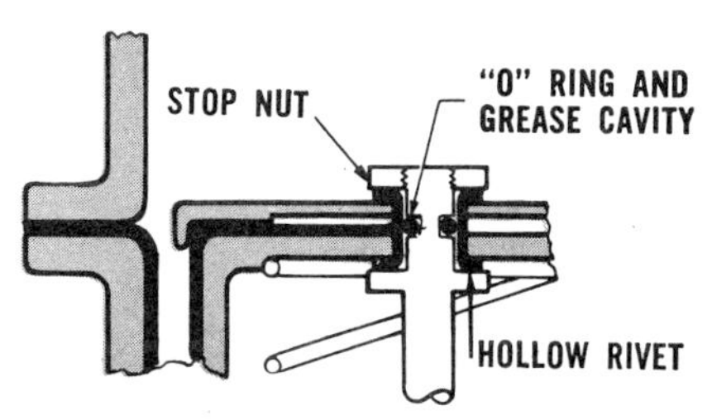

Fig. 10-8. Rotating slip joint is used to hold retainer plate when there is danger of twisting diaphragm during assembly.

Piston Eccentricity: The convolution width is affected by the piston and cylinder eccentricity. This eccentricity should not exceed 10% of convolution width, and some means of guiding the piston shaft must be used in most applications.

A return spring is required in many applications. Very rarely does the force exerted by

Table 10-5. Flex Life as Related to Axial and Free Circumferential Elongation of Fabric

AXIAL ELONGATION (PERCENT)	CIRCUMFERENTIAL ELONGATION* (PERCENT)	EXPECTED LIFE FOR 1 IN. BORE AND 0.8 IN. STROKE AT 100 PSI (1,000 CYCLES)
5	5	375
4	10†	2,107
3	20†	4,546

*This elongation is measured in the "straight fabric zone" of the side wall near the head.

†Special fabric formed with low axial stretch and extra elongation in "straight zone" of side wall.

a compression spring parallel the spring axis, so a bending moment is introduced to the head of the piston.

Retainer Plates: The rolling diaphragm type (molded in the shape of a top hat) needs inversion during assembly to form the convolution. In some installations, and in certain operating conditions, its natural resiliency may cause a tendency for it to reinvert to its original position. This will cause what is known as side wall scrubbing (Figure 10-9), which can be eliminated by using retainer plates with curved lips (see Table 10-6).

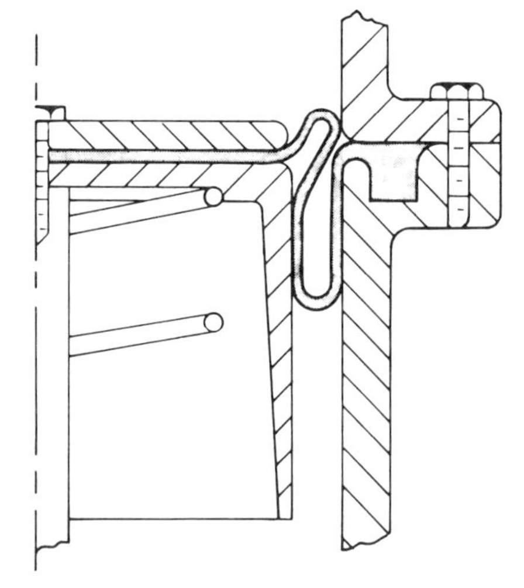

Fig. 10-9. With a straight-sided retainer plate, side wall scrubbing could occur. For all applications except Class 1, the curved-lip plate is recommended.

DIAPHRAGM MATERIALS

The following major material properties are very important:

Table 10-6. Dimensions for Curved Retainer Plate

A = D_p (Piston Dia.)

Clearance Hole For Retainer Cap Hold Down Screw

Full Radius

F

Grad.

B

C

D

E

A

CYLINDER BORE D_c	A	B	C	D	E	F	GRAD.
3.7 to .99	$D_p + 2WSW$	Not required	.015	1/16	1/8	Not required	.025
1.00 to 2.50	$D_p + 2WSW$	.7 D_p	.025	3/32	3.16	.010	.030
2.51 to 4.00	$D_p + 2WSW$	.71 D_p	.030	7/64	7/32	.015	.040
4.01 to 8.00	$D_p + 2WSW$	.7 D_p	.030	1/8	1/4	.015	.060

All dimensions in inches.

D_p = outside diameter piston skirt.

W_{SW} = max. side wall thickness of rolling diaphragm.

Burst Strength

This is the pressure at which a diaphragm will fail. It is affected by two other properties, tensile strength and modulus. Since the ultimate tensile strength is based on the tensile stress required to cause failure for a given sectional area, any reduction of sectional area caused by elongation under pressure progressively lowers the bursting strength of a given sample. Therefore, a high-modulus compound should ordinarily show a higher burst test than would a low-modulus compound, though the ultimate tensile strengths of both are about the same.

Flexibility

This term is usually employed to describe the relative deformation or deflection of a material when it is subjected to a given load. This property is dependent on modulus. "Flexible compounds" describes the low-modulus compounds, and "stiff compounds" describes the high-modulus compounds.

Material Selection

The selection of a material for the diaphragm involves the choice of a fabric and the choice of an elastomer. Listed below are the most widely used fabrics and the more common elastomers.

Fabric

Cotton is the least expensive fabric but very seldom used. Cotton yarn is formed from short fibers, and its tensile strength is low compared to that of continuous-filament yarns. When temperatures are above 200°F, cotton should not be used, or the result will be abrasion or fatigue.

Dacron* is well suited for use in diaphragms, since it can withstand temperatures up to 350°F, and has excellent resistance to abrasion and fatigue. Operation at higher pressure ratings or longer cycle life has resulted from using Dacron yarns in diaphragms.

Nylon* is used in flat diaphragms. Temperatures in excess of 250°F degrade it rapidly. Nylon has an excellent resistance to abrasion and fatigue, but has a tendency to creep in pressure applications and to take a "set" if kept in a fixed position for long periods. The cost of nylon and Dacron is approximately 50% above that of cotton.

Nomex fabric has replaced glass and fluorocarbon fabrics in most diaphragms for high-temperature applications. Its heat resistance follows closely that of the best elastomers, and it can be used for prolonged exposures at 500°F. It has excellent resistance to fatigue and abrasion, in contrast to the poor resistance of glass and fluorocarbons. Nomex does not have the "weave shift" problems of glass and fluorocarbon fabrics.

Elastomers

Buna N is the most commonly used elastomer because it has excellent resistance to solvents, oils, fats, and aromatic hydrocarbons; good flex resistance and abrasion; and a temperature range of −50° to +225°F. It is degraded by ozone, but this problem can be overcome by additives.

Neoprene has good resistance to oxidation and ozone, and performs well in oil and some chemicals. It also has good tensile strength. Neoprene's usable temperature range is −45° to +225°F.

Silicone carries the widest service temperture range of any of the elastomers, −125° to +550°F. It has good ozone and weathering properties. Recently silicones with higher strengths have become available.

Fluorosilicone is used instead of silicone in applications that require a resistance to oil, solvents, and aromatic hydrocarbons. The cost of fluorosilicone is high.

Ethylene propylene, which has an excellent resistance to ozone and steam, has a temperature range of −50° to +275°F, and has good strength.

Viton is used in diaphragms because of its ability to resist many fluids over a wide temperature range, from −10° to +550°F. Its cost is very high.

Also, many elastomers are used for specific

applications. Polyacrylate is used for oil service with temperatures between 200° and 275°F. Polyurethane is used where high pressures are needed. Where impermeability is important, butyl is used; however, a new elastomer, Hydrin 100, has a one-third lower permeability rate, and is also resistant to most fluids. Hypalon is extremely resistant to alkaline and acid solutions.

Some typical pressure switch applications using the rolling diaphragm are shown in Figures 10-11 and 10-12.

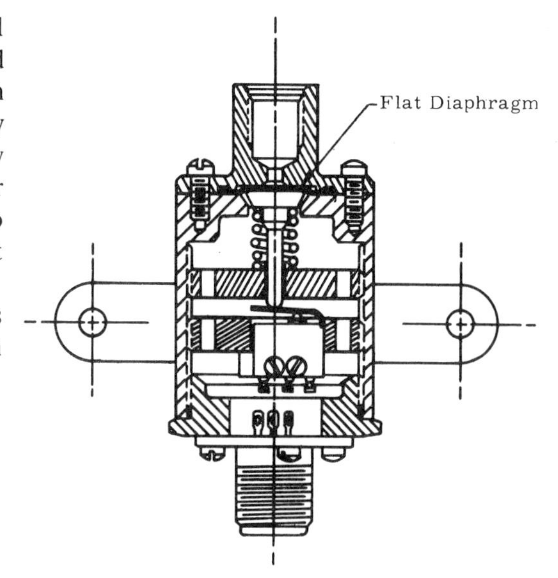

Fig. 10-10. Flat diaphragm. *(Courtesy of Essex Cryogenics Ind., Inc., St. Louis, Missouri.)*

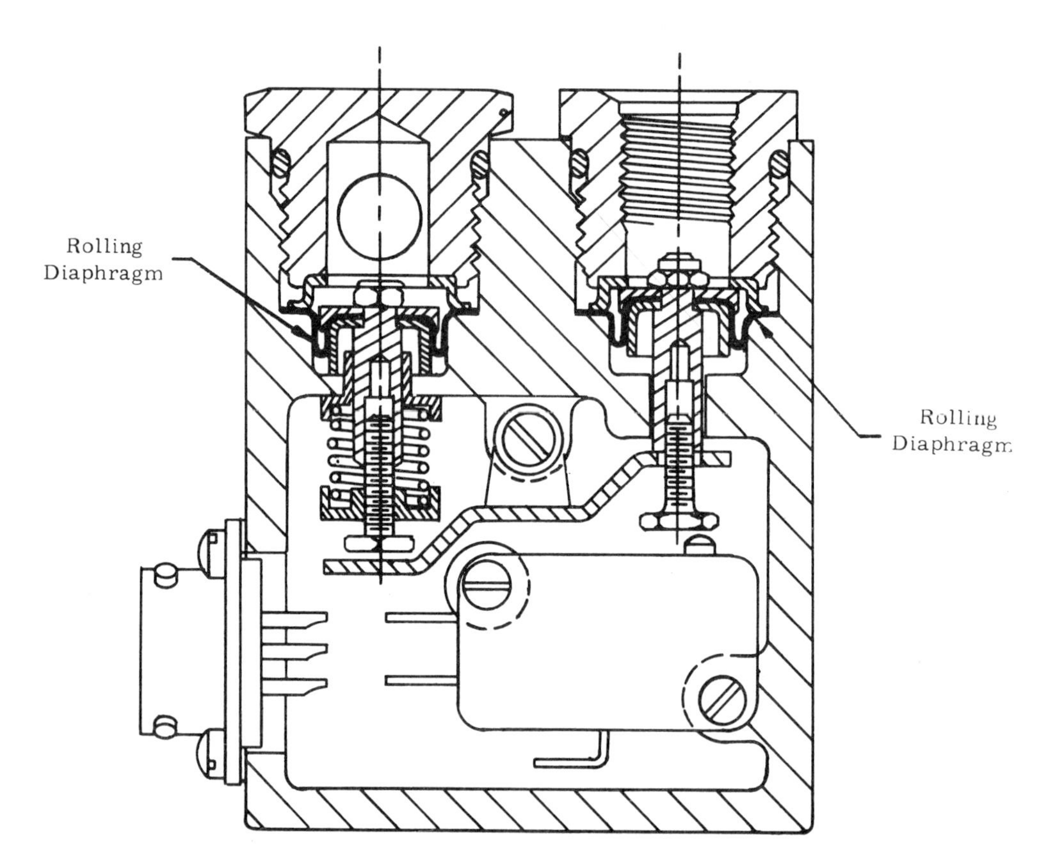

Fig. 10-11. Rolling diaphragm differential pressure switch. *(Courtesy of Essex Cryogenics Ind., Inc., St. Louis, Missouri.)*

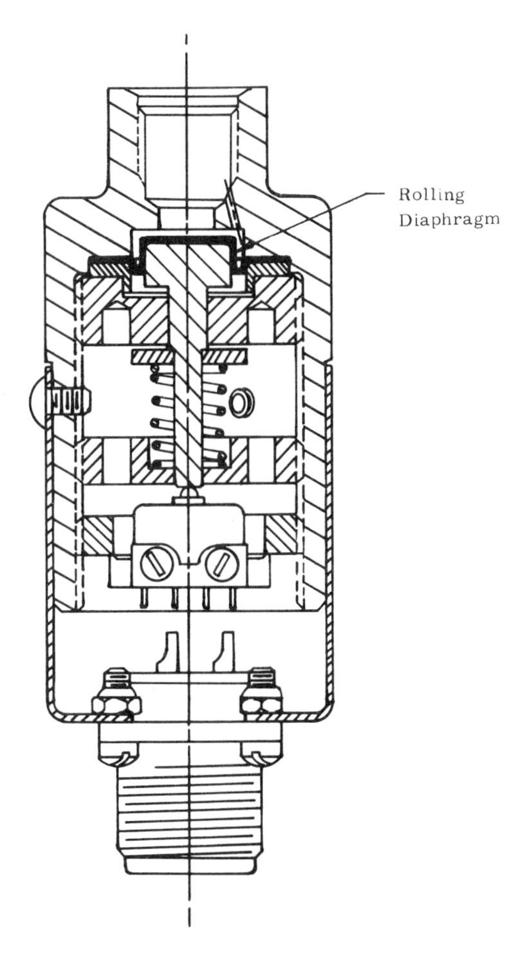

Fig. 10-12. Rolling diaphragm pressure switch. *(Courtesy of Essex Cryogenics Ind., Inc., St. Louis, Missouri.)*

REFERENCES

1. Diaphragm Design Manual, Bellofram Corporation, Burlington, Massachusetts.

B. METALLIC DIAPHRAGMS

INVESTIGATIONS OF THE PROPERTIES OF CORRUGATED DIAPHRAGMS

The pressure-deflection characteristics of corrugated diaphragms are correlated by methods of dimensional analysis. Experimental results for various sizes, materials, thicknesses, and shapes of diaphragms indicate that the performance for diaphragms of any given shape may be computed from a dimensionless formula derived from experimental data on other diaphragms of that shape. Linear shell equations are derived for combined bending and stretching effects with lateral loading terms for rotationally symmetrical shells in appropriate independent and dependent variations suitable for complicated meridional shapes, and with boundary conditions associated with practical diaphragm applications. The method used for solving this system of equations on an electronic digital computer is described, and numerical solutions are presented for a specific diaphragm subjected to uniform pressure loading. Suggestions are presented for future research, both theoretical and experimental, on diaphragm properties and performance.

INTRODUCTION

Metallic diaphragms, singly or in pairs as capsules, find a wide use in pressure-measuring instruments of many types. Their advantages include compactness, simplicity, cheapness, as well as a wide range of load, sensitivity, and deflection characteristics. The use of corrugated disks makes possible larger deflections and affords a control of the shape of the load-deflection curve. There are, however, certain limitations in practice that confront the designer who wishes to make use of diaphragms in any given application. For one thing, there are no generalized design formulas to predict the behavior of corrugated diaphragms. Even for the limiting case of flat disks, where theory has led to the development of load deflection equations (1-8), convenient charts, tables, or other aids to design have only recently become available. It is possible to set theoretical limits to the gain in deflection that can be achieved by corrugating disks (1), but there is no easy way of predicting, except in a general qualitative fashion, the performance to be expected for a given type of corrugation. This practical difficulty is serious because it is always desirable to use the diaphragm best suited to the specified application; the selection of the best form of diaphragm by the trial and modification method may involve considerable development expense.

Diaphragm instruments are subject to various errors arising from the imperfect elastic properties of the diaphragm materials, evident in the phenomena of hysteresis, drift, aftereffect, recovery, and zero shift (or set). For many modern applications, in industry as well as in research, the errors arising from these sources must be less than 0.1% of the maximum deflection. This relatively high standard of performance is achieved by limiting the load to values that will not stress the material to near its yield point, by using those materials with the best elastic properties, by following closely a satisfactory technique of manufacture coupled with selection on the basis of inspection, and, in some measure, by calibrating under conditions corresponding to the manner of use.

There are two ways to determine the properties of corrugated diaphragms: The empirical approach is to make diaphragms of various sizes, shapes, and materials and to load them in various ways and measure their deflection, drift, hysteresis, recovery, and set. From judicious selection of sizes and shapes for experiment it may be possible to deduce design criteria and generally applicable formulas which may be used in design of other diaphragms of similar shape and materials or, by interpolation or extrapolation, of other shapes and other materials.

The second approach is the analytical,

mathematical one which requires that equations be derived relating the load-deflection characteristics to the geometry of the diaphragm and the elastic constants of the material. It appears that predicting the approximate linear characteristics of a "perfect" diaphragm represents the limit of what may be practically obtained from this approach at present, although future progress might make it possible to predict something about the elastic defects or good nonlinear behavior.

With the present imperfect knowledge of the laws governing drift, hysteresis, and aftereffect, experience alone can tell which material, which shape, or which technique of preparation, will result in producing the diaphragms with the most desirable properties.

1. EXPERIMENTAL INVESTIGATIONS

It is only in recent years that partially successful mathematical attacks have been made on the diaphragm problem, so that the history of diaphragm design is primarily a record of cut-and-try methods. Because of the great importance of elastic elements in pressure-measuring systems, the NBS has maintained an interest in diaphragm design for many years and has cooperated with other government agencies and industry in several investigations.

Hersey in 1923 (9) reviewed the experimental work and suggested the dimensional-analysis approach to correlation of test results. In the late 1930s a project for research on diaphragm performance was undertaken on behalf of the National Advisory Committee for Aeronautics, and it was decided that the empirical approach would be the most fruitful in view of the mathematical difficulties and the fact that many of the most important characteristics for practical use relate to the elastic defects, which even now are unlikely to be predicted analytically. An arbitrary diaphragm design was selected, and dies were made in various sizes so that a family of test specimens could be made. It was anticipated that the data could be correlated by dimensional analysis to give general design formulas, each applicable to certain shapes. The results of these investigations were published as Technical Notes, by the National Advisory Committee for Aeronautics (10 and 11) and received limited distribution. The essential conclusions of these investigations will be summarized herein, and the pertinent figures reproduced.

In the early NBS work, a large number of corrugated diaphragms of beryllium copper, phosphor bronze, and Z-nickel, having geometrically similar outlines (NBS Shape 1) but of various diameters and thicknesses, were formed by hydraulic pressing. The diaphragms were made with geometrically similar outlines in order that the data on various sizes could be easily used in developing design formulas by dimensional analysis. The load-deflection data experimentally obtained were correlated by dimensional analysis to obtain a general formula that applies to the particular corrugation outline used.

Less complete data available on diaphragms with other corrugation shapes were also correlated in this manner to obtain some indication of the effect of changes in shape. Some work was done on the deflection characteristics of diaphragms subjected to concentrated control loads.

The shape of the family of diaphragms covered in this early investigation was as shown in the cross-sectional view of the clamping and forming dies, Figure 10-1. It will be noted that the outline consists of circular arcs except at the center and that the outer corrugation joins the rim at a 90° angle. Four dies were used differing in the values of the effective diameter D, which were 1½, 2, 2½, and 3 in. The forming dies and clamping base are shown in Figure 10-2. Diaphragms were formed from commercially available sheet materials from 0.013 down to 0.002 in. in thickness. Some measurements were made on still thinner diaphragms formed by etching, so that data were obtained over a ratio of thickness to diameter over the range 3.6×10^{-4} to 80×10^{-4}. The central flat area of most of the diaphragms was reinforced by means of a stiffening disk soldered in place.

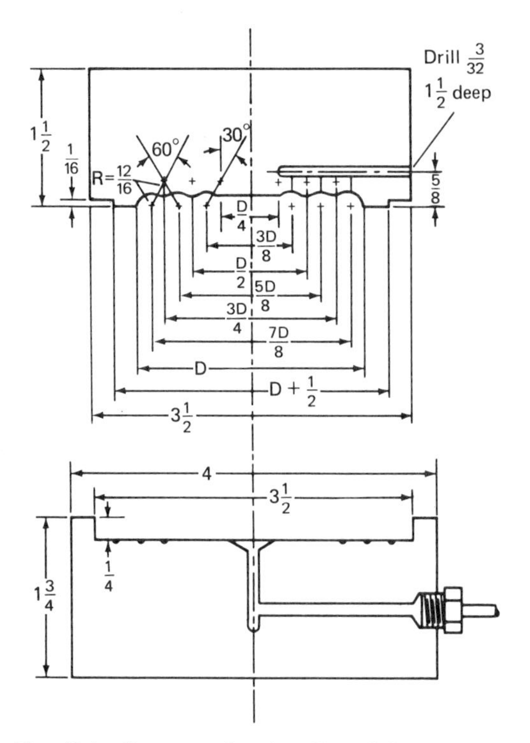

Fig. 10-1. Cross-sectional outline of die and pressure band used in forming diaphragms having NBS Shape 1. (The four dies used had geometrically similar outlines, differing only in values of *D*, which were 1½, 2, 2½, and 3 in.)

Fig. 10-2. Photograph of dies, 1 to 4, and base for dies.

Test data were obtained on these diaphragms by pneumatic loading, with measurements of the central deflection obtained by use of a screw micrometer at each pressure step. Data were obtained with both increasing and decreasing pressures over a number of cycles in order to determine hysteresis, drift, aftereffect, recovery, and zero shift.

We define these elastic defects as follows:

Hysteresis. The difference between the deflections of the diaphragm at a given load, for decreasing and for increasing loads.

Drift. The increase of deflection with time under a constant load.

Aftereffect. The deflection remaining immediately after removal of the load, that is, hysteresis at no load.

Recovery. The decrease of aftereffect with time under no load. (The term also may be applied to the time decrease of hysteresis at a constant load but was not studied in this sense.)

Zero Shift. The permanent deformation, that is, the difference in no-load deflection before loading and sufficiently long after unloading for recovery to occur; or the difference between aftereffect and recovery.

Zero shift appears to be a fairly absolute measure of the useful range of deflection; thus, if each loading causes permanent deformation, it is apparent that the loadings cannot be repeated indefinitely. Even this criterion is, however, to be scrutinized closely. Diaphragms will often show some zero shift initially after each of a number of loadings. Afterward their performance (up to that load) becomes stabilized and they are said to be seasoned. It is characteristic that imposition of a higher load will again cause a zero shift, which may decrease with succeeding loadings and stability may again be obtained. Heat-treatments for stress relief will often reduce the number of workings necessary to attain stability. There finally comes a point, however, in which the total deformation incident to the working is so great that the characteristics of the diaphragm over the low-pressure range are entirely different from the original. Alternatively the diaphragm may break during seasoning before stability is attained.

Presumably diaphragms like other mechanical elements will exhibit fatigue failures after repeated working, even well within their elastic limits. However, no definitive experimental results are available on this phase, at least for measuring instrument diaphragms.

Low-Deflection Characteristics

The shape of the diaphragms of NBS Shape 1 was such that the central deflections were nearly proportional to the differential pressures up to deflections of 2 to 3% of the diameter. Figure 10-3 shows the pressure-deflection curves for several reinforced-center beryllium-copper diaphragms. Dotted lines are extensions of straight lines through the 2% deflection points. None of the diaphragms had a perfectly linear characteristic over their total useful range. Finer-scale information on the exact nature of the load-deflection curves is shown better in graphs of deviation from linearity such as Figure 10-4. At larger deflections all such deviation-from-linearity curves will of course turn downward, i.e., become more stiff than over the approximately linear range.

There appears to be a correlation between the thickness-to-diameter ratio *t/D* and the phenomenon of increasing flexibility over part of the range. For diaphragms with *t/D* values below 2×10^{-3} the flexibility was found in general to increase above the initial value throughout part of the range.

The limiting useful deflections, investigated for several different materials (11), were found to differ markedly with different materials. For beryllium-copper hardened by appropriate heat-treatment after forming, deflections as great as 7% of the diameter could be obtained without apparent overstressing of the material. For some of the other materials excessive elastic defects became apparent when deflections exceeded 2 or 3% of the diameter. There appeared to be

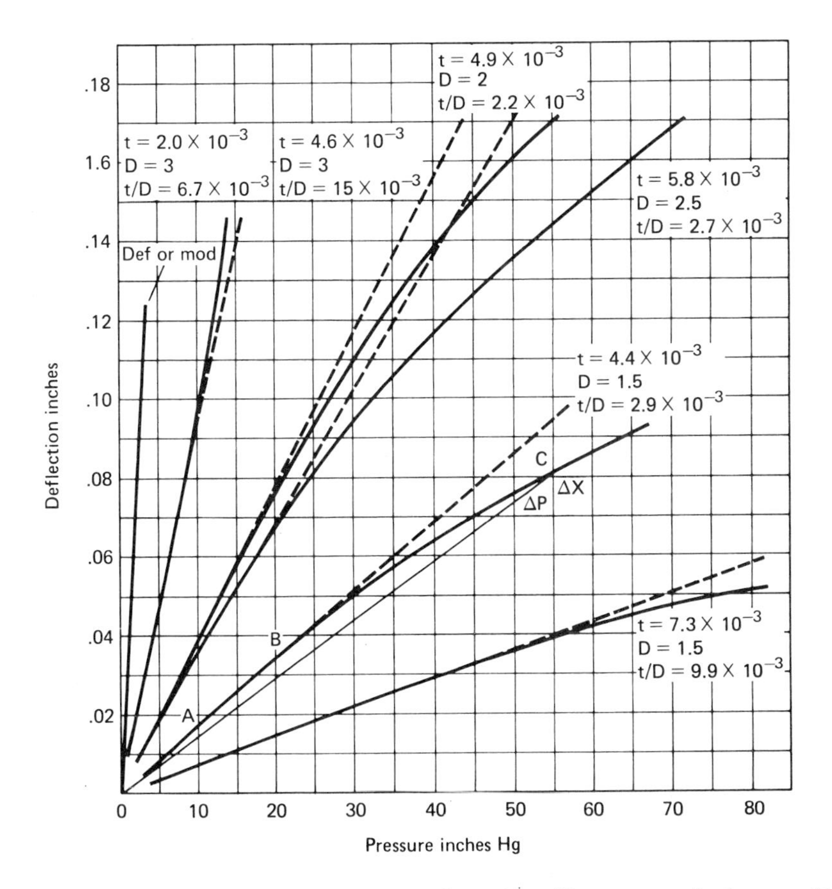

Fig. 10-3. Pressure-deflection curves for several reinforced beryllium-copper diaphragms. (Dotted lines are extensions of the straight lines through the $X = 0.02D$ points.)

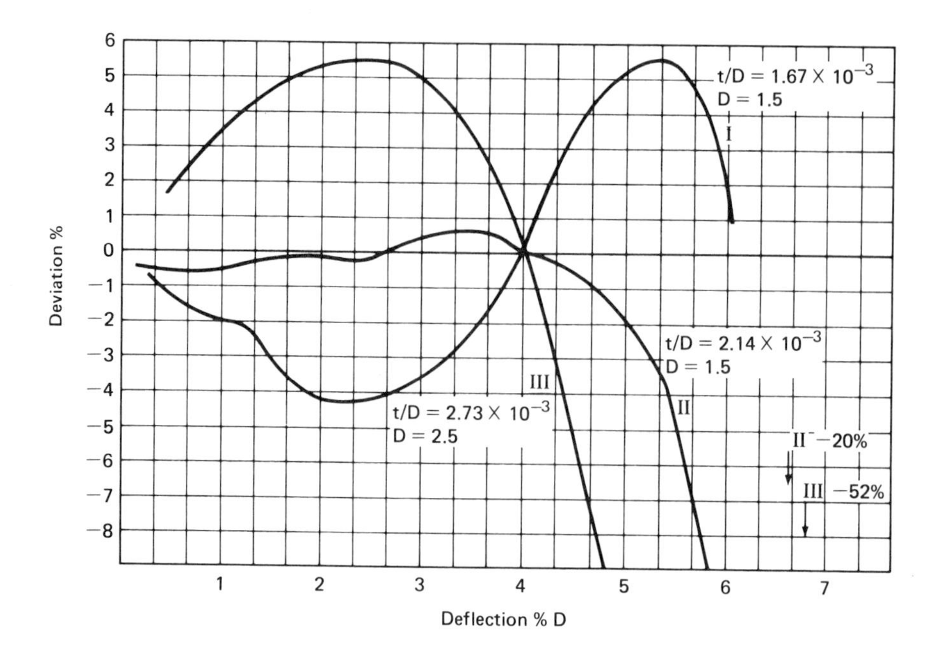

Fig. 10-4. Deviations from linearity of pressure-deflection curves for reinforced beryllium-copper diaphragms. (Deviations from straight lines passing through origin and load-deflection curve at $X = 0.04D$ are plotted as percentages of $0.04D$.)

a fairly good positive correlation between hardness of the material and the limiting deflection. However, if the softer materials could have been deflected without further yielding, they would presumably have shown somewhat similar load-deflection curves, as these are determined primarily by the shape of the diaphragm rather than its material.

Analysis of Experimental Results

In order to compare diaphragms of different materials and of different sizes to determine to what extent the character of the pressure-deflection curve is determined by shape factors, it is necessary to simplify the problem by combining some of the variables.

The method of dimensional analysis, developed by Buckingham (12), is useful in this connection. It is of interest that Buckingham published his epochal paper on engineering applications of dimensional analysis just forty years ago in the *ASME Transactions.* The method is based on Buckingham's famous π-theorem which states that a functional relationship will exist between the various nondimensional products which can be formed by combinations of the physical quantities which may vary in the given situation. The number of such nondimensional products is less than the number of physical variables by the number of independent fundamental units needed to specify the physical quantities (i.e., by three, in purely mechanical problems, since mass, length, and time suffice to define dimensionally any mechanical quantity).

The deflection of a diaphragm is influenced by the pressure applied, the diameter, the thickness, the shape of corrugation outline, and the elastic constants of the material.

Let X = deflection, D = diameter, t = thickness, P = applied pressure, E = Young's modulus, F = plate modulus, $E/(1 - \nu^2)$, and ν = Poisson's ratio. For all diaphragms of the same corrugation shape (dimensionally similar diaphragms)

$$X = f(D, t, P, E, \nu) \tag{1}$$

The dimensionless ratios formed by combinations of the parameters are also related, so that

$$\frac{X}{D} = \phi\left(\frac{t}{D}, \frac{P}{E}, \nu\right) \qquad (2)$$

The contour shape of the experimental diaphragms was such that the pressure-deflection curves (Figure 10-3) were very nearly straight lines over a considerable range. The existence of this linear relationship permits a simplification. Since $X/D = kP/E$ for a given diaphragm, then, in general, for diaphragms of similar shape

$$\frac{X}{D} = \frac{P}{E} f\left(\frac{t}{D}, \nu\right) \qquad (3)$$

In the theory of elastic deflection of flat plates, ν enters mainly in a factor, $(1 - \nu^2)$, that may be combined with E. Its effect, other than this, is less than 2% for a variation of ν from 0.25 to 0.3 (reference 5). This fact does not prove, but does suggest, that the influence of ν on performance of a corrugated diaphragm may be approximately the same.

In view of these considerations, the equation might be expected to apply fairly accurately to diaphragms of these different materials when written as

$$\frac{X}{D} = \frac{P}{E/(1 - \nu^2)} f(t/D) \qquad (4)$$

The modulus $E/(1 - \nu^2)$, the plate modulus, is designated by the symbol F.

Experimental observations on many diaphragms of the given shape show that the pressure-deflection relation is fairly linear over the range of deflections up to $X/D = 0.02$. For this range of deflections and over the range of $1000t/D$ from 0.6 to 6.0 approximately, the following equation was found to hold for all the diaphragms tested:

$$\frac{FX}{PD} = 2.25 \times 10^5\, (t/D \times 10^3)^{-1.52} \qquad (5)$$

The absolute magnitude of the exponent increases beyond each end of this range of $1000\, t/D$.

The general curve developed as Figure 9 in reference 10 from experimental results on diaphragms of this shape over the range of X/D up to 0.02 is reproduced here as Figure 10-5. The points plotted were determined from data for reinforced beryllium-copper diaphragms, loaded on the convex side. The effective values of F for other materials were so chosen that the points for other diaphragms fall on the same curve. This curve defines the stiffness of the diaphragms P/X (or flexibility, X/P) over the given range of deflections as a function of D, effective thickness t, and the "effective" plate modulus of the material F. For beryllium copper, the value of the plate modulus was taken as 18.9×10^6 psi. The corresponding effective values for unreinforced diaphragms of beryllium copper were 17.0, phosphor bronze, 15.1, A-nickel, 31.7, B-Monel, 22.2, K-Monel, 24.5, Inconel,

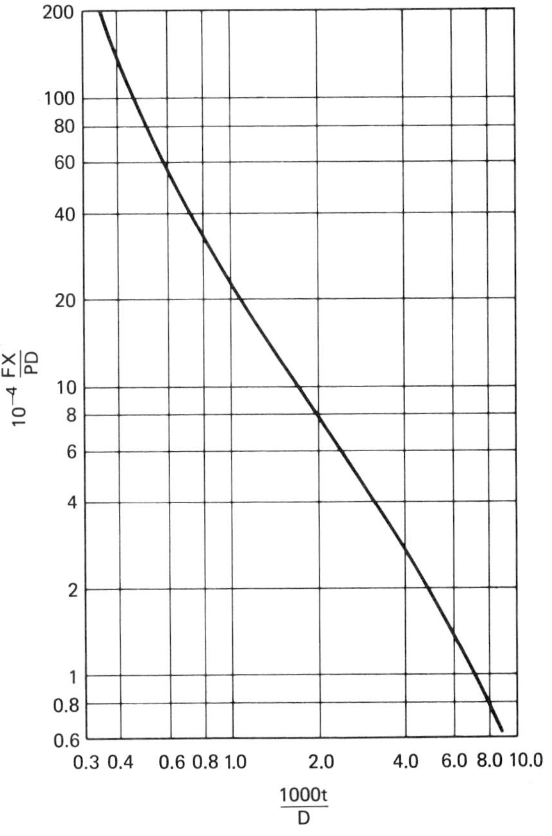

Fig. 10-5. Smooth curve shows relation between dimensionless ratios $(FX)/(PD)$ and t/D for a set of beryllium-copper diaphragms of NBS Shape 1.

26.4 (all $\times 10^6$), effective values being defined as those values for which the deflection characteristics of the diaphragm would coincide with the curve shown, obtained for reinforced beryllium-copper diaphragms.

The experimental points for beryllium copper fall on the curve with an average deviation of not more than 5%. Average experimental results for other materials also fitted the curve over the range of $(t/D \times 10^3)$ from 0.8 to 4 but with somewhat greater scatter.

It should be noted that although this nondimensional chart relating the initial flexibility (or the average flexibility over the working range) furnishes a very valuable design aid, a similar chart for comparing families of diaphragms having quadratic or logarithmic load-deflection characteristics might also be quite useful. For this one might choose as a "nondimensional flexibility"

$$\frac{X}{D} \log \frac{F}{P} \text{ or } \left(\frac{F}{P}\right)^{1/2} \left(\frac{X}{D}\right)$$

Other functions representative of other types of load-deflection characteristics of particular interest might also be used. However, experimentation would be required to determine which function, if any, would fit the facts for any particular diaphragm shape.

During World War II many people who needed to make diaphragms for various purposes utilized the NBS Shape 1, probably because it was about the only shape on which design data were available. Fortunately, this shape had turned out to have a fairly linear characteristic over a fairly good range, and results were generally satisfactory. In several cases, diaphragms were made somewhat larger than those on which experimental data had been obtained at NBS, but it was found that the nondimensional chart applied fairly well even for the larger diaphragms. Several years ago the H. A. Wilson Company exhibited a large diaphragm capsule at several trade shows in connection with its advertising of strip metals and alloys. Through the courtesy of this company we have recently been able to borrow and test this large diaphragm capsule, which is made of Z-nickel, about 0.050 in. thick, and is approximately 30 in. in effective diameter. This giant diaphragm capsule had been made by spinning, with a shape not greatly different from the NBS Shape 1. The only difference of possible significance appeared to be that the central flat area was about 29% of the diameter rather than 25% of the diameter as in NBS Shape 1, with correspondingly slightly smaller corrugations.

A 1/2-in.-thick nickel reinforcing disk was soldered in the center of one diaphragm of the capsule, and the pressure-deflection curve of each side measured by reducing the internal pressure. Each of the diaphragms was found to be quite linear, with sensitivities of 0.093 in./psi and 0.120 in./psi for the reinforced and unreinforced sides, respectively, for deflections up to at least 1% of the diameter. These data are represented by a single point for the reinforced diaphragm on the nondimensional graph Figure 10-6. The plate modulus was taken as 33.2×10^8 psi for Z-nickel corresponding to the published value.

While the very close coincidence of the point for the reinforced center diaphragm with the graph for NBS Shape 1 may be considered slightly fortuitous in view of uncertainties in effective thickness, plate modulus, and effect of edge fastening, the proximity of the points certainly indicates that the dimensionless relation is valid for diaphragms of these large dimensions.

The calibration curves showed hysteresis as large as $1\frac{1}{4}$% for the unreinforced diaphragm, but no other unusual elastic defects were found in the limited tests made.

The flexibilites of thin diaphragms (small t/D value) may be such that they will not support atmospheric pressure, and may not be used in the usual fashion for evacuated capsules. The advantage of sensitivity over reduced-pressure range may be retained by a capsule formed of two nesting diaphragms, as described by Brombacher, Goerke, and Cordero (13).

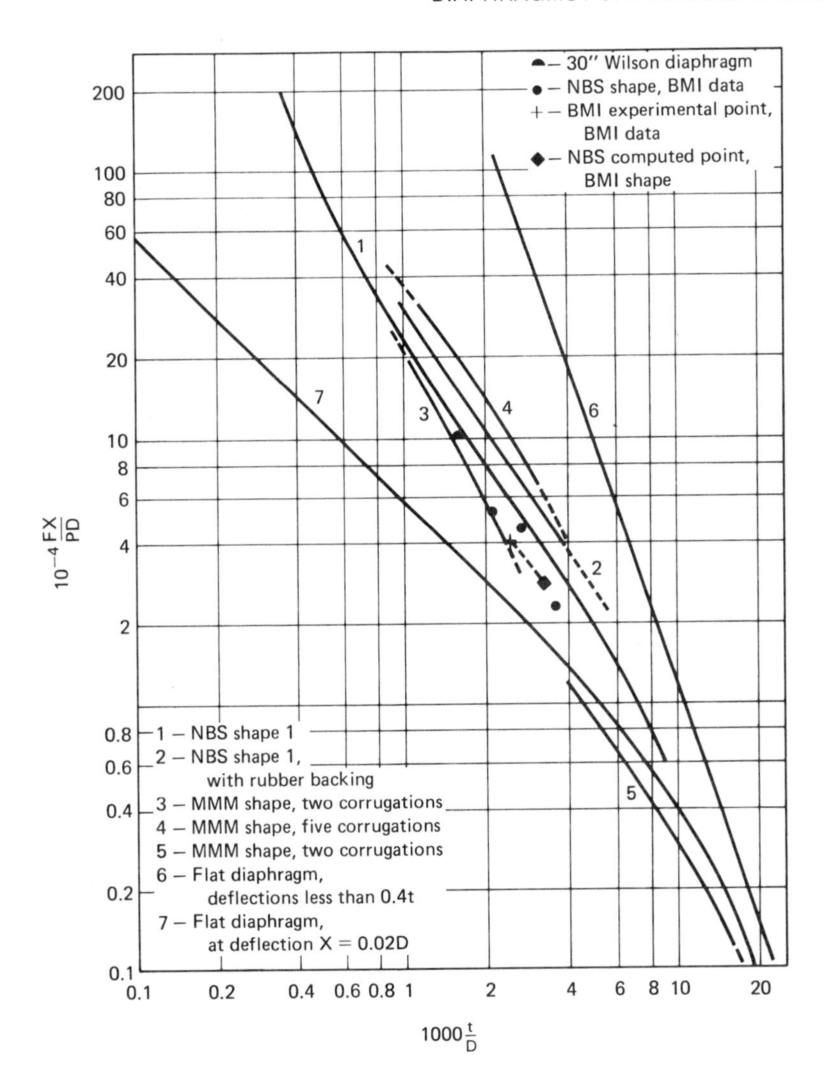

Fig. 10-6. Comparison of performance of diaphragms of different shapes. Curves show relation between dimensionless ratios (*FX*)/(*PD*) and *t/D* for various types of diaphragms. Additional points shown are those obtained using Battelle data (four points), data on 30-in. capsule (one point), and data obtained by SEAC computation (one point).

Effect of Variation in Shape

At the same time that these early investigations were being made on the single-diaphragm shape, plans were made to extend the investigation to other shapes. Diaphragm dies were made with various shapes as shown in Figure 10-7 and listed in Table 10-1 (NBS Shape 1 is shown as No. B-3). Because the pressure of war work made it impossible to continue this investigation at NBS, no test diaphragms were made from these dies for some years. In 1946, the Battelle Memorial Institute undertook an investigation on diaphragm-capsule performance for the Signal Corps, and the NBS made available the sets of dies for use at Battelle in making families of various shapes. Battelle apparently did not use the NBS dies directly but made other sets using male and female dies to form diaphragms with the NBS corrugation shape. Because the test data were obtained at Battelle

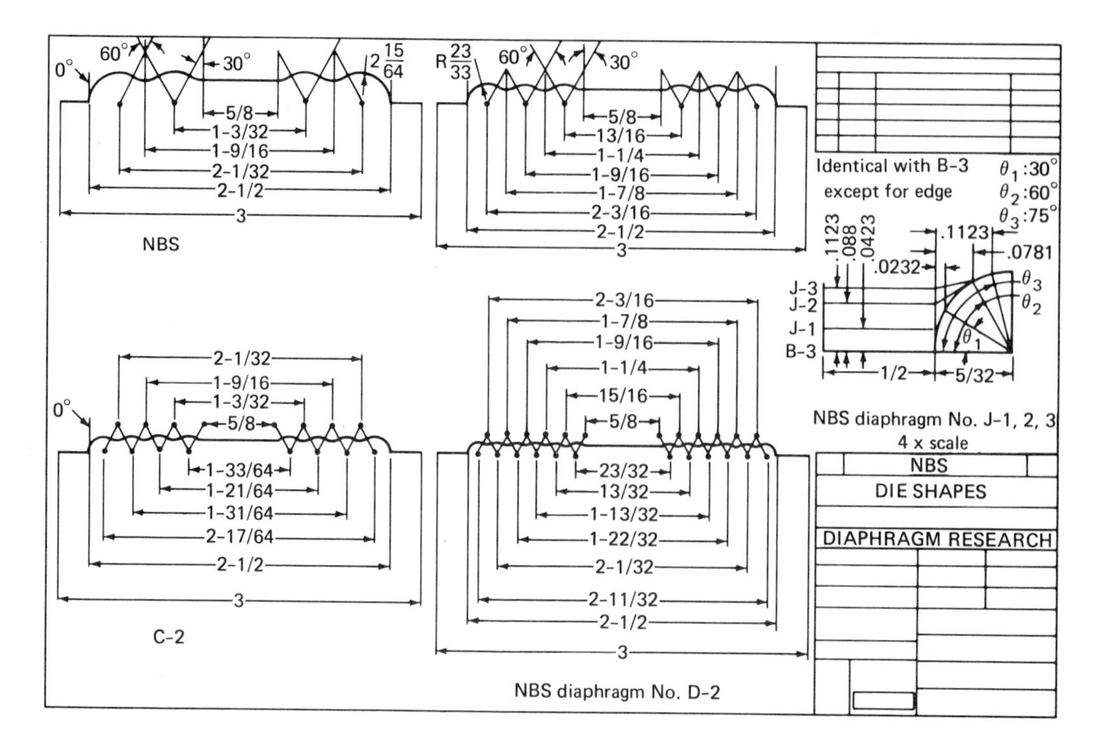

Fig. 10-7. NBS die shapes. NBS Shape 1 is shown here as diaphragm B-3.

Table 10-1. Shape of Dies

NBS DIE NUMBER	1 EFFECTIVE DIAMETER *D*, IN.	2 NUMBER OF CORRUGATIONS	3 CORRUGATION ARC ANGLE, DEGREES	4 CORRUGATION RADIUS IN.	5 EDGE ANGLE, DEGREES
A-2	2½	2	60	3/32 D	0
B-3	2½	3	60	1/16D	0
C-2	2½	4	60	3/64 D	0
D-2	2½	6	60	1/32 D	0
J-1	2½	3	60	1/16 D	30
J-2	2½	3	60	1/16 D	60
J-3	2½	3	60	1/16 D	75

Notes:

1. Outer diameter of corrugation nearest the rim. Add 1/2 in. for total diameter (1/4-in. rim).
2. Number of ridges with outermost corrugation above rim (facing down).
3. Arc of each semicorrugation.
4. Radius of corrugation arcs.
5. Angle between perpendicular to rim and tangent to outer corrugation at point of contact between outer corrugation and rim.

only after diaphragms had been joined together to form capsules, their load-deflection data are not exactly comparable to those obtained at NBS on single diaphragms with clamped edges. However, when approximate corrections are made for the boundary conditions, the Battelle data are in fair agreement with NBS data on the NBS Shape 1. (Based on SEAC computations on the Battelle shape, the clamped edge increases the stiffness by about 5% above that of the capsule with no flange.) By applying the same correction to the other shapes it is possible to make some comparisons of performance of capsules of other shapes with similar single diaphragms. Results of some of these comparisons are shown as newly added points on the non-dimensional chart, Figure 10-6.

Pressure-deflection data from four capsules made by Battelle are plotted toward the lower right in Figure 10-6. Three of these capsules were fabricated on BMI dies with outlines corresponding to NBS Shape 1, and a fourth capsule with a special BMI shape (shown below in Figure 10-12). This is the special BMI shape for which Grover and Bell computed the stress distribution (14). Corrections made to the BMI data involved a 10% reduction of the nominal thickness to approximate an effective thickness, and an added percentage of the value of the ratio FX/PD equal to twice the percent "departure from linearity" as determined by Battelle. These departures from linearity ranged from 4 to 9%. Also, the initial value of FX/PD was reduced by 5% to correct for the difference in stiffness between a capsule and a single diaphragm.

The results of computations at NBS (described in Part 2 of this section) for the same special BMI shape with a clamped edge are also shown as an added point on Figures 10-6 and 10-10. The fact that the NBS computations and Battelle experimental results lie approximately on the same line parallel to the general curve for NBS Shape 1 is encouraging. In Figure 10-6, curve 1 is the experimental curve previously shown in Figure 10-5.

Curve 2 is for diaphragms made on the same dies as NBS Shape 1, but with one thin sheet of rubber between. Curves 3, 4, and 5 represent approximate performance data obtained through the courtesy of Maxwell, Manning, and Moore on diaphragms of their manufacture. These diaphragms were of beryllium-copper 2½ in. in effective diameter with a reinforced flat in the center of 5/16 in. diameter. The corrugations were circular arcs of approximately 60° with suitable radii for forming two or five complete corrugations. The corrugations themselves were symmetrical about the plane of the rim; that is, the outline was not offset or concave as was the NBS outline. Curves 3 and 4 are for all values of deflections up to $0.2D$, while curve 5 is only for the value $0.03D$, representing an average stiffness to that deflection. The load-deflection characteristics of these diaphragms were not as nearly linear as the NBS Shape 1. As an indication of the limits of flexibility, curves 6 and 7 are shown. Curve 6 represents the initial flexibility for flat diaphragms. It is valid only for relatively small deflections up to about 0.4 of the thickness. Curve 7 represents a computed approximation to the deflection-pressure ratio for flat diaphragms deflected to $X = 0.02D$ and represents the average flexibility up to that deflection. The initial flexibility of a flat sheet is presumably the limiting flexibility for any stable diaphragms of the same t/D ratio. It may be noted that none of the diaphragms tested approaches the initial flexibility of the flat sheet, the most flexible (12 corrugations, 90° arcs) being still about eight times as stiff.

More recently, with support from Wright Air Development Center, the Bureau made some single diaphragms using the NBS dies (Figure 10-7), which, in the meantime, had been returned by Battelle, and using still other shapes as shown in Figure 10-8 and listed in Table 10-2. These experimental results, on load-diaphragm characteristics, were transmitted to the Air Force in NBS reports prepared by Wexler, Garland, Garfinkel, and Fairbanks but have not been published. Some results will be summarized here.

Figure 10-9 shows that the effect of increasing the number of corrugations is to increase the initial flexibility as well as the average flexibility over the usable range, although the diaphragms with more corrugations have nonlinear characteristics and become progressively stiffer for larger deflections. The differential flexibility, given by the slope of the curve at any point, will decrease and may even become less than that for the diaphragms of fewer corrugations at larger deflections.

Confirming earlier results, it was found

Table 10-2. Forming Data and Geometric Parameters

DIE NO.	CLAMPING FORCE, TONS	FORMING PRESSURE, PSI	GEOMETRIC PARAMETERS (SEE FIG. 10-1) A IN.	B IN.	C IN.	D IN.	E IN.	F IN.	G IN.	θ DEG	ϕ DEG	NO. CORRUG.	CORRUG SHAPE
01	10	1500	0.3125	0.380	1.250	1.500	0.018	0.036	0.134	60	60	4	Circular
02	24	3600	0.3125	0.333	1.250	1.500	0.009	0.018	0.041	45	90	12	Circular
03	10	1500	0.3125	0.380	1.250	1.500	0.027	0.054	0.134	45	90	4	Circular
04	24	3600	0.3125	0.344	1.250	1.500	0.009	0.018	0.063	60	60	8	Circular
05	24	3600	0.3125	0.344	1.250	1.500	0.014	0.028	0.063	45	90	8	Circular
06	10	1500	0.3125	0.380	1.250	1.500	0.009	0.018	0.134	75	30	4	Circular
07	28	4200	0.3125	0.333	1.250	1.500	0.006	0.012	0.041	60	60	12	Circular
L1	32	4800	0.3125	0.380	1.250	1.500	0.018	0.036	0.134	75	. .	4	Triangle
K1	28	4200	0.3125	0.380	1.250	1.500	0.018	0.036	0.134	60	. .	4	Trapezoid

A—radius of central disk
B—radius to crest of first corrugation
C—radius to inside of clamping edge
D—radius to outside of clamping edge
E—semicorrugation depth
F—corrugation depth
G—width of semicorrugation
θ—tangent angle
ϕ—arc angle

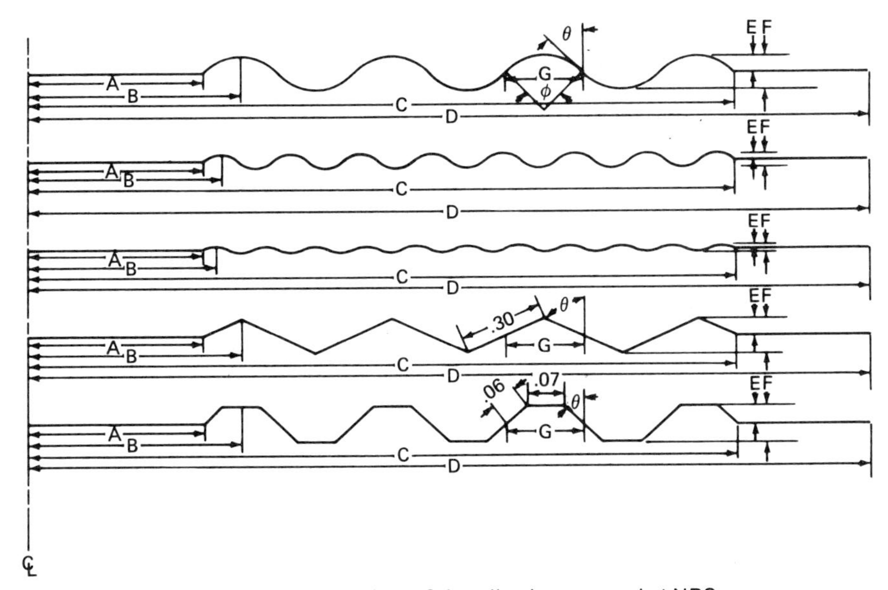

Fig. 10-8. Die profiles. Other die shapes used at NBS.

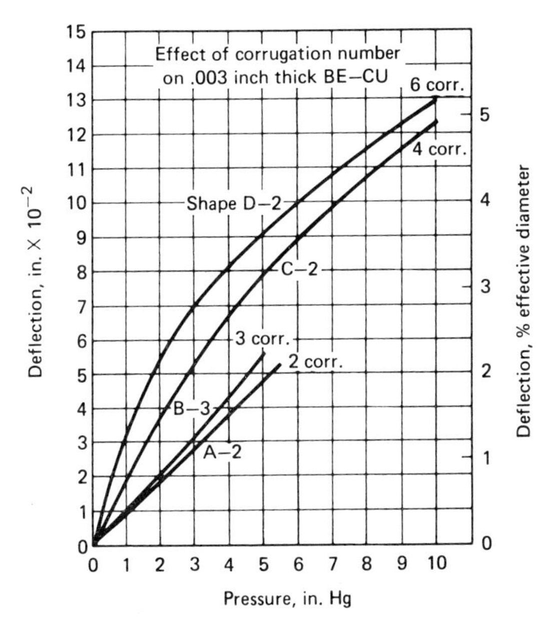

Fig. 10-9. Effect of corrugation number on deflection characteristics of 0.003-in.-thick beryllium-copper diaphragms.

that diaphragms with shallower corrugations, obtained by forming the blank against the dies with varying numbers of thin rubber sheets between the blank and the die, gave better initial flexibilities within limits but at the expense of decreasing the linear range. Some of the diaphragms thus formed gave deflections closely proportional to the square root of the pressure over quite a large range. In particular, the 2½-in.-diameter diaphragm with 0.003 in. material with four corrugations (die shape C-2) formed with two thicknesses of 0.008-in.-thick dental rubber dam gave a slope of 0.46 on a log-log plot, thus roughly approximating the square root characteristic. Also of interest is another nonlinear characteristic in which the deflection is proportional to the logarithm of the pressure. The 2½-in.-diameter diaphragms of 0.003-in.-thick material made on die Shape 02 (Table 10-2) with 12 corrugations closely approximated this characteristic over the upper two thirds of its useful deflection range.

The diaphragms made with the triangular and trapezoidal shapes were quite linear up to at least 2% of the diameter, with the trapezoidal outline being slightly stiffer and the triangular slightly more flexible than the corresponding diaphragm with the same number of circular corrugations (die Shape 01).

The initial flexibilities of the diaphragms of these different shapes are plotted in nondimensional form in Figure 10-10, in which a portion of the curve for NBS Shape 1 is reproduced. For only a few of the shapes were tests made with material of more than one thickness so that for most of the data only one point is available. However, this point represents the average of test results on

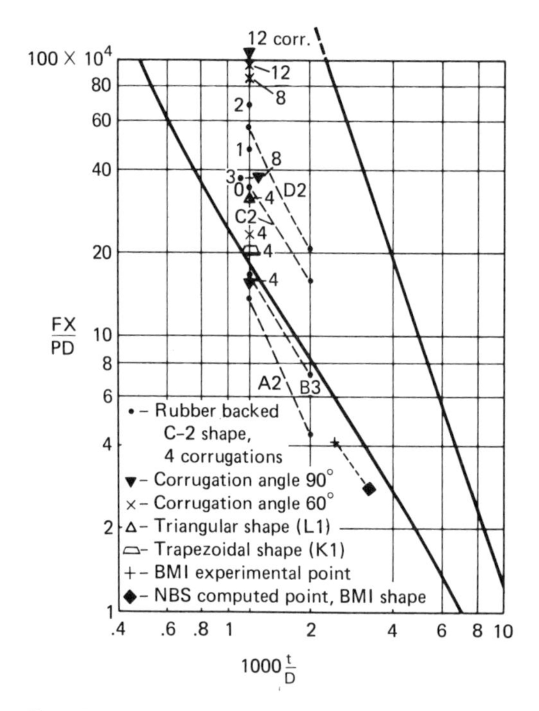

Fig. 10-10. Expanded plot of $(FX)/(PD)$ versus t/D showing data obtained on diaphragms of different shapes.

at least three diaphragms. In view of the general finding that the characteristic curves for other shapes run somewhat parallel to the curve for NBS Shape 1 over some range of values of t/D, it is reasonable to assume that these points may serve to locate roughly the curves for their corresponding series.

At the time the empirical program was planned in the late 1930s, it was contemplated that eventually studies might be made of various other shapes involving radial corrugations, conical segments, spiral corrugations, twisted cones, etc., but no specific designs were evolved. In 1951 the Engineering Research Associates (now a subsidiary of Remington-Rand) undertook on contract to the Air Force a general study of diaphragm performance and design and made interesting experiments and analyses of pleated cones and tubes, which responded to pressure variations by twisting, and on noncircular capsules, some with radial pleats or corrugations, etc. The results of this project are significant for guiding further work but cannot be compared or applied directly to the other work reported here.

2. THEORETICAL

Mathematical Analysis Using Shell Equations

A corrugated diaphragm is precisely an elastic shell with rotational symmetry; thus a mathematical analysis of stresses and displacements in diaphragms logically should be based upon shell theory. This has hitherto not been attempted for capsules of arbitrary meridian shape because of the complexity of the differential equations and the inability to find satisfactory solutions either in closed form or by approximations. Our threefold purpose here is (a) to derive the linear shell equations for combined bending and stretching effects with lateral loading terms for rotationally symmetric shells in appropriate independent and dependent variables suitable for complicated meridional shapes, and with the boundary conditions associated with practical diaphragm applications; (b) to describe the method used for solving this system on an electronic digital computer; and (c) to present numerical solutions for a specific diaphragm subject to air-pressure loading.

Several analyses have been published previously using simplified physical assumptions to obtain somewhat rougher information about diaphragm behavior, instead of using full elastic shell theory. In this connection, the analyses by Griffith (1), Pfeiffer (17), Charron (18), and Haringx (15) should be mentioned. In particular, the ingenious approach by Haringx essentially refers the diaphragm behavior back to flat-plate behavior, by smearing out the effects of the corrugations over a wavelength. This leads to consideration of a factitious "equivalent" flat plate with variable elastic qualities that, on the average, approximate the elastic characteristics of the diaphragm. Haringx was able to derive a curve that would predict the performance of the NBS series (Shape No. 1) when adjusted on the basis of one empirical coefficient derived from the NBS experimental work. We attempted to deduce from the Battelle and NBS data, on capsules of the various other NBS shapes, some data that can be used to test the applicability of the Haringx formulation to other shapes, but this effort has not been carried to a definite conclusion. Although such approaches will furnish approximate stiffness information, they cannot be expected to yield the rapidly varying stress concentrations in every local region. For data requiring this order of accuracy, one must have recourse to full shell theory.

One publication has already appeared in which a diaphragm problem has actually been based upon results from shell theory. This is the excellent work by Grover and Bell (14), in which known approximate solutions to the shell equations for elementary meridional shapes having constant curvature were pieced together. Since approximate solutions are available for circular arc and conical shells, a diaphragm consisting piecewise of such portions can be handled in this way. These authors chose a specific meridional shape consisting of six circular arcs and six straight lines as the basis for their computations. By adjusting the arbitrary parameters occurring in the general solutions for each piece to produce a continuous fit for the stresses, this procedure required the solution of a system of 44 algebraic equations. Their results gave information for stresses and moments but not for displacements. To serve as a mutual check on results, our present preliminary investigation has chosen the specific shape used by Gover and Bell, although our method applies equally well for any shape and does not simplify by restricting the analysis to shapes consisting merely of circles and straight lines.

The equations in Timoshenko (8) for rotationally symmetric shells subject to combined bending and stretching cannot be used directly in the present computations. The independent variable ϕ used in reference 8 (also ϕ in our Figure 10-11) would lead to multiple-valued functions for complicated shapes; also the present investigation requires a formulation containing lateral loading terms. Furthermore the choice made in reference 8 of the angle of meridional rotation and the modified shear as dependent variables is not suitable for shells having horizontal sections, and the use of these unknowns also would require an extra machine integration in order to obtain the displacements. For these reasons, one chooses here as independent variable the meridional arc length s, and as dependent variables the tangential and normal displacement components $\bar{v}$ and w and the shear Q (Figure 10-11). The resultant N_ϕ and N_θ are stretching forces per unit length; the resultant Q is a shear force per unit length contributing to bending; and the resultant bending mo-

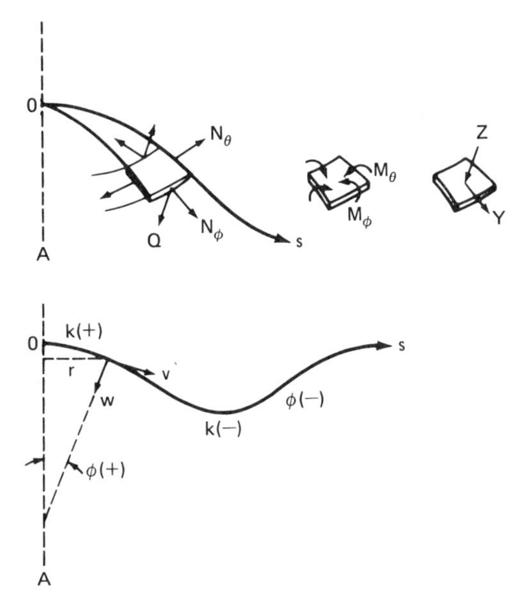

Fig. 10-11. Equilibrium forces and moments acting on a shell element.

ments per unit length are M_ϕ and M_θ. The quantities Y and Z are the components of the external applied lateral load in the tangential and normal directions, respectively, and have dimensions of force per unit area. All quantities are taken in the positive sense as indicated in Figure 10-11. Other quantities are absent because of the symmetry. Young's modulus is E, Poisson's ratio is $\bar{\nu}$, and the thickness of the shell is t. The meridional line OS is a plane curve which when rotated about the axis OA generates the middle surface of the diaphragm. Let $k(s)$ be a quantity having the same absolute value as the curvature of the meridional curve. The variable signs of k and ϕ are chosen as indicated in the figure. The arc length s and the radial distance r are always taken with positive sign. With s as independent variable, considerations of strain and curvature change for the middle surface and the stress-strain equations lead to the relations

$$\left.\begin{aligned}
M_\phi &= -\frac{Et^3}{12(1-\nu^2)}\left[k'\bar{v} + kv' + w'' + \frac{\nu\cos\phi}{r}(k\bar{v} + w')\right]\\
M_\theta &= -\frac{Et^3}{12(1-\nu^2)}\left[\frac{\cos\phi}{r}(k\bar{v} + w') + \nu(k'\bar{v} + kv' + w'')\right]\\
N_\phi &= \frac{Et}{(1-\nu^2)}\left[v' - kw + \nu\left(\frac{\cos\phi}{r}\bar{v} - \frac{\sin\phi}{r}w\right)\right]\\
N_\theta &= \frac{Et}{(1-\nu^2)}\left[\frac{\cos\phi}{r}\bar{v} - \frac{\sin\phi}{r}w + \nu(v' - kw)\right]
\end{aligned}\right\} \quad (6)$$

where the primes indicate differentiation with respect to s. When these expressions are substituted into the three equations for force equilibrium in the tangential and normal directions and moment equilibrium for the shell element, using the relations $\phi' = k$ and $r' = \cos\phi$, one obtains the system of three equations for v, w, and Q:

$$\left.\begin{aligned}
&(rk + \nu\sin\phi)(w' - \left(\frac{\sin\phi\cos\phi}{r} - k\cos\phi - rk'\right)w - rv'' - (\cos\phi)v' +\\
&\left(\frac{\cos^2\phi}{r} + \nu k\sin\phi\right)\bar{v} + \left(\frac{rk}{C}\right)Q = \left(\frac{r}{C}\right)Y\\
&\left(\frac{\sin^2\phi}{r} + rk^2 + 2\nu k\sin\phi\right)w - (rk + \nu\sin\phi)v' - \left(\frac{\sin\phi\cos\phi}{r} + \nu k\cos\phi\right)\\
&\bar{v} - \left(\frac{r}{C}\right)Q' - \left(\frac{\cos\phi}{C}\right)Q = \left(\frac{r}{C}\right)Z\\
&rw''' + (\cos\phi)w'' - \left(\frac{\cos^2\phi}{r} + \nu k\sin\phi\right)w' + rkv'' + (k\cos\phi + 2rk')v' +\\
&\left(k'\cos\phi + k''r - \frac{k\cos^2\phi}{r} - \nu k^2\sin\phi\right)\bar{v} + \left(\frac{r}{D}\right)Q = O
\end{aligned}\right\} \quad (7)$$

where $C = Et/(1-\nu^2)$ and $D = Et^3/12\,(1-\nu^2)$.

By including the three auxiliary equations $v' - P = O$, $w' - R = O$, and $R' - T = O$, and substituting the new quantities P, R, and T, thus defined, into equations (7), one obtains a system of six equations, each of first order, equivalent to system (7). The equations are linear but have highly variable coefficients and two nonhomogeneous terms. It is in this first-order form that the system of equations is actually integrated to obtain solutions for the six unknowns, $\bar{v}$, w, Q, v', w', and w''.

For air-pressure loading of a diaphragm we take $Y = O$ and $Z =$ air pressure. The equations are then nonhomogeneous only because of the term $(r/C)\,Z$ in the equation expressing normal force equilibrium.

The shape actually considered here is defined in Figure 10-12 where distances are given in inches and angles in radians. This investigation studies the same diaphragm as chosen by Grover and Bell for their calculations (which is assumed to approximate closely the shape of the BMI capsules on which they obtained experimental data). Also following Grover and Bell, we take $E = 15 \times 10^6$ psi, $\nu = 1/3$, $t = 0.007$ in., and $Z = 15$ psi, acting in a normal direction on the top face of the diaphragm. It is assumed that there is no central stiffening disk present, and that the diaphragm has constant thickness and is not presented. Also the gravity effect due to diaphragm weight is neglected. The choice of $Z = 15$ is obviously not significant, since all quantities in the linear theory are directly proportional to Z and can be obtained in this manner for any positive or negative value of Z from our results.

For the flat central AOA of the capsule, Grover and Bell used the solution to the nonlinear equations for a flat circular plate obtained by Way (5), but used results from linear shell theory for the rest of the capsule. Instead of mixing nonlinear and linear effects, this analysis employs equations that are linear over the entire diaphragm. Furthermore, the solution employed by Grover and Bell for the ring-shell sections is the approximate one obtained by Stange (19), which is valid only for a ring shell having a radius of its central circle that differs just slightly from the radii of its bounding circles. Thus their use of Stange's solution for the corrugations close to the center O necessarily introduces some error. For these reasons, one should not expect exact agreement between the Grover and Bell results and the present results obtained by direct integration of the linear system (7) over the entire diaphragm radius.

In the portions where the shell is flat and horizontal, the stretching and bending effects separate, and equations (7) simplify considerably to become, respectively, the plane-stress equation:

$$v'' + \frac{1}{r}\,v' - \frac{1}{r^2}\,\bar{v} = -\,\frac{Y}{C} \tag{8}$$

and the pure bending equations:

$$Q' + \frac{1}{r}\,Q = -\,Z \tag{9}$$

$$w''' + \frac{1}{r}\,w'' - \frac{1}{r^2}\,w' = -\,\frac{Q}{D} \tag{10}$$

For $Y = 0$, the general solutions that remain finite and satisfy the three boundary conditions $\bar{v} = w' = Q = 0$ at the center $r = 0$ are:

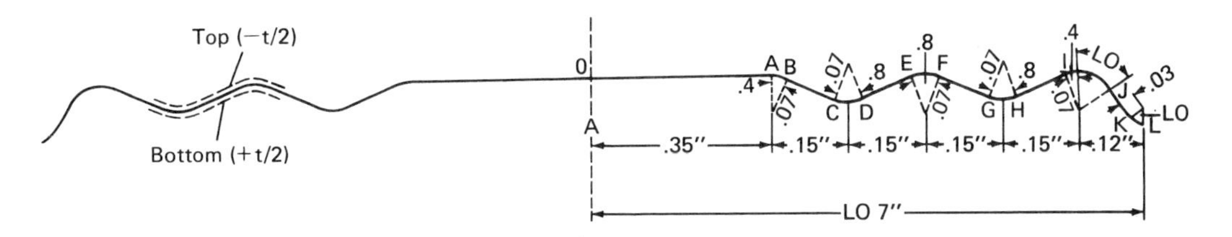

Fig. 10-12. Shape and dimensions of BMI diaphragm.

$$\overline{v} = c_1^s \tag{11}$$

$$Q = 0 - \frac{Z}{2}s \tag{12}$$

$$w = c_2 + c_3 s^2 + \frac{Z}{64D}s^4 \tag{13}$$

$$P = c_1 \tag{14}$$

$$R = 2c_3 s + \frac{Z}{16D}s^3 \tag{15}$$

$$T = 2c_3 + \frac{3Z}{16D}s^2 \tag{16}$$

where c_1, c_2, and c_3 are arbitrary constants and the terms involving Z are the nonhomogeneous particular solution. Three more conditions are at our disposal to define a specific diaphragm problem; however, these must be imposed at the outer boundary (point L). The solutions (11) to (16) hold from O to A; beyond A these solutions must be extended to L by numerical integration of the six curved-shell equations equivalent to system (7).

Numerical integration by electronic digital computer of a system of differential equations can be accomplished by several alternative methods. A finite-difference scheme which imposes the desired boundary conditions at both ends of the interval could be used; this then would require the inversion of a large matrix for each problem. Computationally, for practical reasons of time and accuracy, it is far preferable, however, to perform the numerical integration of the system step by step, proceeding as an initial value problem where all six unknowns must be specified at the starting point. To adapt this method to a boundary-value problem, it is necessary first to obtain the general solution by solving numerically for three independent solutions to the homogenous equation, each of which is an extension beyond point A of the solutions (11) to (16), by assigning three arbitrary sets of values to the three unspecified constants in equations (11) to (16), and likewise by solving for one particular nonhomogeneous solution. One obtains three independent homogeneous vector solutions $\overline{G_1}$, $\overline{G_2}$, and $\overline{G_3}$, and a nonhomogeneous solution $\overline{G_p}$, defined by the choices

	c_1	c_2	c_3	Z
$\overline{G_1}$	1	0	0	0
$\overline{G_2}$	0	1	0	0
$\overline{G_3}$	0	0	1	0
$\overline{G_p}$	0	0	0	15

which thereby produce initial values at the point A for numerical integration.

The two specific problems for which results are included in the present preliminary report were solved on the electronic digital computer SEAC at the National Bureau of Standards. This machine has a high-speed internal memory of 1,024 addresses. Because of the many variable and complicated coefficients occurring in shell equations, it was necessary also to use an external magnetic-tape unit to augment the storage capacity. Although this necessarily increased the computing time, it was still reasonably short, so that the use of SEAC for solving diaphragm problems can be considered to be altogether feasible economically.

The procedure actually used to solve each of the four initial value problems $\overline{G}_{1,2,3,p}$ was the Runge-Kutta fourth-order integration method. This can be described briefly as follows: Let the s interval be divided by a mesh having a basic interval size Δs. Also let each interval (s_n, s_{n+1}) of length Δs be bisected by a point $s_{n+1/2}$. For an unknown vector $\overline{y}(s)$ defined by a system of ordinary differential equations $\overline{\dot{y}} = A(s)\,\overline{y} + \overline{f}(s)$, where $A(s)$ is the matrix of coefficients, this method obtains the value at the $(n+1)$ mesh point by the relation

$$\overline{y}_{n+1} = \overline{y}_n + \frac{1}{6}(\overline{K_1} + 2\overline{K_2} + 2\overline{K_3} + \overline{K_4})$$

with an accuracy of the fourth order in Δs, where

$$\overline{K}_1 = \Delta s\ [A(s_n)\overline{Y}_n + \overline{f}(s_n)]$$

$$\overline{K}_2 = \Delta s\left[A(s_{n+1/2})\left(\overline{Y}_n + \frac{1}{2}\overline{K}_1\right) + \overline{f}(s_{n+1/2})\right]$$

$$\overline{K}_3 = \Delta s\left[A(s_{n+1/2})\left(\overline{Y}_n + \frac{1}{2}\overline{K}_2\right) + \overline{f}(s_{n+1/2})\right]$$

$$\overline{K}_4 = \Delta s[A(s_{n+1})(\overline{Y}_n + \overline{K}_3) + \overline{f}(s_{n+1})]$$

The total machine program consisted essentially of four parts: (a) a code to calculate all the variable coefficients in $A(s)$ at each mesh point over the range; (b) a code for the Runge-Kutta method applied four times for $\overline{G}_{1,2,3,P}$; (c) a code to form the appropriate linear combination of these solutions for the quantities v, w, Q, P, R, T to solve the specific boundary-value problem defined by the outer edge conditions; and (d) a code to compute the membrane forces N_ϕ, N_θ and the bending moments M_ϕ, M_θ from the solutions obtained for the six unknowns. A variety of different boundary-value problems can be solved quickly for a diaphragm of fixed geometry using only codes (c) and (d) after the four basic solutions have been obtained from codes (a) and (b).

The integration of the problem for the shape chosen (Figure 10-12) used about 1,000 words of the internal memory and about 3,000 words on auxiliary magnetic tape, and required about 1,000 words of data from an auxiliary magnetic-wire input. Except in 11 small intervals about the transition points $A, \ldots, \ldots, K$, the interval $(s_n, s_{n+1/2})$ was taken as 0.00175 in., making a basic Δs size of 0.0035 in. At the center of each straight-line section, a smaller compensating interval also was used in order that a mesh point could fall exactly on each transition point.

A special complication arises at each transition point $A, \ldots, K$. Not only the curvature $k(s)$ enters the shell equations (7), but also its first and second derivatives k' and k''. Since this diaphragm shape consists of sections having different constant curvatures joined at the transition points $A, \ldots, K$, $k(s)$ therefore has a finite jump discontinuity at each such point. These jumps could be handled easily in the numerical integration without special considerations, but the quantity k' puts a delta function into the coefficients, and k'' inserts a derivative of the delta function (a double delta function).

The presence of these infinite singularities in the coefficients of the differential equations will produce singularities in some of the six unknowns at each transition point. The quantities $\overline{v}$, w, and Q must remain continuous, but v' and w' will have finite jump discontinuities, and w'' will behave as a delta function plus a finite jump. Our computed results for these quantities clearly tend toward this limiting behavior. In order to overcome this computational difficulty inherent in the nonsmooth diaphragm shape chosen, it was found necessary to insert suitable smoothing functions over small intervals of length $4\Delta s$ centered about each of the points $A, \ldots, K$.

A typical set is shown qualitatively in Figure 10-13; this set was used at the points D and H. A linear sawtooth curve for k'' was chosen to approximate the derivative of the appropriate delta function. Each curve is obtained from the one below it by integration. The approximate curves for k'', k', and k were chosen so that the integrals would have the same required behavior as the integrals of the singular functions which they replaced. When solving the shell equations on SEAC through these 11 transition regions of rapid change, the mesh had to be considerably reduced in size throughout each such interval in order to retain accuracy in the numerical solutions. Therefore a much finer mesh which had length $\Delta^* s = 0.0005$ in. was used at such places. The sizes actually chosen for the large mesh and the special fine mesh were determined in the following manner:

When the shell has no lateral loading, the shear Q must be exactly zero at each point where the meridian has a horizontal tangent (e.g., midway between points C and D), by considerations of static equilibrium of the vertical force components acting on that

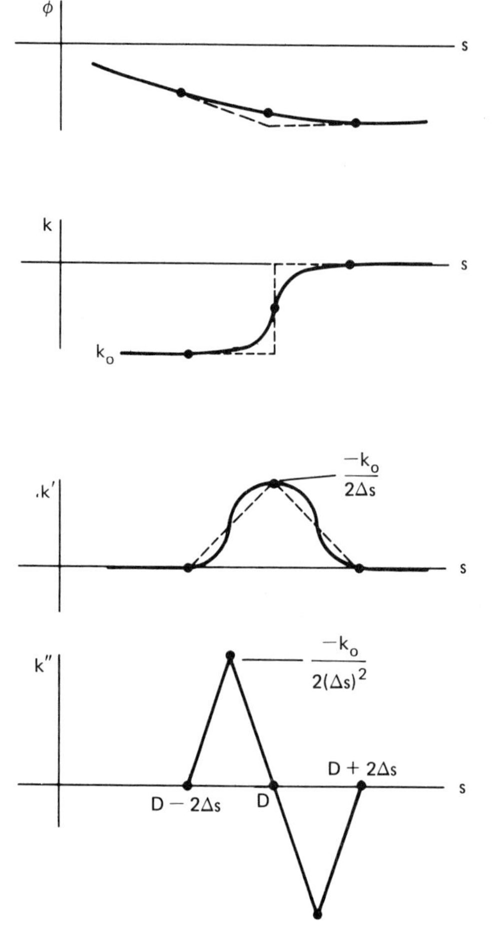

Fig. 10-13. Typical set of smoothing functions.

portion of the shell. Likewise when loading Z is present, at such horizontal points one has the exact relation $2\pi rQ = -\pi r^2 Z$ to determine Q. Various test problems were computed with decreasing mesh sizes until the numerical results differed from these exact values of Q at the horizontal points by 3% or less. The first mesh sizes that attained this accuracy were the ones actually chosen for the full diaphragm problem.

SEAC is basically a fixed-point binary digital computer; in this application, however, it was used with a "floating decimal-point" routine, giving eight significant decimal figures. Over the range of integration from $s = 0.343$ to $s = 1.155$ in., about 190 large-size Δs intervals were used, plus over 300 of the smaller $\Delta^* s$ intervals in the transition zones, thus making a total number (including the $s_{n+1/2}$ midpoints) of almost 1,000 mesh points over the full integration range. The computing time required to solve for one solution such as $\overline{G}_1$ was about 1½ hour; hence 6 hours were needed to obtain the required four basic solutions. Then 1/2 hour was needed to compute the appropriate linear combination of these, plus another 1/2 hour to compute the bending moments and membrane forces from the displacements.

Two boundary-value problems have so far been solved for this-size shape. When two diaphragms are placed together to form a capsule, the condition at the outer edge for each half is characterized by having no normal deflection, no meridional rotation, and no radial force (Figure 10-14). Although such a

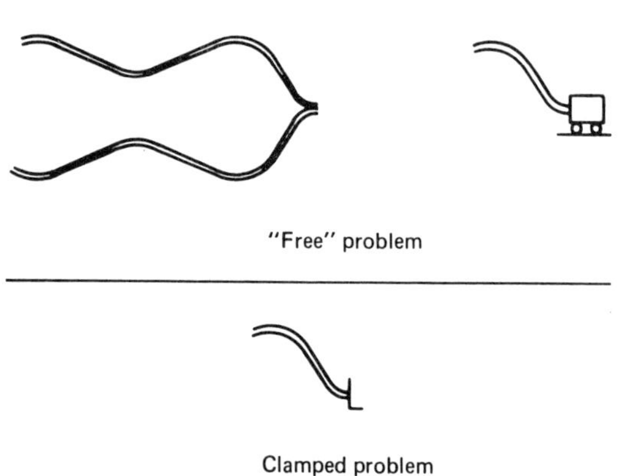

Fig. 10-14. Two boundary-value problems.

boundary is free only with respect to the radial force, this case will be termed here the "free" problem. Using the expression $w' + k\overline{v}$ for the angle of rotation of a meridional element and the expression for N_ϕ in equations (6), the boundary conditions at L for the free case become:

$$\begin{Bmatrix} w = 0 \\ w' + k\overline{v} = 0 \\ v' + \frac{\nu}{r}\overline{v} = 0 \end{Bmatrix} \quad (17)$$

When the diaphragm is clamped at its outer boundary, the conditions at L are:

$$\begin{Bmatrix} \bar{v} = 0 \\ w = 0 \\ w' = 0 \end{Bmatrix} \tag{18}$$

Using the appropriate components of the general solution

$$\overline{G} = a_1\overline{G_1} + a_2\overline{G_2} + a_3\overline{G_3} + \overline{G_p} \tag{19}$$

to express the three relations in equations (17) or (18) at L, one then solves the resulting algebraic system for the constants a_1, a_2, and a_3 and inserts these values in equation (19) to obtain the solution for $\bar{v}$, w, Q, v', w', w'' for the free or clamped problem.

Figure 10-15 presents our results for $\bar{v}$, w, and Q. The clamped case is only slightly stiffer than the free, with a different pattern occurring only near the outer boundary. The difference between the two curves for Q throughout the corrugation range is too small to appear on the scale used except where indicated; the Q-values are identical in the central flat-plate range. The exact values for Q at the horizontal positions are indicated by circles, and lie on the extension of the straight line that defines Q in the range OA. The greatest error for Q in the computed results at these various horizontal positions occurs at the point between I and J ($r = 0.95$) and is about 3% for both problems. Further reduction in the error by employing finer meshes was not considered necessary in view of the

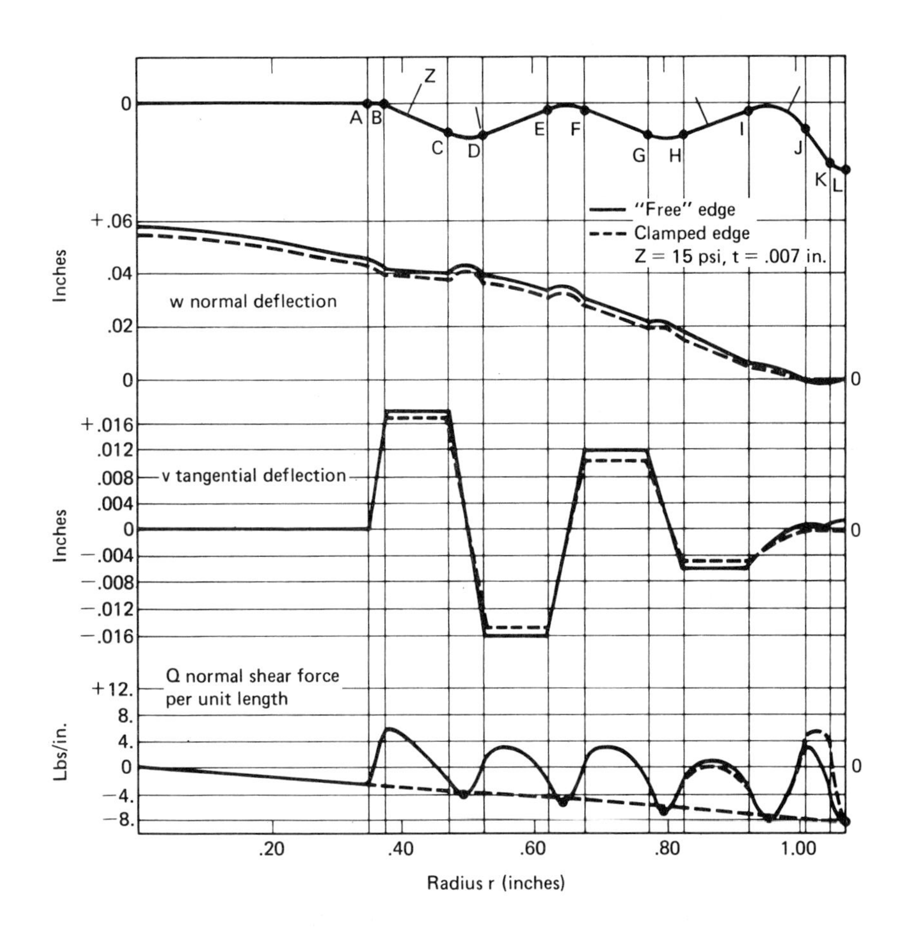

Fig. 10-15. Computed deflections and shear.

variations normally existing among commercial diaphragms of the same type. The resultant Q is produced by the shear stress $\tau_{\phi z}$. This stress vanishes at the top and bottom faces, and has, according to shell theory, a parabolic distribution through the thickness of the shell. Its maximum value $\tau_{\phi z}(c)$ occurs on the central surface of the shell ($z = 0$) and can be obtained by the relation $\tau_{\phi z}(c) = 3Q/2t$. This shear stress has therefore its maximum positive value at $r = 0.38$ and its maximum negative value at the outer boundary $r = 1.07$, but is negligible compared with some of the other stresses.

The vertical deflection W and the horizontal deflection V can be obtained easily by computing the appropriate combintion of w and $\bar{v}$ with $\sin\phi$ and $\cos\phi$ at each point. The behavior of V and W is illustrated in Figure 10-16. W is taken positive in a downward direction, and V is positive when the displacement is outward from the center O.

The $(FX)/(PD)$ ratio derived from the computed deflection is shown in Figures 10-6 and 10-10 as an added point. The good agreement with experiment has already been mentioned.

Behavior of All Local Stresses

The normal stress ς_z is altogether negligible compared with some other stresses, since it decreases through the thickness (as a known third-degree polynomial function, according to shell theory) from its value + 15 psi on the top face of the diaphragm to its zero value at the bottom face.

The shear stress $\tau_{\phi z}$, as mentioned previously, has its greatest positive value at $r = 0.38$ in. and its greatest negative value at the outer boundary $r = 1.07$ in. The maximum stresses (for both the clamped and the free problems) corresponding to these positions are about + 1,300 and − 1,720 psi, respectively, and occur midway between the two faces. This shearing stress is likewise negligible compared with some of the other stresses developed in the diaphragm.

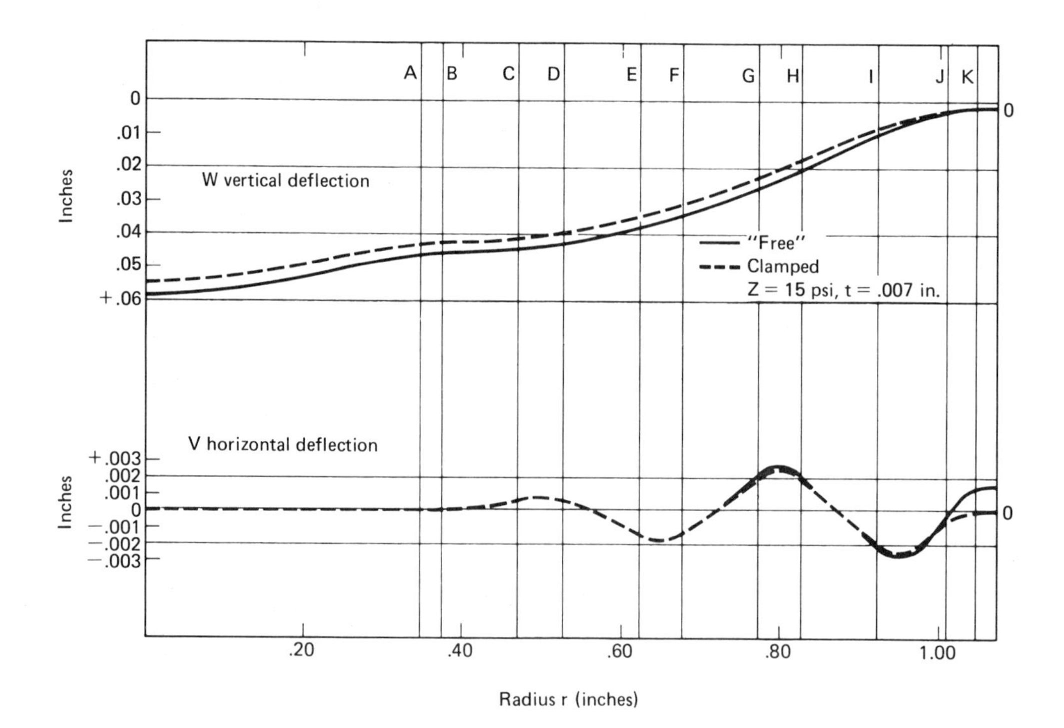

Fig. 10-16. Computed vertical and horizontal deflections.

The radial stress ς_ϕ, which acts along the variable tangential direction s, and the circumferential stress ς_θ each vary linearly, according to shell theory, through the shell thickness between their values at the top face ($z = -t/2$) and the bottom face ($z = +t/2$). The resultants are related by

$$N_\phi = \int_{-t/2}^{t/2} \varsigma_\phi dz \text{ and } M_\phi = \int_{-t/2}^{t/2} \varsigma_\phi z dz$$

with analogous relations for N_θ and M_θ. Decomposing each such linear function into its even and odd components with respect to the thickness variable z, we write:

$$\varsigma_\phi = \overset{e}{\varsigma}_\phi + \overset{o}{\varsigma}_\phi$$

and:

$$\varsigma_\theta = \overset{e}{\varsigma}_\theta + \overset{o}{\varsigma}_\theta$$

The even components $\overset{e}{\varsigma}_\phi$ and $\overset{e}{\varsigma}_\theta$ are thus constant through the thickness, and the other components $\overset{e}{\varsigma}_\phi$ and $\overset{o}{\varsigma}_\theta$ are odd linear functions with respect to z. The constant quantities $\overset{e}{\varsigma}_\phi$ and $\overset{e}{\varsigma}_\theta$ produce the resultant stretching (membrane) forces N_ϕ and N_θ, respectively, but produce no resultant bending moments. The odd contributions $\overset{o}{\varsigma}_\phi$ and $\overset{o}{\varsigma}_\theta$ produce the resultant bending moments M_ϕ and M_θ, respectively, but produce no membrane-force resultants.

For the part of the stress that causes the membrane effects, one has the relations

$$\overset{e}{\varsigma}_\phi = N_\phi/t \text{ and } \overset{e}{\varsigma}_\theta = N_\theta/t$$

These membrane stresses are presented in the first and third graphs of Figure 10-17.

The top graph in Figure 10-17 shows the behavior of $\overset{e}{\varsigma}_\phi$ for both the clamped and free cases. This radial bending stress is so much smaller than the other stresses shown in Figure 10-17 that a magnified vertical scale has been used for it. This stress is constant in the flat central section. Results for the clamped and free problems differ everywhere, but the difference becomes significant enough to graph as two distinct curves only in a narrow region near the outer edge. At the outer edge $\overset{e}{\varsigma}_\theta = -1{,}560$ psi for the clamped case.

The third graph in Figure 10-17 presents the behavior of the other membrane stress $\overset{e}{\varsigma}_\theta$. This is constant in the central section and identical with $\overset{e}{\varsigma}_\phi$ there. Results for $\overset{e}{\varsigma}_\theta$ for the clamped and free problems differ everywhere, but again the only significant differences occur near the outer edge. The computed value of $\overset{e}{\varsigma}_\theta$ for the clamped case is -520 psi at the outer edge, which is exactly one third of our result mentioned previously for $\overset{e}{\varsigma}_\phi$ there, as is required by the value of $\nu = 1/3$. The maxima for $\overset{e}{\varsigma}_\theta$ occur at $r = 0.8$ in., and are about $+50{,}600$ psi and $+48{,}800$ psi for the free and clamped cases, respectively.

The second graph in Figure 10-17 presents the quantity $\overset{o}{\varsigma}_\phi(t/2)$, which is the part of the radial stress causing bending, evaluated at the bottom face ($z = +t/2$). The value of the linear function $\overset{o}{\varsigma}_\phi$ at the top face is then obtainable by $\overset{o}{\varsigma}_\phi(-t/2) = -\overset{o}{\varsigma}_\phi(t/2)$. The resultant radial bending moment is related by

$$\overset{o}{\varsigma}_\phi(-t/2) = 6M_\phi/t^2$$

Again, the free and clamped results differ everywhere, but the only significant differences occur beyond $r = 0.85$ in. The peak values on the bottom face are $-54{,}200$ psi at $r = 0.995$ in. for the clamped case, and $-48{,}000$ psi at $r = 0.998$ in. for the free case. One thus observes that the peak values for the radial bending stress and for the circumferential membrane stress are almost equal, but the bending peak occurs closer to the outer edge.

The bottom graph in Figure 10-17 shows the behavior of $\overset{o}{\varsigma}_\theta(t/2)$, which is the circumferential bending stress, evaluated at the bottom face. Again, the only significant differences between the clamped and free cases occur near the outer edge. The relation for the corresponding resultant bending moment is

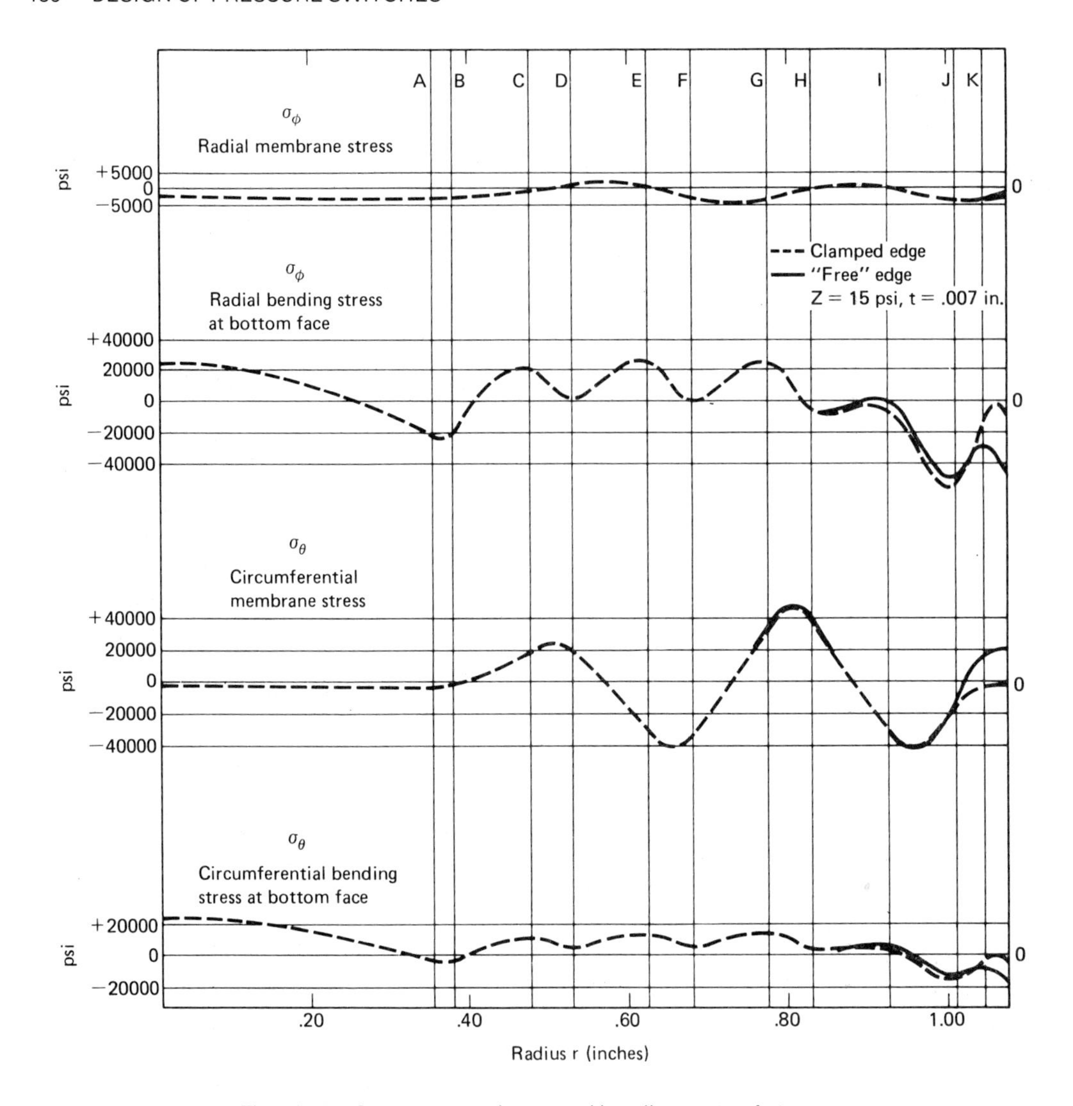

Fig. 10-17. Separate membrane and bending parts of stresses.

$$\overset{0}{\varsigma}_\theta(t/2) = 6M_\theta/t^2$$

Of the foregoing four stresses, the two membrane stresses exhibit the same frequency of oscillation as the frequency of the corrugation shape, although with different phase relationships. The two bending stresses, however, possess a double-frequency behavior.

The actual (total) stress components at the top and bottom faces can be obtained by adding or subtracting the membrane and bending parts as follows:

$$\varsigma_\phi(t/2) = N_\phi/t + 6M_\phi/t^2$$

$$\varsigma_\phi(-t/2) = N_\phi/t - 6M_\phi/t^2$$

$$\varsigma_\theta(t/2) = N_\theta/t + 6M_\theta/t^2$$

$$\varsigma_\theta(-t/2) = N_\theta/t - 6M_\theta/t^2$$

The total radial stresses on each face are shown in Figure 10-18. The curves on the upper axes give results for the free problem, and the clamped problem is given below. At

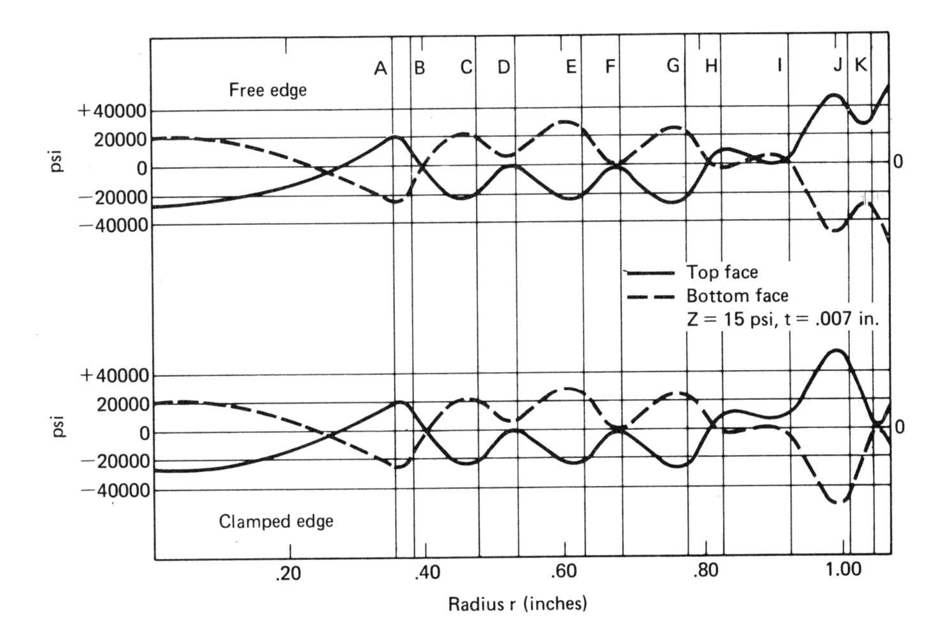

Fig. 10-18. Total radial stress at each face.

any fixed distance $r = r_0$ from the center, the actual radial stress at any point inside the diaphragm is immediately obtainable by linear interpolation between the values shown on the curves at r_0 for the two faces.

The total circumferential stresses on each face are likewise presented in Figure 10-19, with the free problem given previously and the clamped problem given later. The largest stress component occurring in each problem is shown in Figure 10-19 on the curves for the circumferential stress at the bottom face. This peak tension occurs at $r = 0.79$ in. and is about +62,100 psi for the free problem and +60,300 psi for the clamped case. The peak value in the free case represents a stress amplification of about 4,140 times the applied lateral pressure.

Finally, a comparison can be made with the results obtained by Grover and Bell (14), but only for the stresses and for the free problem, which are the only results given by them. Two typical stresses are compared in Figure 10-20. The vertical scale has here been doubled compared with previous figures in order to show the differences more clearly. The upper pair of curves shows the total circumferential stress at the top face, and the lower pair shows the total radial stress at the bottom face. The large differences occurring in the flat central section are clearly attributable to the use of nonlinear equations in reference 14. It is interesting to observe that these large differences in the center have only a small effect on the quantities in the corrugation zone, where our results show merely a slight increase in the peak values over the Grover and Bell results. The greater differences occurring near the outer boundary, however, are possibly due primarily to the approximate nature of the functions employed by them in the corrugation zone, rather than due to the nonlinear effects at the center.

It is expected that the procedure and the machine codes now available will be applied to solving other diaphragm problems for different shapes and thicknesses and for cases in which a central stiffening disk is present.

3. NEEDS FOR FUTURE WORK

For the future it is hoped that it will be possible to have experimental studies made, at NBS

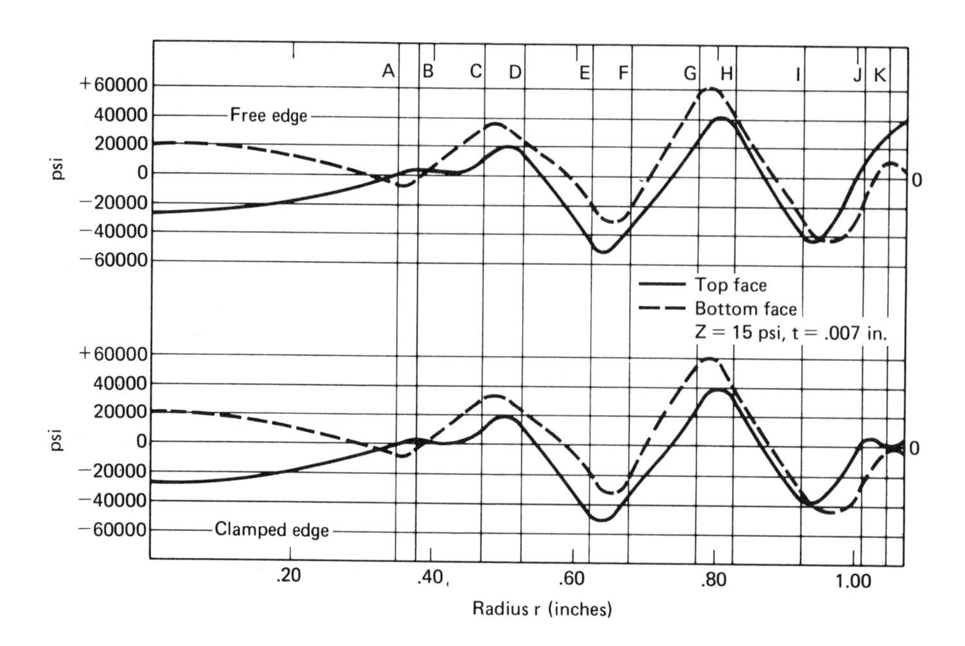

Fig. 10-19. Total circumferential stress at each face.

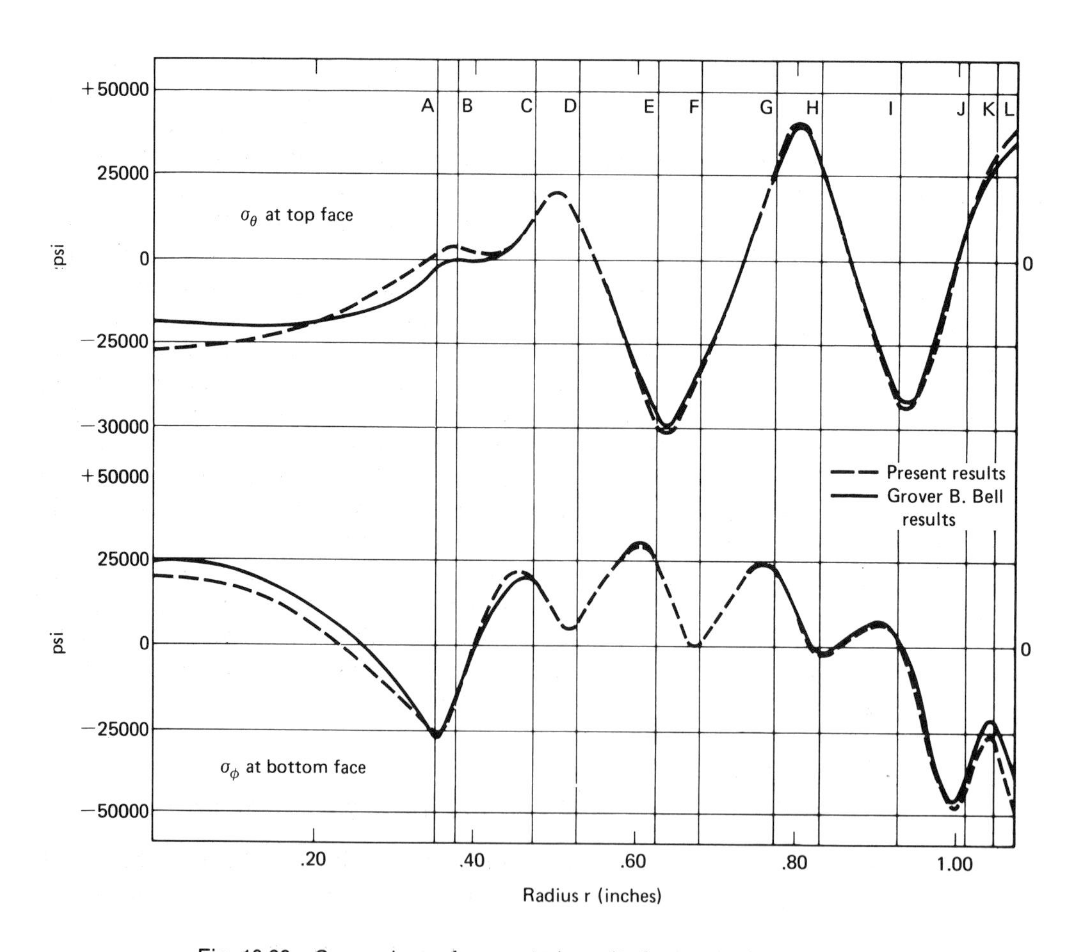

Fig. 10-20. Comparison of computed results for two typical stresses.

or elsewhere, on a wider scope of the variables on which some investigations have been made, and also on many other facets of the general problem, in conjunction with an exhaustive computational program.

In connection with the problem of shape, it is clearly desirable to extend investigations to extremes of corrugation depth, corrugation number, corrugation shape, i.e., extension of work on square wave, triangle wave, etc., so that approximate results may be obtained by interpolation for other shapes. The effect of superposing a built-in deflection or dishing, positive or negative, or a combination of small corrugations on larger corrugations also should be studied. Radial corrugations, spiral corrugations, and noncircular edges also should be studied further.

In the field of materials, further studies should certainly be made on high-performance alloys, titanium, beryllium, and materials such as plastic, glass, quartz, glass-reinforced plastic, and so on. Possible advantages of special techniques of manufacturing diaphragms, including evaporation, casting, hot-forming, machining, spinning, electroforming, etc., may warrant further investigation. Probably the most important objective to be sought by improved techniques and materials is the reduction of elastic defects, although increased useful deflection range may also be attained. Presumably, hysteresis is the important criterion. Since the magnitude of the other elastic defects appears to be related to the hysteresis, aftereffect is often the practical measurement to make.

One result found in the early work at NBS was that the percentage hysteresis (hysteresis/deflection) increased markedly for extremely small deflections. This effect should be studied very carefully, since it may have a very important bearing on the ultimate attainable performance. This effect may have been peculiar to the method of clamping, since it has not been noted by others working on free-edge capsules, but this appears to be an open question for the present. The ultimate sensitivity of diaphragm instruments depends, of course, on the transducer used to measure the deflection. Since it is now possible to measure deflections with a sensitivity of 10^{-7} in., a more thorough investigation of the fine-scale hysteresis would be possible. Inseparably connected with the problem of materials is the problem of treatment of materials—heat-treatment, seasoning by cold-working, overstressing, and so on—and the effect of temperature, or temperature gradients, on the performance of the element. These factors, while important, can probably be evaluated by straightforward and painstaking experimentation.

The question of maximum useful deflection is important in most design problems, and this is affected not only by the choice of material but also by the geometry of the diaphragm shape as it affects the concentration of stresses. Direct measurement of stress in thin diaphragms is experimentally quite difficult, although bonded strain gage, brittle lacquer, and other techniques have been used to indicate the regions of maximum stress in loaded diaphragms. Photoelasticity is another method that probably should be tried. Presumably these techniques could help to determine which shapes were most favorable in minimizing stress concentrations. However, the same general conclusions can be arrived at from the maximum deflections experimentally attainable without excessive drift or hysteresis or, in some cases, from stress-pattern computations of the type we have described.

In our experimental work we found that a central force of about 0.4 of the distributed pressure load would result in the same central deflection for NBS Shape 1. For a flat diaphragm this ratio is 0.25 over the linear range. For each diaphragm shape, this factor should be determined for all deflections, since it is of importance in determining the overall elastic constant of the diaphragm when coupled to mechanically loaded members. Also, the natural frequency of the diaphragm is not easily determinable from static measurement, and should be measured. This may require testing in evacuated chambers, particularly for thinner diaphragms, to eliminate air damping. Resonant flexing also might be used to permit rapid fatigue testing, which may be of importance for knowledge of reliability. Simple approximate formulas for computing natural frequency might be derived which would be quite useful. Exact values can be obtained as solutions to a type of eigenvalue problem that is easily handled by digital computers.

The effective thickness was determined rather crudely in the NBS work, and some attention might well be given to better definition and determination of this quantity. It appears that each element of the diaphragm should be weighted according to its actual thickness and the stresses to which it is subjected. This suggests that computed values of stress might be combined with experimental measurements of thickness and experimental or computed values of deflection to arrive at more precise values of the effective thickness.

As mentioned previously, it may be useful to relate the performance of diaphragms with nonlinear load-deflection curves on nondimensional design charts by appropriately chosen nondimensional flexibility functions. Assuming that the linear range may soon be well in hand through high-speed computation, attention will focus on further theoretical work on the nonlinear problem such as Haringx (15 and 16) is attempting. Perhaps an exact formulation of the nonlinear problem that is also suitable for machine computation will be difficult to obtain. Even so, approximations will be most useful, with the linear solutions now becoming available for checking limiting conditions.

The computation of stress distribution may now be used to guide the modification of shapes to achieve better stress distribution and presumably larger deflection ranges. It may well be that an optimum shape will be found to have corrugations of varying depth, width, and even of outline along the radius! Computed deflection distributions will also provide a potent guide to design interpolation and fine-scale adjustment of the shape of load-deflection curves. Looking even further, and perhaps more optimistically, it may be hoped that the growing knowledge of inelastic behavior of materials coupled with the knowledge of stress distribution may some day make possible the computed prediction of hysteresis.

While it is possible, and desirable, to obtain computational results on idealized diaphragms to accuracies of better than 1%, it should be realized that variations or uncertainties in material properties such as modulus, Poisson's ratio, effective thickness, or residual stresses, make it unlikely that diaphragm manufacture will become "exact" to anywhere near this precision. Variations of several percent between deflection curves of supposedly identical diaphragms are still to be expected.

In the experimental testing and evaluation, there is a real need for applying better instrumentation to the measurement of diaphragm performance. Not only is high precision required in measuring load-deflection characteristics, in order to obtain data on elastic deflections by subtraction, but means should also be devised to permit rapid testing through many cycles, with automatic recording, reduction, and analysis of the data. Adequate techniques are now available but so far have not been combined and used in this problem.

Electronic counting (and subdividing) of optical interference fringes appears to offer an attractive method of adequately accurate and rapid measurement of deflection. This method can be easily tied into automatic data-handling equipment, curve plotters, and computers. For precise drift measurements other methods, such as the use of the capacitance gage, may be better. Automatic plotting of reduced data—such as the deviation of the load-deflection curve from the desired shape, or drift, or hysteresis—simultaneously with the experimental testing would greatly facilitate routine evaluation of larger numbers of test specimens, and also make possible on-the-spot changes in test work to investigate newly discovered effects.

There is undoubtedly a wealth of empirical data in the files and in the minds of experienced diaphragm designers. One of the most productive research projects would be the collection, coordination, and dissemination of such "knowledge of the art." We suggest this to the ASME Research Committee on Mechanical Pressure Elements as a worthy task, and at the same time we commend the Committee for its leadership in arranging this symposium on diaphragm research, and express our thanks for this opportunity to review the past and look to the future.

BIBLIOGRAPHY

1. The Theory of Pressure Capsules, A. A. Griffith, Great Britain Aeronautical Research Council, R & M No. 1136, 1928.
2. On the Depression of the Centre of a Thin Circular Disc of Steel Under Normal Pressure, Stanley Smith, *Transactions of the Royal Society of Canada,* 3rd series, vol. 16, section 3, May 1922, pp. 217–225.
3. *Elastische Platten,* A. Nadai, Julius Springer, Berlin, Germany, 1925.
4. Uber die Berechnung der dünnen Kreisplatte mit Grosser Ausbiegung, Karl Federhofer, *Eisenbau,* vol. 9, 1918, pp. 152–166.
5. Bending of Circular Plates With Large Deflection, Stewart Way, *Trans. ASME,* vol. 56, paper APM-56-12, 1934.
6. Zur Berechnung der dünnen Kreisplatte mit Grosser Ausbiegung, Karl Federhofer, *Forschung auf dem Gebicic des Ingenieurwesens,* Ausg. B. bd. 7, hoft 3, May–June, 1936, pp. 148–151.
7. Über die Randwertaufgabe der Gleichmässig belastoton Kreisplatte, J. Barta, *Zeitschrift für angewandie Mathematik und Mechanik,* bd. 16, heft 5, October, 1936, pp. 311–314.
8. *Theory of Plates and Shells,* S. Timoshenko, McGraw-Hill Book Company, Inc., New York, N.Y., and London, England, 1940.
9. Diaphragms for Aeronautic Instruments, M.D. Hersey, NACA TR No. 165, 1923.
10. Corrugated Metal Diaphragms for Aircraft Pressure-Measuring Instruments, W. A. Wildhack and V. H. Goerke, NACA TN No. 738, 1939.
11. The Limiting Useful Deflections of Corrugated Metal Diaphragms, W. A. Wildhack and V. H. Goerke, NACA TN No. 876, 1942.
12. Model Experiments and the Forms of Empirical Equations, E. Buckingham, *Trans. ASME,* vol. 37, 1915, pp. 263–288.
13. Sensitive Aneroid Diaphragm Capsule With No Deflection Above A Selected Pressure, W. G. Brombacher, V. H. Goerke, and F. Cordero, *National Bureau of Standards Journal of Research,* vol. 24, no. 31, 1940.
14. Some Evaluation of Stresses in Aneroid Capsules, H. J. Grover and J. C. Bell, *Proceedings of the Society of Experimental Stress Analysis,* vol. 5, 1948, p. 125.
15. The Rigidity of Corrugated Diaphragms, J. A. Haringx, *Applied Scientific Research,* The Netherlands, vol. 2, series A, 1950.
16. Stresses in Corrugated Diaphragms, J. A. Haringx, *The Anniversary Volume—Applied Mechanics,* dedicated to C. B. Biezeno, Haarlem, Antwerpen, Djakarta, N.V. De Technische Uitgeverij H. Stam., 1953, pp. 199–213.
17. A Note on the Theory of Corrugated Diaphragms for Pressure-Measuring Instruments, A. Pfeiffer, *Review of Scientific Instruments,* September, 1947, pp. 660–664.
18. *Étude des Capsules Aneroides,* F. Charron, Ministère de l'Air, Paris, France, 1940.
19. Der Spannungszustand einer Kreisringschall, by K. Stange, *Ingenieur-Archiv,* vol. 2, 1931.

DISCUSSION

J. A. HARINGX.[7] The writer was greatly interested in reading an attractive paper, since it contains extensive information on experimental results with various types of corrugated diaphragms, a new mathematical approach taking advantage of the possibilities offered by modern electronic computers, and a number of valuable considerations on work to be done in the future.

Referring to the writer's own work on this subject, it is agreed that from a strictly mathematical point of view the principle of smearing out the effects of the corrugations, which has led to the introduction of a factitious equivalent flat plate,[8] is only justifiable for a diaphragm with a large number of small corrugations. However, there exist various methods of concentrating the elasticity of structural parts into a number of elastic hinges. It has been shown that by this method very satisfactory results are obtained, even for cases where the number of these hinges is limited to two or three. The writer's experience has convinced him of the same reliability of the smearing-out method. He refers to the elastic stability of helical compressions springs[9] and of flat spiral springs.[10] Apart from the approximation based on the smearing-out principle, a more exact method of calculation was derived for these cases, so that the results of the approximation could be checked. Even when the number of spring coils is small, the approximation is surprisingly satisfactory.

Nevertheless, the writer's belief in the smearing-out method when applied to diaphragms having a small number of corrugations has been more or less speculative. He was therefore very glad that in Figures 10-15 and 10-16 the authors published the full deformation of the diaphragm that was earlier investigated by Grover and Bell. Although the profile of this diaphragm is rather unsuitable to be treated in the way dealt with in the writer's paper,[8] he thought it worthwhile to check the vertical deflection W derived on a sound mathematical basis with the deflection y given in equations (31) and (32) of the said paper.[8]

For the region with uniform corrugations, the quantity q is 8.78 (see reference 2 in the writer's paper[8]) and as shown in Figure 10-21 the curves for the vertical deflection are nearly the same, even for this "unsuitable" diaphragm.

The difference in nature of the curves results from the different ways of dealing with the central part of the diaphragm. In the approximation the corrugations are tacitly extended to the center where they are so stiff that this region can more or less be regarded as a rigid flat plate. The authors, on the other hand, treat the central region as an elastic flat plate. However, the deflection of this region is about 0.015 in. so that, for a sheet thickness of 0.007 in., the assumption that the deflection is proportional to the load is certainly not valid. Here we see that this more exact solution also has its limitations. It is thus still a matter of uncertainty whether or not in practice the value of the maximum deflection calculated by the authors is really better than the writer's approximation, which requires no more than a few minutes of calculation.

The writer must object against the statement by the authors, that the curve showing the agreement between his approximation and the experiments by Wildhack and Goerke (ef. Fig. 7 in the writer's paper[8]), was found "when adjusted on the basis of one empirical coefficient derived from the NBS experimental work." This is certainly not true and must be caused by some regrettable misinterpretation of his work. As shown in the writer's paper[8] the formulas given lead directly to a definite deflection, whether right or wrong but without any adjustment.

L.E. Wood.[11] The paper has been read with great interest, and the following comments, while perhaps departing to some degree from the general trend of the paper, are thought to be applicable in a broad sense.

The Friez Instrument Division of the writer's company has manufactured a large number of diaphragms, many of which are, in general, somewhat similar to the National Bureau of Standards Shape 1. Over a period of approximately twenty years, our diaphragms have been made of three principal materials, namely, phosphor bronze, which was replaced during World War II by beryllium copper, owing to its superior elastic properties, and which, in turn, in recent years, has been largely replaced by Nickel Alloy Span "C," primarily because of its superior thermoelastic properties. During this period, somewhere between one and two million diaphragms have been produced, and those made today are greatly superior from the standpoint of performance and uniformity over those made ten to twenty years ago. However, the improvements lie mostly in the realm of better materials and closer process

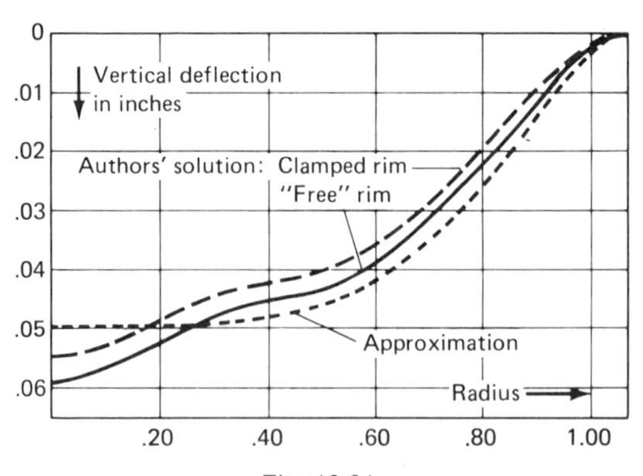

Fig. 10-21.

control, rather than in improvements in fundamental knowledge as to the design of diaphragms.

Based on ideas gathered from reading the National Bureau of Standards paper and our own experience, it is thought that perhaps diaphragm-development work could be simplified a little if the following generalizations are acceptable:

1. A diaphragm is a combined spring and fluid barrier.
2. The properties required as a satisfactory fluid barrier may be taken for granted insofar as the material of the diaphragm itself is concerned.
3. The physical properties of the material affecting its use as a diaphragm are the same as those affecting its use in any spring application.
4. The properties which determine the excellence of the material as a spring, such as chemical composition, modulus of elasticity, thermoelastic modulus, hardness, microstructure, etc., are subject to reasonably exact determination and analysis.
5. There appears to be no reason, assuming a material can be formed into the necessary shape for a diaphragm, that its performance as a diaphragm would not bear the same relationship to its physical properties as its performance as a conventional spring.
6. In other words, diaphragms having identical geometric configurations should behave in relation to one another in quite direct relationships to the readily measured physical properties of the materials.
7. Consequently, significant research on the diaphragm shapes should be possible, using one material having known and carefully controlled characteristics.

We do not have detailed and extensive laboratory data, verifying the foregoing, since as a result of practical necessity it has been necessary usually to vary several factors simultaneously during the transition from one material to another, but would like to suggest it for what it may be worth.

In general, our experience has agreed with that found at the Bureau of Standards in relation to the amount of motion that can be obtained from diaphragms and the general characteristics of behavior. The references in the paper, with regard to center loading and free versus clamped diaphragms, apply to areas where we feel further work is needed as well as on the more broad aspects of the diaphragm problem.

AUTHORS' CLOSURE

Dr. Haringx has again contributed valuable new data in Figure 10-21 showing the closeness with which his approach can predict deflections of diaphragms having shapes admittedly somewhat unsuitable to his approach. In the present paper of Haringx the results that he obtained in this case for stresses as well as for deflection are surprisingly good. These results again confirm the usefulness of his formulations in diaphragm design.

We explained in the third and fourth paragraphs following equation (7) that we computed the deflection for a loading of 15 psi in order to compare results with those of Grover and Bell, but we pointed out that the entire mathematical solution was limited to the linear range. Some confusion appears to be inescapable, since our (linear) calculations cannot apply to the central part of the diaphragm for loading of 15 psi. However, if both our results and those of Grover and Bell were scaled down linearly, by a factor of, say, 5 or 10, then the comparisons would be equally valid in the outer parts of the diaphragm, but Grover and Bell's curve would be in error in the central portion because of applying a linear scaling down to a nonlinear solution. Since Haringx' approximation tacitly implies a "more or less" flat plate in the central region, a better comparison would be obtained if horizontal lines were drawn to all curves of Figure 10-21, intersecting them at a radius of 0.35 in. We may agree that it is yet to be proved by further comparison of experiment and calculations whether the computation is more accurate than Haringx' approximation, but to be meaningful the comparisons must be made on truly comparable conditions. (Incidentally, if the "exact" calculation were made on the basis of a rigidly reinforced central disk, slight changes would result in the deflection of the outer part as a result of the changed boundary condition. This might be true also, to some extent, for Haringx' approximation.)

The authors regret the statement objected to by Haringx. The sentence should have read: "Haringx was able to derive a curve which would predict the performance of the NBS Series (Shape No. 1)," The "adjustment" referred to the following sentence on the tentative and inconclusive attempts, at NBS, to test Haringx' formulation for other shapes.

Mr. Wood's generalizations appear quite reasonable, and we certainly concur with the suggestion that further work is needed with respect to central loading and "free" versus "clamped" diaphragms, since these factors are of great importance in practical design work.

NOTES

[1] The experimental phases of the NBS work reported herein were sponsored at different times by the National Advisory Committee for Aeronautics and Wright Air Development Center of Air Research and Development Command. The research in Part 2 was supported by the United States Air Force through the Office of Scientific Research of the Air Research and Development Command.

[2] Chief, Office of Basic Instrumentation, National Bureau of Standards, U.S. Department of Commerce.

[3] Chief, Mathematical Physics Section, National Bureau of Standards, U.S. Department of Commerce.

[4] Chief, Mechanical Instruments Section, National Bureau of Standards, U.S. Department of Commerce. Mem. ASME.

[5] Numbers in parentheses refer to the Bibliography at the end of the chapter.

[6] Part 2 was prepared by R. F. Dressler. This research was supported by the United States Air Force, through the Office of Scientific Research of the Air Research and Development Command.

[7] Chief Engineer, Philips Research Laboratories, Eindhoven, The Netherlands. Mem. ASME.

[8] Design of Corrugated Diaphragms, by J. A. Haringx, published in this issue of the *Transactions,* pp. 55–64.

[9] On Highly Compressible Helical Springs and Rubber Rods, and Their Application for Vibration-Free Mountings, by J. A. Haringx, thesis, Technological University at Delft, The Netherlands, 1947; *Philips Research Reports,* vol. 3, 1948, pp. 401–449.

[10] Elastic Stability of Flat Spiral Springs, by J. A. Haringx, *Applied Scientific Research,* vol. 2, series A, 1949, pp. 9–30.

[11] Chief Engineer, Bendix Aviation Corporation, Baltimore, Md.

Contributed by the Diaphragm Research Subcommittee of the Research Committee on Mechanical Pressure Elements and presented at the Diamond Jubilee Annual Meeting, Chicago, Ill., November 12–18, 1955, of the American Society of Mechanical Engineers.

Statements and opinions advanced in papers are to be understood as individual expressions of their authors and not those of the Society. Manuscript received at ASME Headquarters, September 27, 1955. Paper No. 55-A-181.

This material was extracted from the *Transactions of the American Society of Mechanical Engineers (ASME),* January, 1957, Investigations of the Properties of Corrugated Diaphragms, written by W. A. Wildhack, R. F. Dressler, and E. C. Lloyd.

PART THREE

RELATED DEVICES

XI Pressure Transducers

SOME BASIC POINTERS ON USING STRAIN-GAUGE PRESSURE TRANSDUCERS†

Strain-gauge pressure transducers provide a convenient and reliable means of measuring the pressures of gases and liquids. They are especially suited for use in systems containing problem fluids such as slurries and corrosives. A transducer is a device that converts one physical quantity (such as force, temperature, flow, or displacement) into another physical quantity with more usable output characteristics. Bonded strain-gauge pressure transducers convert the displacement caused by the force of system pressure into an analogous electrical output signal.

Operating Principle. Bonded strain-gauge pressure transducers have an etched metal foil strain gauge intimately bonded to a sensing element; the strain gauge is fashioned into a full, four-arm Wheatstone bridge. When pressure causes the sensing element to deflect, the variable arm of the bridge is stressed. Stress deformation causes a change in electrical resistance, producing a change in electrical output signal that corresponds to the change in pressure.

Advantages of Pressure Transducers. Strain-gauge pressure transducers are unitized, hermetically sealed units that are installed by threaded connection directly at the point of measurement, Figure 11-1. They are offered with a variety of input and output connection options, Figure 11-2. Because the units convert fluid pressure into an electrical signal directly at the point of measurement, no capillary elements or instrument piping is required. Signal transmission from transducer to point of readout is by electrical wiring, eliminating the response time lag encountered when fluid is used as the signal-transmission medium. Elimination of instrument piping is also an asset if the measured fluid is toxic or corrosive.

Pressure transducers are designed so that the fluid contacts either a stiff diaphragm or a small-volume pressure-sensing tube. Pressure changes on these elements cause only a small change in cavity volume, and the sensing element presents minimum restriction to fluid flow. There are no moving parts to wear out.

The Importance of Proper Application. Although pressure transducers are rugged and reliable and require virtually no maintenance, they are subject to misapplication. Proper application not only assures that measurements will be of high integrity; but it also ensures long transducer life.

The Effects of Temperature. All strain-gauge pressure transducers are compensated for

†Article written by Bernard H. Shapiro, Vice-President, Engineering, BLH Electronics, Waltham, Massachusetts. Courtesy of *Plant Engineering Magazine,* Technical Publishing Company, Barrington, Illinois 60010.

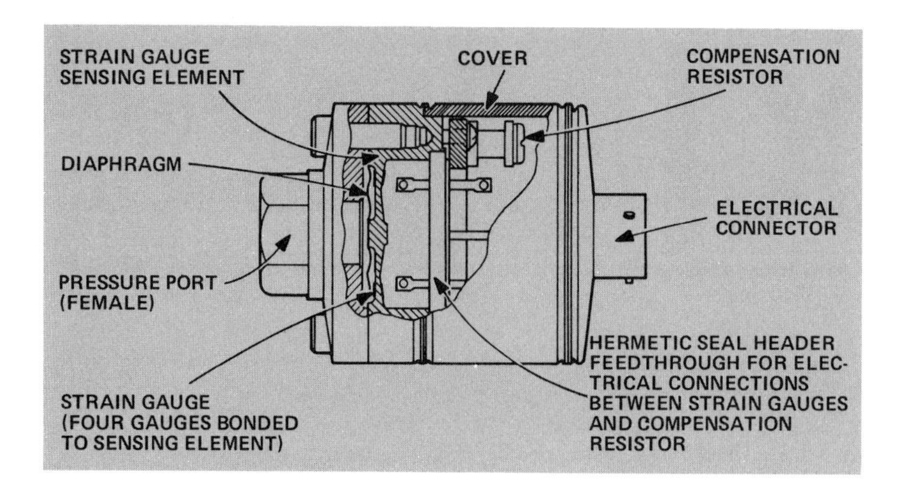

Fig. 11-1. Strain-gauge pressure transducers are hermetically sealed units designed for installation directly at the point of measurement without the need for instrument tubing. Units are of stainless steel, ideally suiting them for measuring pressure in systems containing corrosive gases or fluids. Force of fluid pressure causes deflection of the sensing element, stressing the variable arm of the strain-gauge bridge. The resulting change in resistance of the bridge arm produces a change in the output electrical signal that is proportional to the change in pressure. Because of the very minute deflection of the sensing element, there is virtually no change in cavity volume of the vessel containing the measured fluid, and there are no moving parts.

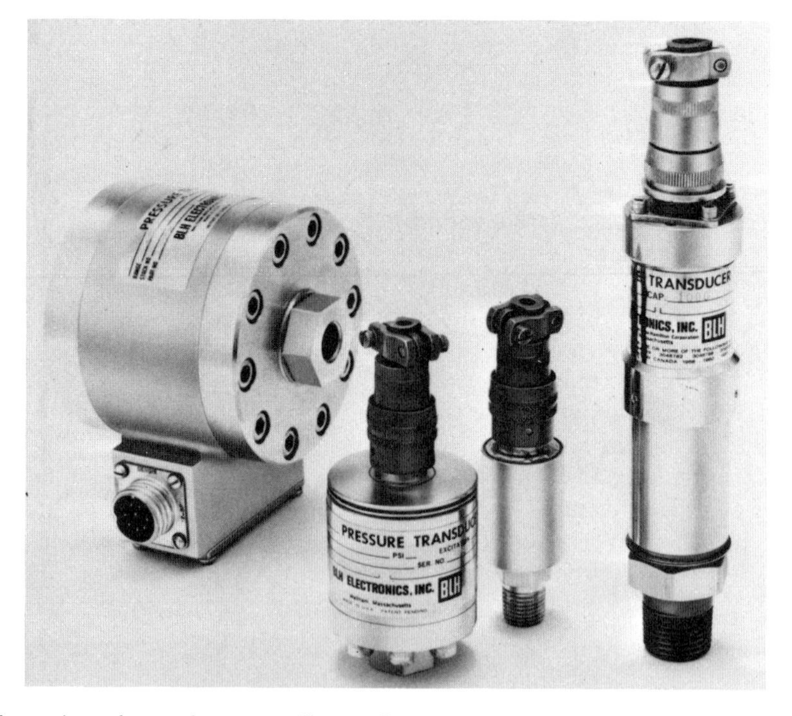

Fig. 11-2. A number of manufacturers offer strain-gauge pressure transducers in a wide variety of styles, shapes, pressure ranges, and stainless steel. Transducers can be obtained for any application, and in constructions that are compatible with any fluid or ambient environment.

temperature errors at steady-state conditions. This compensation is precise and very effective, and error resulting from thermal gradients is seldom a problem. However, errors can be introduced by rapidly changing temperatures in either the ambient environment or the measured fluid. Magnitude of error is a function of the temperature differential across the transducer.

If pressure measurements must be precise, and there are wide, frequent changes in ambient temperature, thermal protection should be provided to minimize the temperature differential across the transducer. Thermal protection can be provided by insulating materials or by baffles or heat shields. If there are significant transient temperature changes in the measured fluid, the effects of these temperature changes can usually be stabilized by baffles or multi-path ports.

Media Compatibility. Most pressure transducers are made of corrosion-resistant stainless steels that are compatible with most fluids. The element of the transducer that contacts the fluid is usually made of stainless steels such as 17-4 PH, 15-5 PH, or 17-7 PH. The outer shell or protective cover is usually of a series 300 stainless.

Corrosion within a pressure transducer is more than a cosmetic problem, and if there is doubt as to which material is best for a particular application, the transducer manufacturer's application engineers should be consulted. Corrosion of the pressure-sensing tube or pressure-sensing diaphragm can change the transducer's sensitivity and affect its structural strength.

Shock and Vibration. Generally, strain-gauge transducers are constructed to withstand relatively high shock and vibration without damage. Nevertheless, any measuring device should be placed in an area that is relatively free of vibration and shock. If severe vibration and shock may cause a problem, isolation, in the form of a flexible connection or other means, should be provided to decrease the forces on the transducer.

Frequency Response. There is sometimes a need to make dynamic measurements of rapidly fluctuating pressures; in some cases, static pressure measurements have dynamic components. Finding the proper transducer to measure dynamic pressure in a particular application can be challenging and difficult. Response time of the transducer is basically affected by the unit's natural frequency, and most transducer manufacturers can provide information on the natural frequency of their transducers.

The interface between the transducer's sensing element and the measured fluid will also affect frequency response. For optimum frequency response, the transducer should be coupled as closely as possible to the point at which the measurement will be taken. Interposing tubing between the point of measurement and the transducer will adversely affect frequency response. Best results are obtained with a transducer with a high natural frequency of flush-diaphragm construction, directly connected at the point where the measurement is to be taken.

Dynamic Overloads. High-pressure spikes superimposed on a steady-state pressure can damage transducers. Water hammer caused by the rapid opening or closing of valves can cause such dynamic transient overloads. Damage can be prevented by opening and closing valves very slowly, or by installing pressure snubbers in the connection to the transducer. Such snubbers are offered by a number of manufacturers.

Handling and Installation. A pressure transducer is a precision measuring device and should be treated accordingly. Although transducers are ruggedly built, they can be damaged if dropped or installed improperly. Accuracy, in some cases, may be affected by deformation of the exterior housing, and zero shifts in the electrical output might result from the torque applied to input pressure fittings during installation. Manufacturer's recommendations for proper torque should always be observed. Zero shift can be compensated for in post-installation calibration of the readout device.

If fittings used to install the transducer have different thermal coefficients of expansion than the metal of the transducer, measurement errors might also result. Fittings, therefore, should be thermally compatible with the metal in the transducer.

Normally, pressure transducers are not mounting-position sensitive. However, some types—usually in the low-capacity range—are sensitive to mounting position. Restrictions on mounting position will be included in the manufacturer's installation instructions.

Readout Instrumentation. A basic pressure-measuring system using transducers consists of the transducer, a power supply, one or more amplifiers, and a readout device, Figure 11-3. If digital readout is desired, Figure 11-4, an analog/digital converter must also be interposed in the system. Transducer output levels usually range from 0 to 20 millivolts and 0 to 5 volts. Transducers with output voltage ranges of from 0 to 1 volt or more usually have integral amplifiers.

Electrical noise imposed on the output circuitry can affect accuracy, especially with

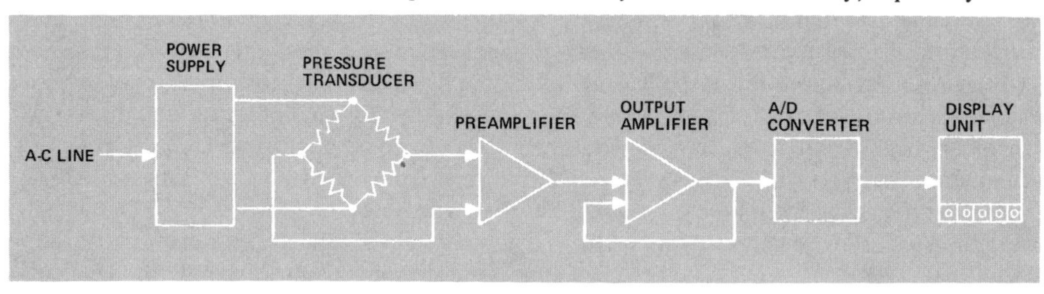

Fig. 11-3. Basic pressure measurement system employing strain-gauge pressure transducers consists of the transducer, power supply, one or more amplifiers, readout device, and associated wiring. The readout instrument can be dial, pointer, chart recorder, or other type of display; analog/digital converter is required only if digital display is used. In some transducers, the amplifier is integral with the transducer.

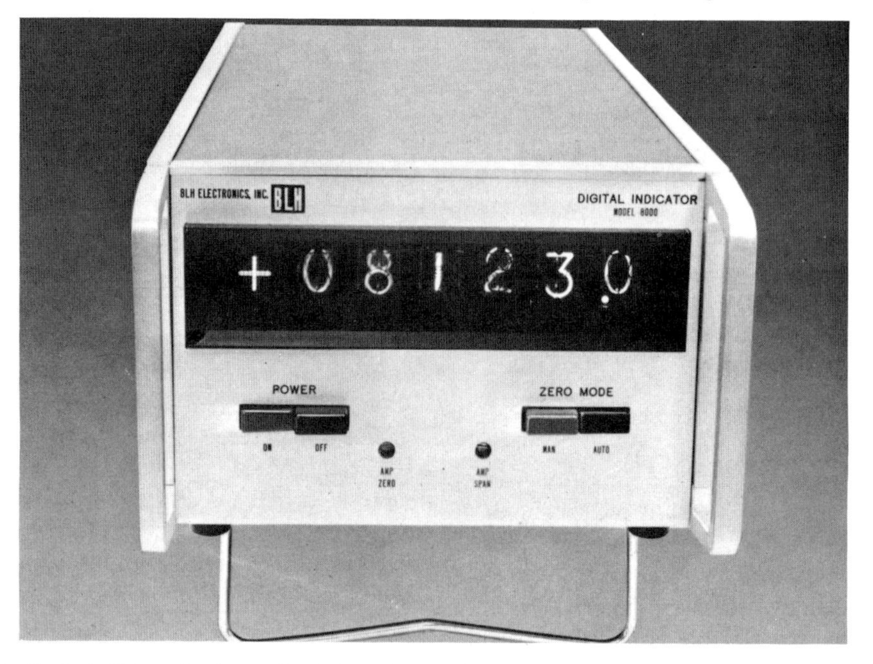

Fig. 11-4. Typical digital display unit designed specifically for system using strain-gauge pressure transducers has integral adjustments for zeroing strain-gauge bridge and for adjusting the amplifier output. Conventional pointer displays, chart recorders, video displays, and other types of displays can also be used. Transducer output can also be used as input for automated process control.

transducers designed for low-level output signals operating at the low end of their pressure range. Cable runs between components of the system should be properly shielded and grounded, and all connections should be of high integrity. Cable between-conductor leakage, and leakage between conductors and shield, should be held within the limits specified by the transducer manufacturer.

The low-profile solid state pressure transducer that weighs less than an ounce and is shown in Figure 11-5 offers an accuracy of 1%. The pressure transducer is temperature-compensated, produces an amplified signal, and requires no recalibration, the Honeywell division said.

A flexible diaphragm in the center of a silicon "chip" only a tenth of an inch square lies at the heart of the solid state device. Chemical etching of the diaphragm and unique ion implementation of four resistors in the chip have helped create a highly accurate and repeatable transducer that converts changes in pressure into proportional electronic signals, the manufacturer said. These signals are applicable to better utilization of engine controls, home appliances, medical instruments, environmental control systems, pneumatic controls, and a variety of other control systems.

The pressure transducer is designed to operate from 0 to 15 psi. It provides a ripple-free linear dc output with a non-zero null and a span output up to 85% of the supply voltage. The chip is said to be drift-free over life. Sensitivity of the transducer is 333 mV psi. Combined linearity and hysteresis do not exceed ±1% of full scale output (FSO). Operating temperature range is −40° to +125°C (−40° to +257°F).

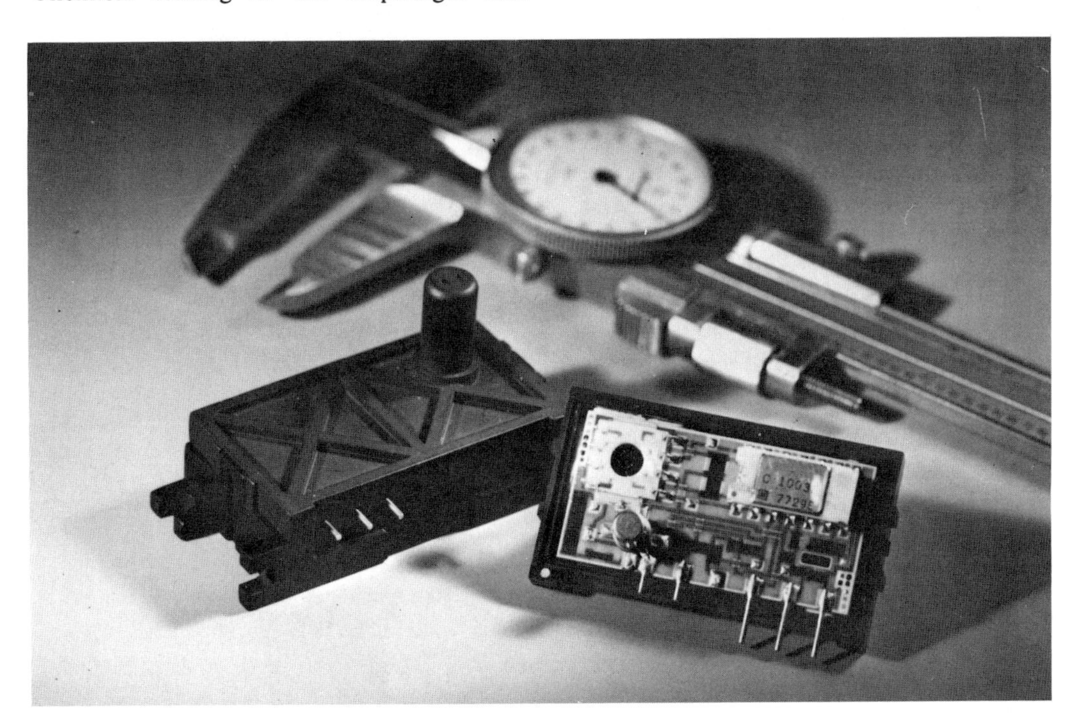

Fig. 11-5. Low profile, high accuracy describes this pressure transducer. A silicon chip (inside circle, center), 0.1 inch square, flexes as pressure changes, inducing proportional electronic signals through resistors implanted in the chip. The 110PC pressure transducer is temperature compensated, requires no recalibration and produces an amplified signal. *(Courtesy of Micro Switch, Freeport, Illinois.)*

XII Temperature Switches

Figures 12-1 and 12-2 show two types of temperature switch. Their operation is described in the figure legends.

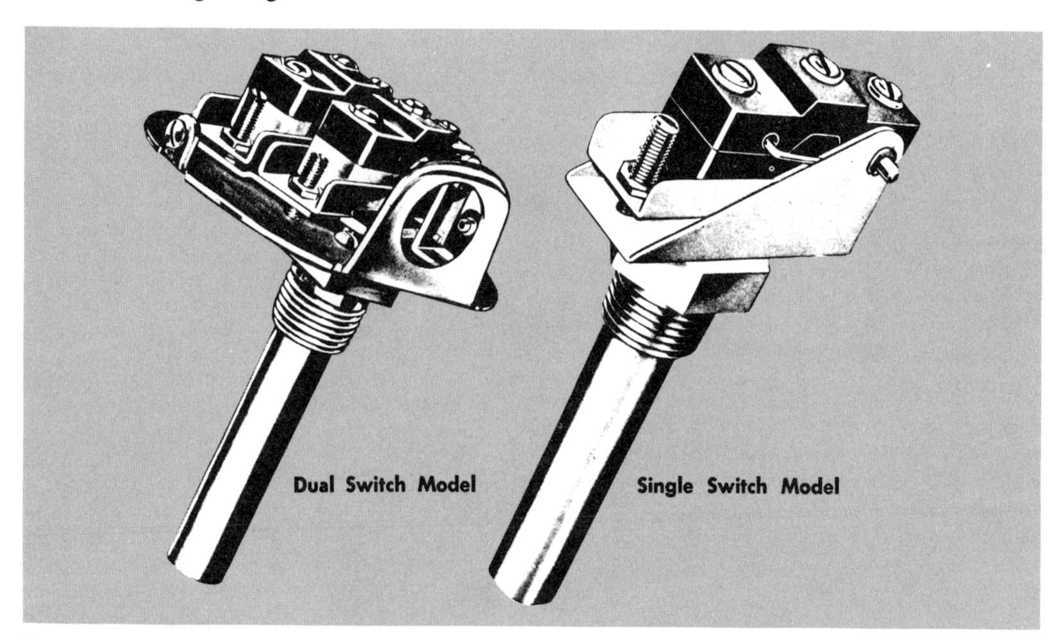

Fig. 12-1. Temperature actuated control switch: Control action is provided by an expandable liquid acting on a bellows assembly. Bellows motion, created by volume changes of the liquid, actuates the switch contacts through a push rod. *(Courtesy of Fenwal, Incorporated, Ashland, Massachusetts.)*

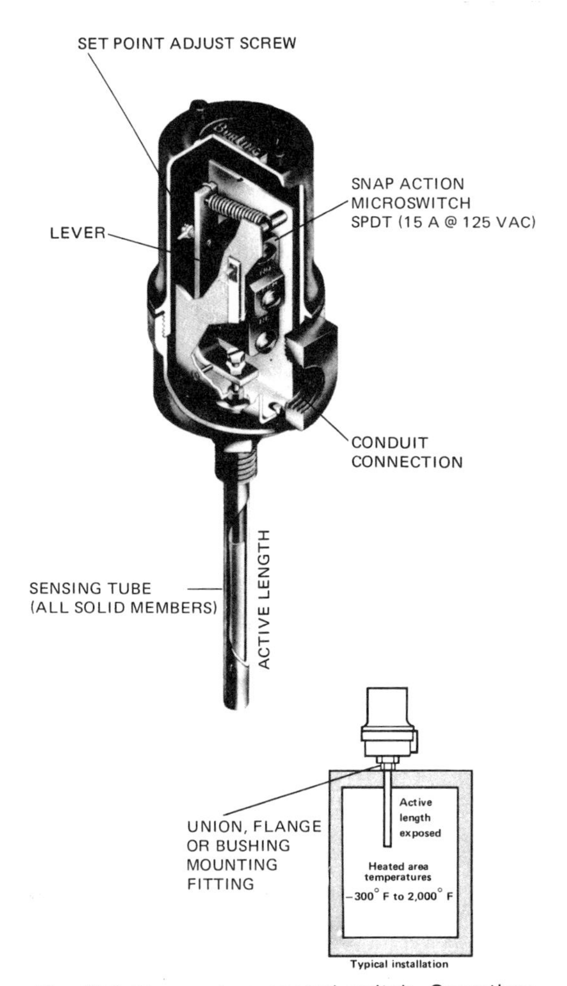

Fig. 12-2. Temperature control switch. Operation: The principle of differential expansion of solids has been adapted in this unit. The outer tube has a higher coefficient of linear expansion than the inner one. Consequently an increase in temperature causes the outer tube to expand more than the inner one, thus causing the spring to move the lever toward the switch. *(Courtesy of Burling Instrument Co., Chatham, New Jersey.)*

XIII Flow Switches

In this chapter we shall briefly discuss a flow switch. Flow switches, in most cases, will differ from the basic pressure switch only from the standpoint that the pressure which operates them is generally considered dynamic pressure, whereas the basic pressure switch is considered to be that of static pressure.

Mathematically defined for simplicity, a static pressure would consist of the formula: $P \times A$ (pressure × area); whereas the dynamic pressure switch formula is equal to $1/2\, \varrho v^2 A/2g$, which for all practical purposes is derived from the kinetic energy equation.

The basic flow switch, in general, is tripped by the flow of the liquid which moves a metal target barrier or shield that mechanically operates a basic switch. (See Figures 13-1a,b,c.)

Other types of flow switches have also been devised in the past, such as arrangements using the physical principle of the bourdon tube effect, and again the pressure × area coupled with an amplifier version.

Typical applications will be shown in the applications section of the book (see Chapter 16 A).

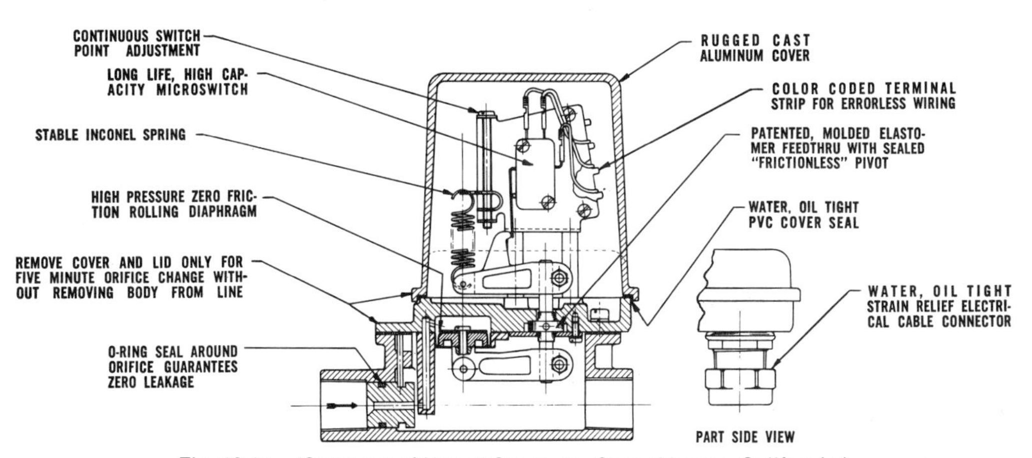

Fig. 13-1a. *(Courtesy of Harwil Company, Santa Monica, California.)*

0.12 TO 8.0 GPM FOR ½″ PIPES

4 TO 32 GPM FOR 1″ PIPES

5 - 85,000 GPM & UP FOR PIPES 1″ - 48″ & UP

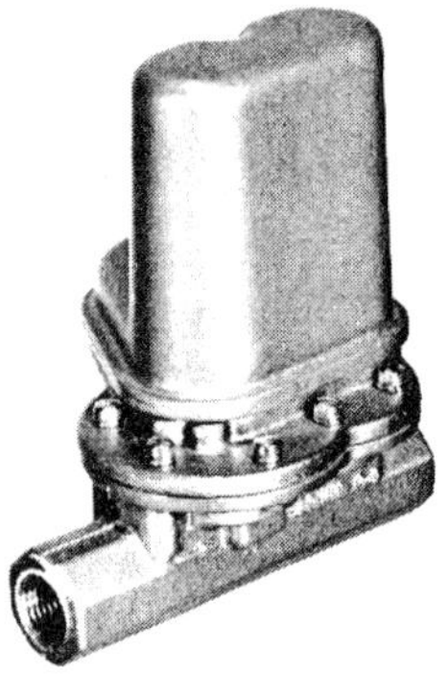

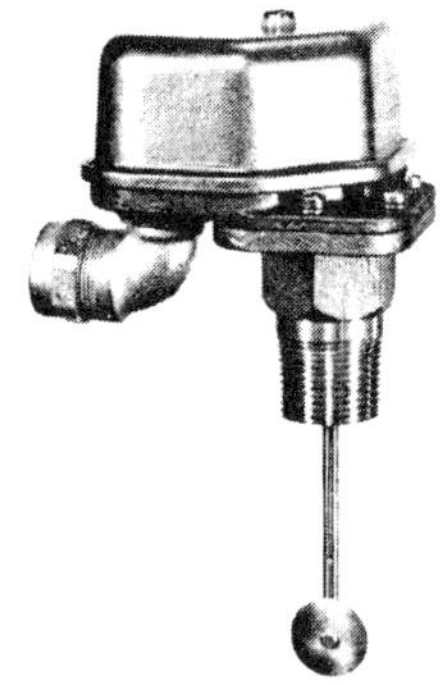

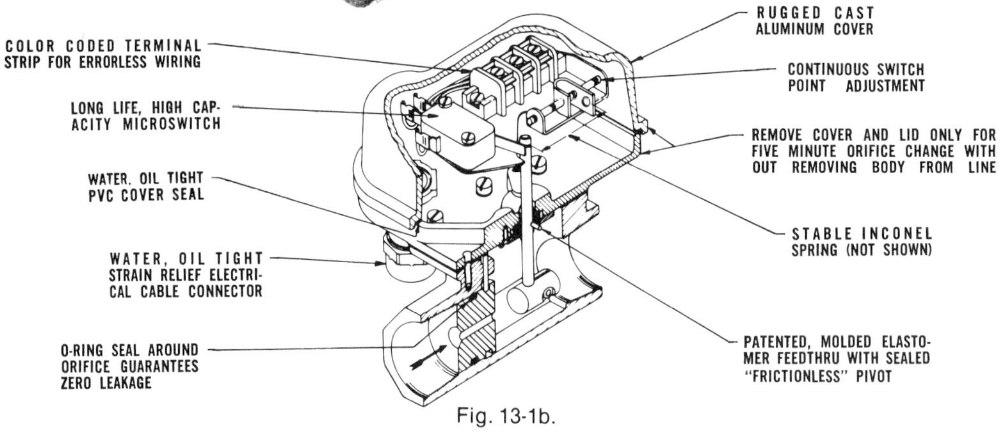

COLOR CODED TERMINAL STRIP FOR ERRORLESS WIRING

LONG LIFE, HIGH CAPACITY MICROSWITCH

WATER, OIL TIGHT PVC COVER SEAL

WATER, OIL TIGHT STRAIN RELIEF ELECTRICAL CABLE CONNECTOR

O-RING SEAL AROUND ORIFICE GUARANTEES ZERO LEAKAGE

RUGGED CAST ALUMINUM COVER

CONTINUOUS SWITCH POINT ADJUSTMENT

REMOVE COVER AND LID ONLY FOR FIVE MINUTE ORIFICE CHANGE WITH OUT REMOVING BODY FROM LINE

STABLE INCONEL SPRING (NOT SHOWN)

PATENTED, MOLDED ELASTOMER FEEDTHRU WITH SEALED "FRICTIONLESS" PIVOT

Fig. 13-1b.

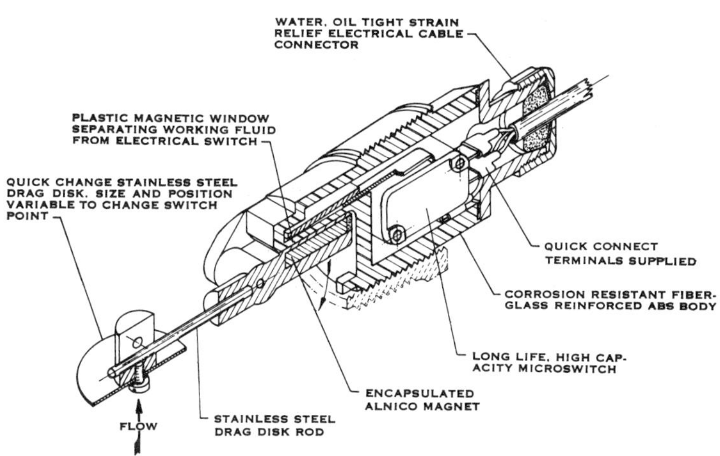

WATER, OIL TIGHT STRAIN RELIEF ELECTRICAL CABLE CONNECTOR
PLASTIC MAGNETIC WINDOW SEPARATING WORKING FLUID FROM ELECTRICAL SWITCH
QUICK CHANGE STAINLESS STEEL DRAG DISK. SIZE AND POSITION VARIABLE TO CHANGE SWITCH POINT
QUICK CONNECT TERMINALS SUPPLIED
CORROSION RESISTANT FIBER-GLASS REINFORCED ABS BODY
LONG LIFE, HIGH CAP-ACITY MICROSWITCH
ENCAPSULATED ALNICO MAGNET
STAINLESS STEEL DRAG DISK ROD
FLOW

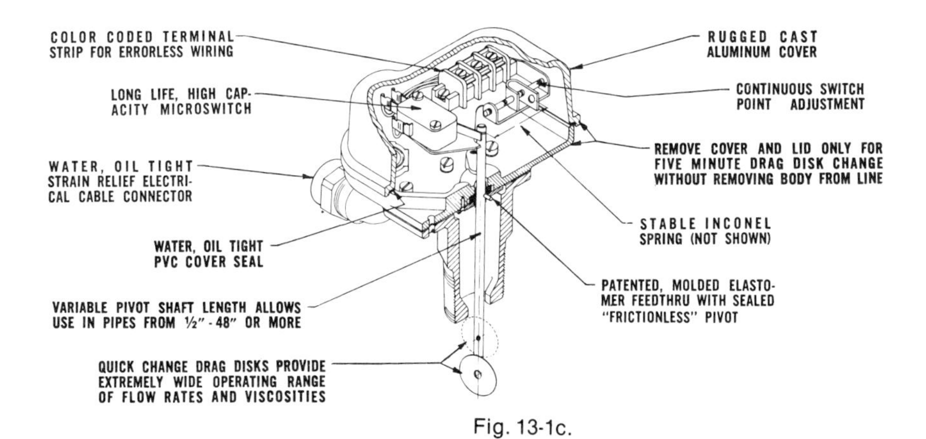

COLOR CODED TERMINAL STRIP FOR ERRORLESS WIRING
LONG LIFE, HIGH CAP-ACITY MICROSWITCH
WATER, OIL TIGHT STRAIN RELIEF ELECTRI-CAL CABLE CONNECTOR
WATER, OIL TIGHT PVC COVER SEAL
VARIABLE PIVOT SHAFT LENGTH ALLOWS USE IN PIPES FROM ½" - 48" OR MORE
QUICK CHANGE DRAG DISKS PROVIDE EXTREMELY WIDE OPERATING RANGE OF FLOW RATES AND VISCOSITIES
RUGGED CAST ALUMINUM COVER
CONTINUOUS SWITCH POINT ADJUSTMENT
REMOVE COVER AND LID ONLY FOR FIVE MINUTE DRAG DISK CHANGE WITHOUT REMOVING BODY FROM LINE
STABLE INCONEL SPRING (NOT SHOWN)
PATENTED, MOLDED ELASTO-MER FEEDTHRU WITH SEALED "FRICTIONLESS" PIVOT

Fig. 13-1c.

XIV Liquid Level Switches

Liquid level can be indicated or controlled by devices based on almost any conceivable operating principle—ranging from buoyancy (see Figure 14-1) to piezoelectricity. Although unique or severe requirements may lead to the use of exotic devices, for most industrial applications only the four major functional groups described herein need be considered.

The four basic types of liquid level switches may be classified in the following manner.

1. Float
   a. Valve in tank
   b. Valve in sump
   c. Float on rod
   d. Rotary gland
   e. Torque tube
   f. Flex tube
   g. Magnetic switch
   h. Armature
   i. Counterweight
2. Probe
   a. Resistance
   b. Capacitance
   c. Capacitance tube
   d. Ultrasonic
   e. Vibrating leaf
   f. Rotating paddle
3. Radiation
   a. Photoelectric
   b. Gamma ray
4. Differential Pressure
   a. Bubbler, unpressurized tank
   b. Bubbler, pressurized tank
   c. Closed tube
   d. Pressure switch
   e. Differential-pressure switch

For applications see Chapter 16B, Figures 16-1 and 16-2.

References: Delaval Turbine, Inc., Barksdale Controls Division, Los Angeles, California, and *Machine Design.*

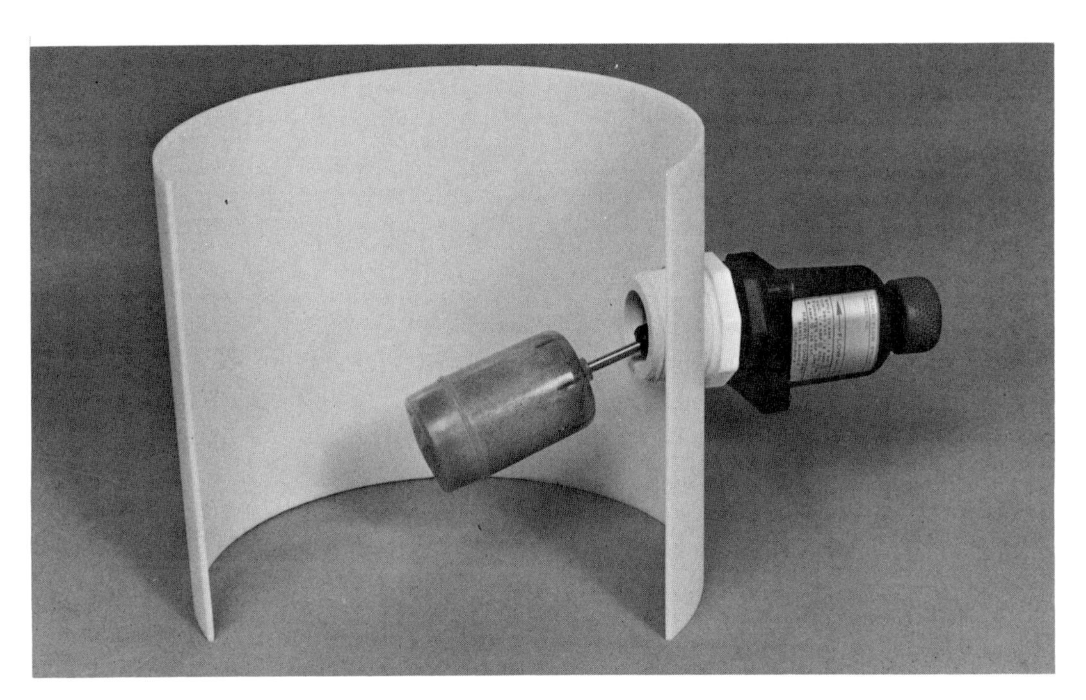

Fig. 14-1. *Courtesy Harwil Co.*

PART FOUR

APPLICATIONS

XV Pressure Switch Application

PRESSURE-DIFFERENTIAL SWITCH SIGNALS FILTER CLOGGING†

In a hydraulic test bench, a pressure-differential switch signals that the filter is clogged when the pressure drop across the hydraulic oil filter exceeds the set-point. A standard 5-micron filter has a rated, maximum pressure drop of 300 psi. The pressure-differential switch, connected across the filter, is set for 150 psi pressure differential.

When the filter element becomes filled with contaminants, the pressure drop across the filter increases. When this pressure drop approaches 150 psi, the set-point of the pressure-differential switch, the electrical circuit closes and an indicator lamp on the test board warns the test bench operator that the filter should be replaced.

Normally, hydraulic oil filters are replaced on a scheduled maintenance basis. The relatively minor expense of a pressure switch and lamp will insure prompt replacement of the oil filter. It also will extend the useful life of the oil filter to the point where the filter is actually clogged before it is replaced.

The method is employed by Barksdale Valves, Los Angeles, California, on their hydraulic test bench.

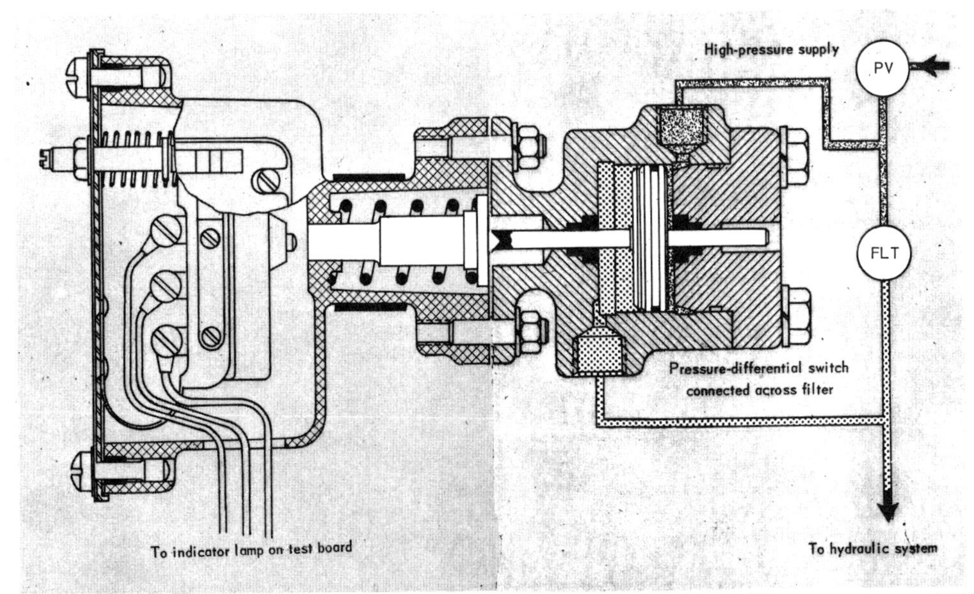

Fig. 15-1. High-pressure switch uses piston movement to measure pressure difference across oil filter.

†By E. W. Schrader.
Reprinted from *Design News* with permission.

PRESSURE SWITCHES FIND USE IN COMPRESSOR START-UP AND UNLOADING PUMP SHUTDOWN

Figures 15-2 and 15-3 show two uses of pressure switches. The first prevents shock to a compressor system. The second application involves automatic shutdown of a pump when a chemical tank car has been emptied.

When the operator wants to open a main compressor intake valve to pump additional gas, he has to pressure the downstream section of pipe up to at least 50 psi differential before he can open the large main valve. This is necessary to prevent shock to the system.

To accomplish this he first opens a small 1″ or 2″ by-pass valve. When the downstream pressure builds up to within 50 psi of the main pressure, the pressure switch opens the main valve. The differential pressure switch keeps the main valve locked closed at all times until the pressure on both sides of the main valve is within 50 psi.

Unloading of acids, caustics, and other dangerous chemicals from tank cars is a relatively simple operation; however, one problem involves automatic shutdown of the pump when the tank car becomes empty.

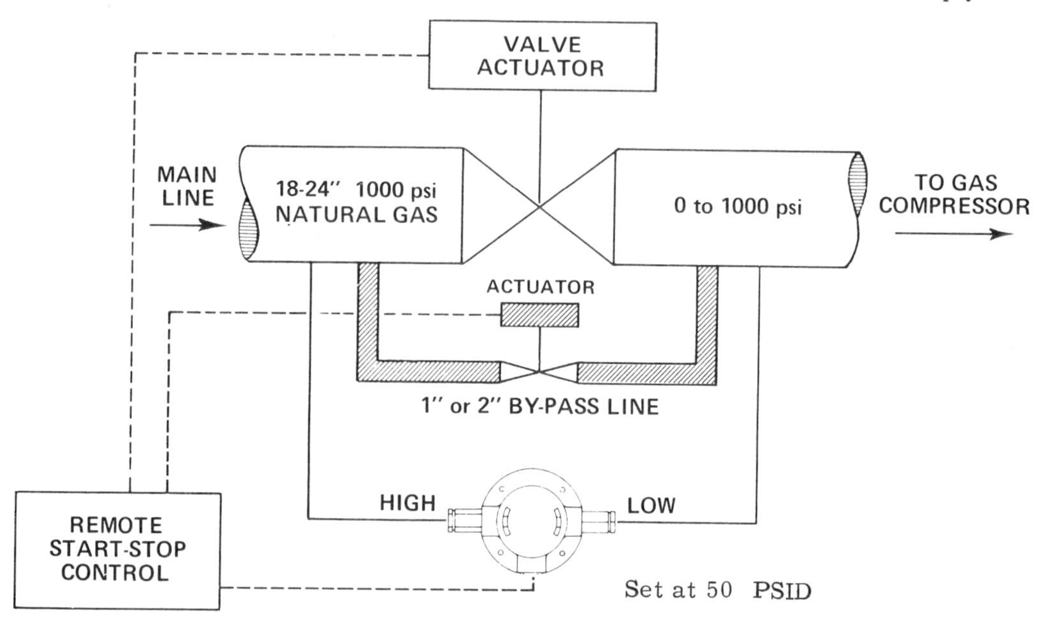

Fig. 15-2. Differential pressure switch to prevent shock on compressor start-up. *(Courtesy of Static-O-Ring Pressure Switch Co., Olathe, Kansas.)*

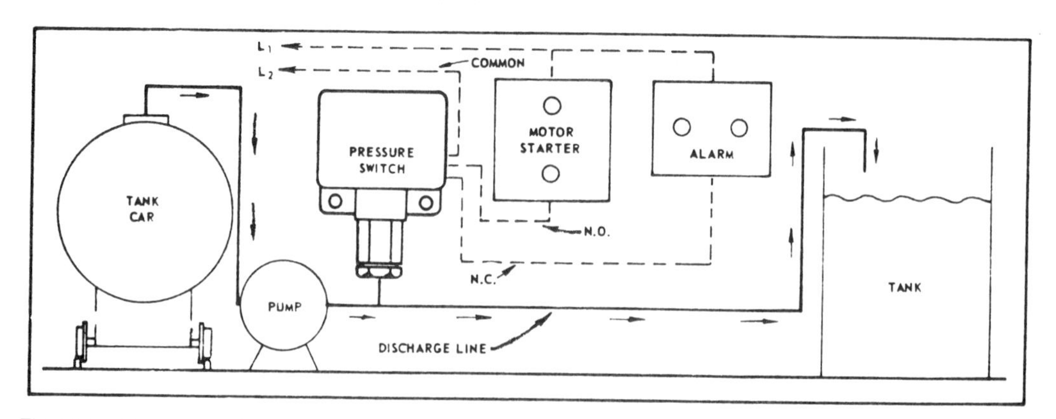

Fig. 15-3. Pressure switch application to unloading pump shutdown. *(Courtesy of Static-O-Ring Pressure Switch Co., Olathe, Kansas.)*

A simple solution involves the application of a pressure switch to the discharge head of the pump. The normally open (N/O) connection of the Micro Switch is wired in series with the pump motor starter; the pressure switch is then set to break contact when discharge head falls to just slightly higher than no-flow discharge pressure (i.e., empty tank car condition).

As long as the liquid is flowing, the discharge head will exceed the set point by a considerable amount. When the tank car is empty, however, flow ceases, and the discharge pressure falls to the head of liquid on the discharge side of the pump.

Immediately, the Micro Switch contact opens and drops out the motor starter. An alarm signal, wired to the normally closed (N/C) circuit, would serve to alert the operator that the tank car is empty.

This releases the operator for other duties without danger of damage to the pump from cavitation, and also permits optimum efficiency of power usage.

Selection of the wetted parts of the pressure switch depends upon the liquid being pumped.

A typical application, now in service, involves a pressure switch applied to sulfuric acid and set to actuate at 25 feet w.c. decreasing head.

When the pump is running, the discharge head is approximately 100 feet w.c., at which pressure the motor starter is in circuit. When flow ceases, the discharge head drops to 20 feet w.c. head, "deactuating" the N/O circuit and dropping out the motor starter to stop the pump.

To re-start, the operator manually actuates the motor starter and holds it in until the discharge head rises to 27 feet w.c., at which time the pressure switch actuates to hold in the motor starter until flow again ceases.

PRESSURE SWITCHES CAN COUNT AND MEASURE FOR CONTROL SYSTEMS†

Sensitive pressure switches, operating on low-pressure shop air, can be an inexpensive means of obtaining an electrical signal for either counting or measuring in automatic control systems. Limit switches are commonplace, but in environments of high humidity, high temperature or fluids, there is some chance of insulation damage.

A low-pressure air line, in the range of 0.5 to 3 psi, of small diameter may be inserted into a fluid to measure shut-off level or may be inserted into a high-temperature drying kiln and into similar environments without damage.

There are two basic ideas, to use low-pressure air as a signal: (1) the transient, or short-period, pressure build-up, when the end of the line is momentarily closed, or (2) the steady-state, or long-period, pressure build-up, when the end of the line is closed for a substantial length of time.

Sensitive pressure switches, with actuation pressure up to 3 psi, can be used for obtaining impulse signals, when the end of the air line is closed for a sufficient length of time to allow pressure to build up in the pipe or tubing. The end of the line may be closed by

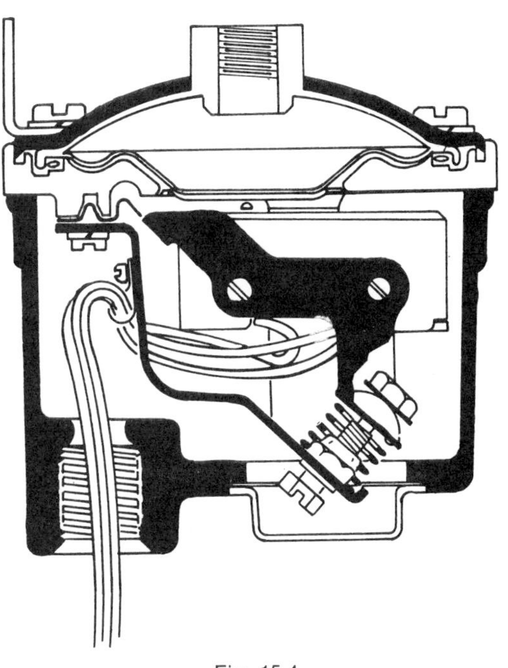

Fig. 15-4

†By E. W. Schrader.

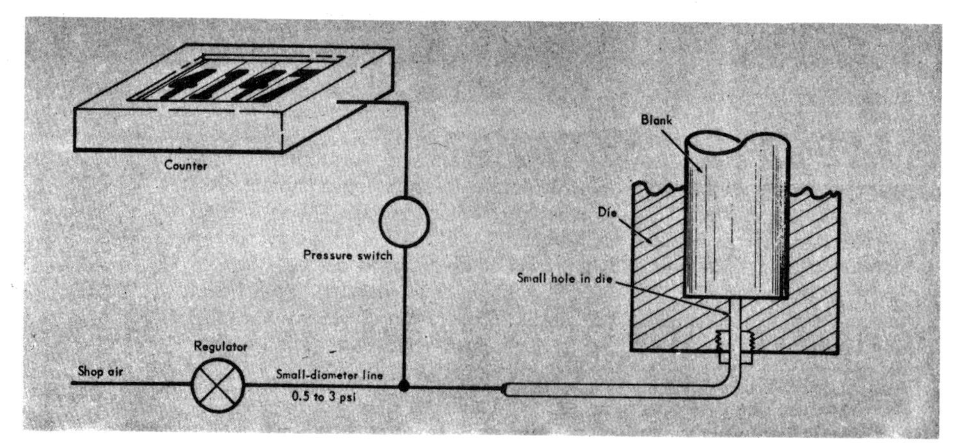

Fig. 15-4. Cold-forming die, with small hole in bottom of die, admits air to assist ejection of formed part. As each blank drops into die, pressure builds up in small-diameter air line and pressure switch trips to operate counter.

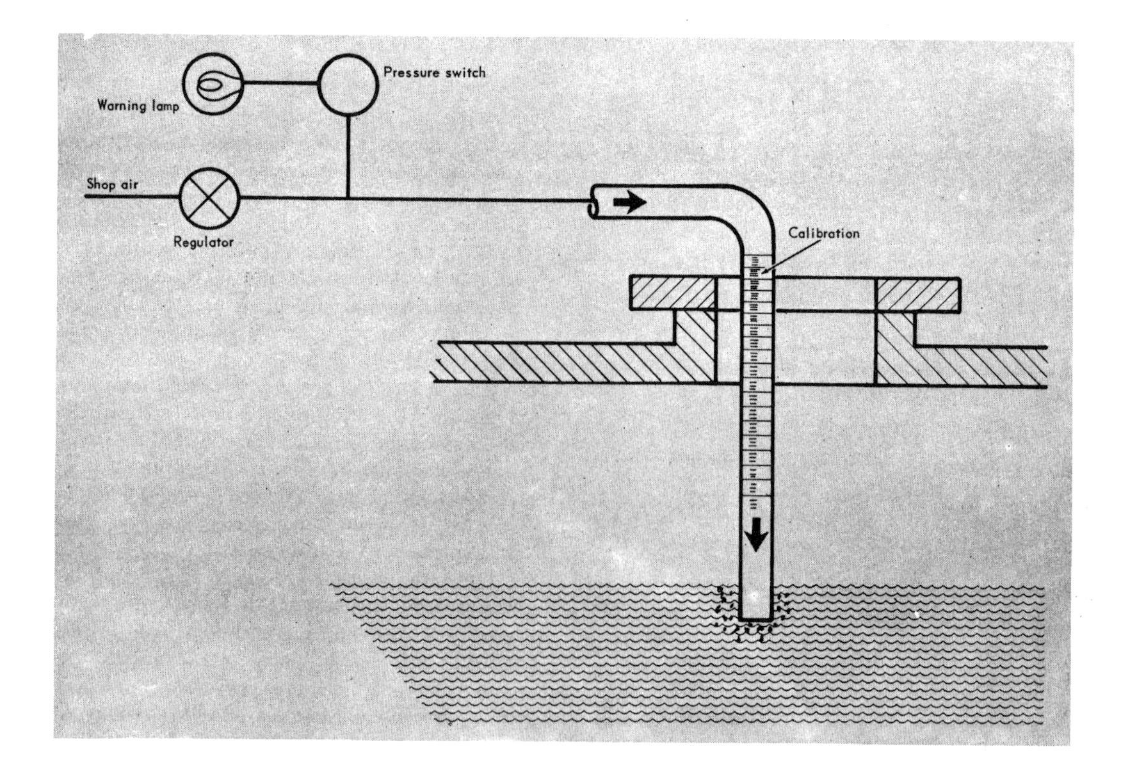

Fig. 15-5. Calibrated pipe is positioned in tank-car opening for appropriate filling level of particular size of tank car. When liquid rises, to close off air supply flowing in pipe, pressure builds up in pipe to trip pressure switch, which, in turn, operates warning lamp or closes electrically operated valve.

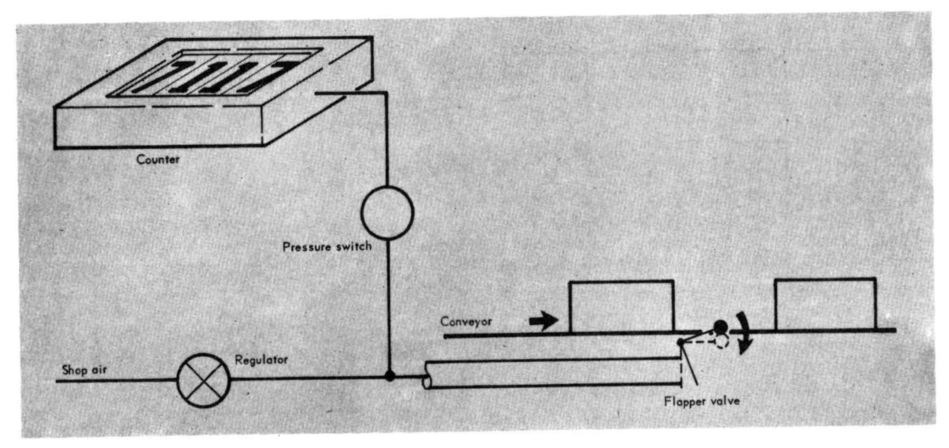

Fig. 15-6. Flapper valve closes end of small-diameter pipe each time object passes along conveyor. Pressure builds up in pipe to trip switch to operate counter or open door to emit object from drying chamber.

a simple flapper valve. It need not be leak-tight, since the regulator will supply a flow of air to make up for the leakage.

Most pressure switches are adjustable in a range for sensing both decreasing and increasing pressures, thus accommodating various installed line lengths.

The application ideas, based on actual installation experience, were supplied by the Pressure Switch Division, Barksdale Valves, Los Angeles, California.

PRESSURE SWITCH CONTROLS LIQUID NITROGEN LEVEL*

Background. High vacuum systems cannot tolerate the presence of vapors, but when liquid nitrogen (LN_2) cold traps are used to remove them, a reliable method of replenishing the traps with LN_2 is necessary. A popular method of controlling the LN_2 level is to take advantage of temperature changes caused by changing liquid levels, but a controller is needed that operates via the pressure developed by the head of LN_2 present in the cold trap.

Design Solution. A controller was designed with a probe inserted into the cold trap to transmit LN_2 pressure changes to a pressure-sensitive diaphragm switch. A relay control in turn energizes a solenoid valve, which introduces gas pressure (dry nitrogen) to the LN_2 reservoir. This pressure forces LN_2 to be transferred from the reservoir to the cold trap.

The cold traps presently being serviced by this system only require liquid level control (differential) from about one to five inches. Narrower differentials are obtainable, but the frequency of transfer is usually high at these narrow differentials. As a result, LN_2 wastage is also high because of the repeated need to cool down the transfer line at the start of each transfer cycle.

The designers report that the dry N_2 pressure supply is not essential to system operation—transfer could be accomplished by blocking only the LN_2 reservoir vent. However, the additional pressure source does speed up the transfer and minimize the time variation in the transfer cycle caused by the falling LN_2 level. According to the designers, the system enjoys a fail-safe feature if dry N_2 pressure is lost.

The automatic level control system was designed by J. E. Swiddle and M. Tomlinson

*1975 Design News Annual.

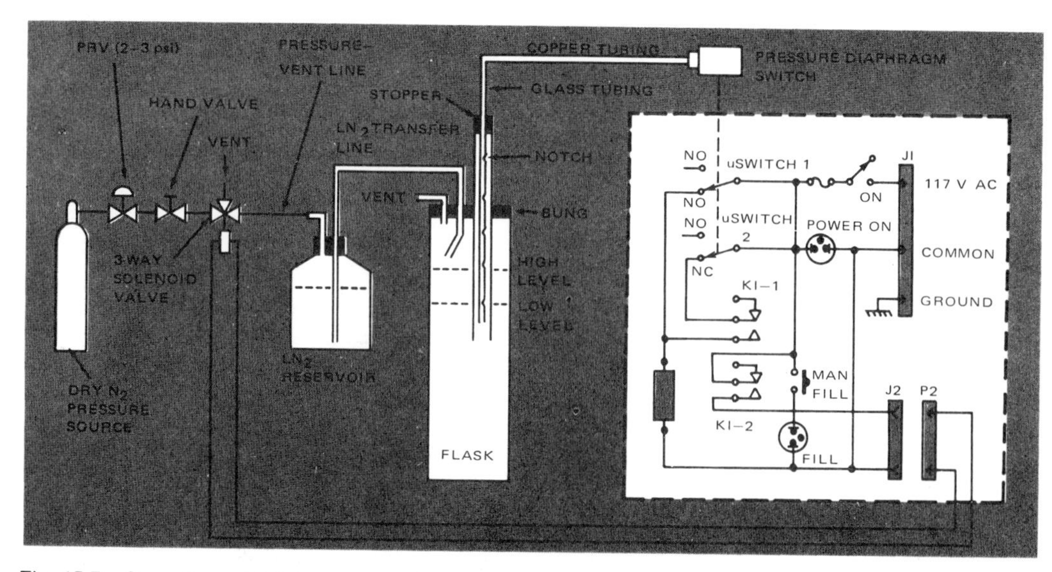

Fig. 15-7. Operation of LN_2 level control. As the liquid level in the cold trap falls, the pressure exerted by the volume of air in the probe decreases. This pressure change is transmitted to the pressure diaphragm switch, which allows the C and NC contacts of Switch 1 to close. Line voltage is then applied to the two-pole relay K1. When K1 energizes, line voltage through contacts C and NC of Switch 2, and latching contacts K1-1, insures that the relay will remain energized after the reactuation point of the low-level switch has been passed while filling.
Line voltage is also applied, by way of contacts, K1-2, to a three-way solenoid valve. In its normally deenergized state, the solenoid valve blocks the line from the dry N_2 pressure source to the LN_2 reservoir. At the same time, the LN_2 reservoir is vented to the atmosphere through the third orifice on the valve. When the solenoid valve is energized, the vent is blocked, and pressurizing gas is fed to the inlet line of the LN_2 reservoir. This forces LN_2 to be transferred from the reservoir to the cold trap.
As the level in the cold trap rises, the increasing head pressure relayed through the probe forces the contacts of Switch 2 to open. Because Switch 1 is open, relay K1 is de-energized. The solenoid valve is also de-energized, blocking the dry N_2 gas line and venting the excess pressure in the LN_2 reservoir to atmosphere.

at Whiteshell Nuclear Research Establishment of Atomic Energy of Canada Ltd., Pinawa, Manitoba. (From *Design News,* June 18, 1973, p. 52.)

PRESSURE SWITCH APPLICATION FOR INDUSTRIAL EQUIPMENT PROTECTION†

Prime objectives of virtually all manufacturing engineering organizations include reducing maintenance costs and increasing the operating effectiveness of industrial equipment. Often, both of these goals can be attained through the use, on such equipment, of relatively simple and inexpensive monitoring devices. The purpose of these devices is to sense unsafe operating conditions and either actuate alarms or shut down the equipment before it fails.

Where pressure is a critical variable, protective devices of this type include single and differential pressure switches. Properly applied, these devices can forestall expensive breakdown repairs and downtime that might otherwise occur. Information provided by pressure switches can also be used to determine optimum preventive maintenance intervals as opposed to shorter intervals which may

†By Lloyd M. Polentz, Consultant, Barksdale Valves, Los Angeles, California.

Reprinted from *Automation,* September, 1963.

have been set arbitrarily or by custom. Savings resulting from longer maintenace intervals include those obtained by postponing expensive overhauls until they are really needed and by using expendable components (filters, for example) over their entire service lives.

Potential sources of savings through the use of pressure switches are numerous and are present in almost every manufacturing plant. Installation of pressure switches should be considered, for example, wherever pressure-lubricated machinery, fluid power systems, gas storage systems, or internal combustion engines are employed.

Although the discussion to follow will cover the previously mentioned areas in particular, the use of pressure switches is by no means confined to such applications. The monitoring and protective actions of the devices, as described in the examples, can be equally effective in safeguarding and improving the operation of many other types of industrial equipment.

Pressure-Lubrication Systems

In pressure-lubricated machinery, high-pressure lines supply lubricant to the bearings and low-pressure lines return the lubricant to a sump. A differential pressure switch connected across a bearing will indicate a normal pressure-difference condition. Should the flow stop, the bearing passage becomes plugged, or the return line to the sump becomes plugged, the differential pressure switch will close an appropriate electrical circuit to cause lighting of a signal lamp or to shut down the machine. For complete protection, this application calls for the use of a double-setting, high-low type of switch that can signal when a minimum preset differential pressure has not been attained and when a maximum preset differential pressure has been exceeded. In multibearing systems, each bearing could be equipped with an individual pressure switch, and a master switch could be provided to monitor pressure at the common lubricating pump outlet. (See Figure 15-8.)

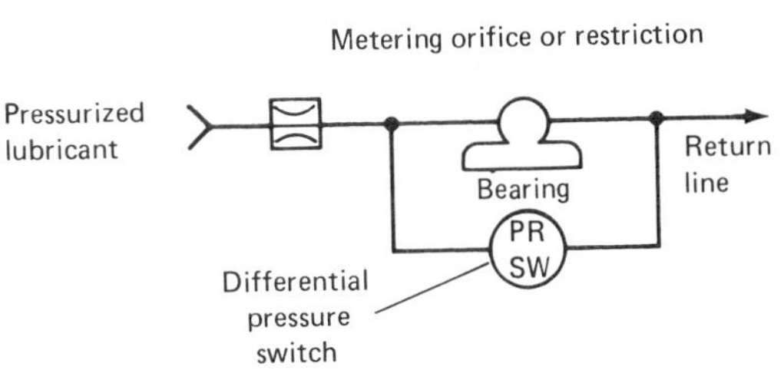

Fig. 15-8.

Hydraulic Systems

Pressure switches can be used to considerable advantage in many types of hydraulic systems. In a hydraulic test bench used to check precision equipment, for example, a differential pressure switch can monitor the condition of the full-flow filter required to guarantee removal of all potentially harmful particles in the system. In such equipment, an adequate and economical filter-cleaning program is almost impossible to schedule due to the many unknowns which are involved. As a result, there is always the danger of collapsing an element. A new filter element may cost seventy dollars or more, whereas the cleaning cost is only about five dollars.

A differential pressure switch, installed as shown in the photograph, is actuated when the element is packed with dirt and requires cleaning. The switch also senses a condition which might result in a surge of dirt into the system due to rupturing of an element that has a high differential pressure across it. A second pressure switch, connected in the pump intake and wired into the drive motor circuit, will prevent the pump from operating with a starved intake. This safeguard can protect against burnout of the pump in case of a failure in the hydraulic fluid supply.

Pneumatic Systems

Compressed air systems can be improved, and maintenance costs reduced, by connecting a differential pressure switch across the air intake filter (see diagram). When the pressure drop across the filter rises to a point where cleaning is required, a signal light can inform the maintenance man of this condition. Such

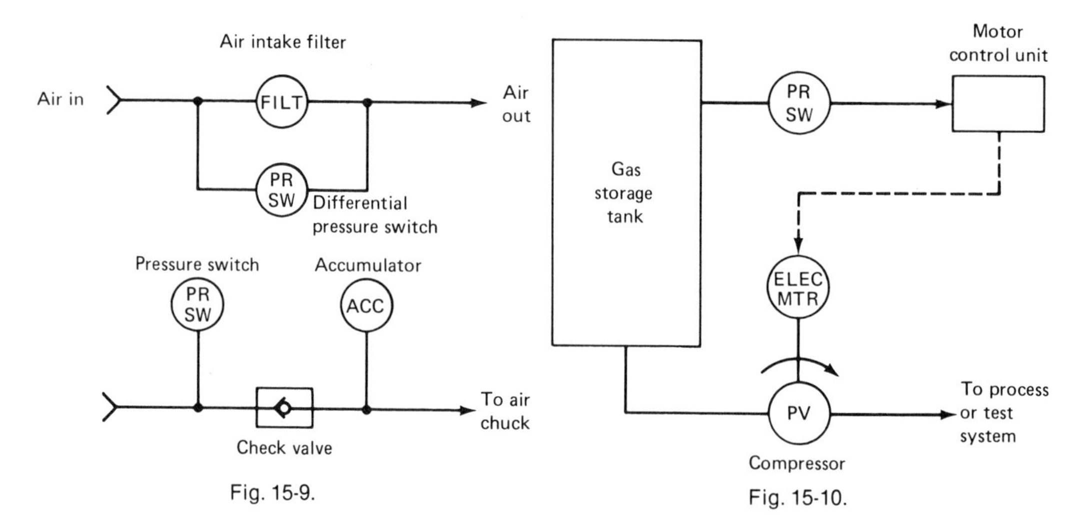

Fig. 15-9.

Fig. 15-10.

a protective device will allow the maximum amount of filter use between cleaning operations and still ensure that the filter does not become so clogged as to restrict the intake of air to the compressor or to be collapsed by atmospheric pressure.

A pressure switch will show a high rate of return on investment when the switch is connected in the air supply system for an air chuck. If the air supply pressure should drop due to excessive use of compressed air elsewhere in the system or due to a broken supply line, the switch will shut down the machine. As shown in the diagram (Figure 15-15), a check valve and an accumulator should also be provided. The function of these elements is to ensure emergency pressure to keep the chuck actuated during deceleration of the machine upon loss of system pressure. Similar principles can be applied in the protection of equipment using air clamps or vacuum plates.

Gas Storage Systems

A storage tank in a closed gas system, such as a helium test system, should be equipped with a pressure switch if there is any possibility that the tank might be exposed to a negative pressure or vacuum. In such a system the gas is stored in a tank and is drawn from the tank by a compressor which supplies pressurized gas for testing. Occasionally, due to leakage or high capacity of systems being tested, the bulk of the gas supply will be withdrawn from the storage tank, and the compressor may pull down the pressure in the tank until it is below atmospheric pressure.

It is a characteristic of such tanks that even a small negative pressure differential can collapse them. For example, a 6-foot diameter steel storage tank with a wall thickness of 1/4 inch will easily hold 125 psig internal pressure, yet may collapse with a negative pressure differential of only 2.5 psi. For safety reasons, all such storage tanks should be equipped with a pressure switch wired into the compressor motor control circuit so as to shut down the compressor in case it ever produces a vacuum in the storage tank. (See Figure 15-10.)

I.C. Engine Power Plants

Differential pressure switches can provide considerable savings in the operation and maintenance of internal combustion engine power plants. For trouble-free operation and long use of such equipment between overhauls, cleanliness of the lubricating oil is mandatory. Usual practice to ensure clean oil is to change the oil filter frequently. This can mean that the filter element is cleaned or replaced well in advance of the time that it becomes unsuitable for satisfactory or safe operation.

Filter element condition can be determined, and maintenance costs reduced, by the use of a differential pressure switch across the filter as previously illustrated. The switch will also increase the safety factor provided by filtering since it will automatically detect any increased dust content of the air that might clog even a new oil filter. A pressure switch placed across the air intake filter will provide the same type of protection for large I.C. engines that was previously described in the case of a hydraulic test bench. It is also good insurance to place a single pressure switch in the pressurized lubrication system so that the engine will be shut down automatically if lubricating oil pressure should drop below a preset minimum value.

SENSE LEVEL WITH PRESSURE SWITCHES†

There is no universally satisfactory solution for one of the oldest of all instrumentation problems—the sensing of liquid levels.

The float-type sensor and controller, the earliest and still most popular method, is far from trouble free. Certain disadvantages and potential sources of trouble inherent in float-type level sensors cannot be eliminated. The float must be located at the desired liquid level and float motion must be transmitted from the inside of the tank or float chamber to the outside by means of a trunnion and stuffing box. This particular difficulty has been overcome through use of mercury switches attached to the float inside the chamber. This has introduced other problems...the mercury switch must be mounted absolutely level and it is subject to unwanted actuations as a result of liquid movement or vibration.

Furthermore, all float-type level devices are dependent upon float movement for actuation of the level-controlling mechanism. Such movement must be controlled by bearings and linkages of one type or another. All mechanical joints and bearings are subject to wear, and after a period of time such wear will cause the setting to drift. Many liquids cause corrosion at the moving points of contact, others cause buildup of precipitates at these joints and liquids of high viscosities seriously hinder satisfactory operation of any float-type control mechanism.

Floatless, liquid-level controllers may overcome these difficulties. Liquid-level sensors utilizing principles of acoustics, supersonics, electric eyes, torque measurement, electrical capacitance, electrical resistance, heat transfer, tank weight and other physical principles have been employed. These methods are relatively complex and require considerable accessory electronic gear. They are expensive, delicate and tricky to operate and maintain. Each type may have applications where it fulfills requirements better than any other type of liquid-level sensor. None satisfies necessary requirements of low first cost, reliability, easy installation and low maintenance necessary for universal industrial use. As a result, none of these special types of liquid-level sensing devices has made notable inroads on use of the float-type liquid-level controller.

Recent advances in accuracy and technology of pressure switches makes it possible to construct simple, dependable liquid-level controls that are less expensive, more accurate, easier to install and more flexible in application. Their use is based on the fact that a head of any liquid will produce a pressure proportional to that head. By measuring this pressure, the head of a known liquid in a tank can be determined. This principle is used to control the level in the tank to a given point or between two pre-selected points. It can also be used to produce a portable liquid-level controller for use in filling or emptying tanks.

Simplest type of application is the indication and control of a single level in a tank vented to the atmosphere, either a low level to prevent emptying or a high level to prevent overfilling. This can be accomplished with one pressure switch as shown in Figure 15-11. The switch need not be located at the controlled level of the liquid. Height of the controlled level may be changed by adjusting the setting of the pressure switch and it is not necessary to change the location of the sensor. The pressure

†By Lloyd M. Polentz, P.E.
Reprinted from *Petro/Chem. Engineer,* December 1964.

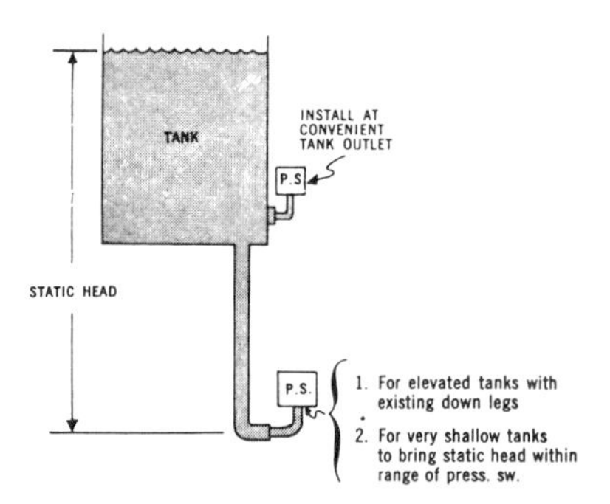

Fig. 15-11. Pressure-switch, liquid-level control installation showing alternate locations of switch.

switch need not be attached to the tank directly, but can be attached to the flow line used to fill or empty the tank (Figure 15-12). It is desirable to locate the pressure tap as

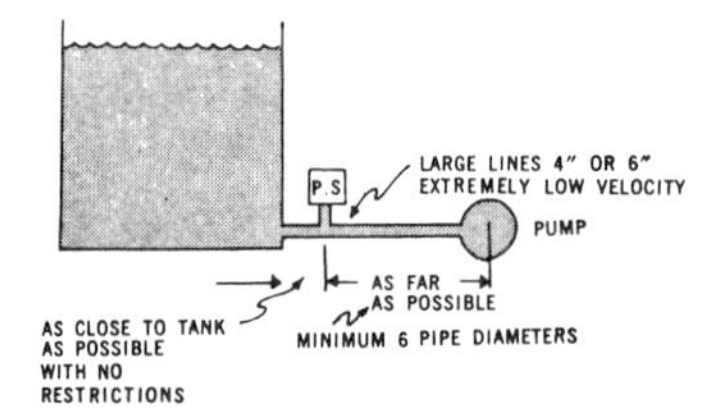

Fig. 15-12. Sensor located in discharge or fill line.

close to the tank as possible preferably in a large diameter flow line.

Pressure-switch, liquid-level controllers can be used to control level of liquid in a tank between two limits. This can be accomplished by installation of two sensors as is done with float-type controllers. Space, time and money can be saved by installing a single, dual setting, high-low, pressure switch (Figure 15-13) with no loss in accuracy or reliability.

Pressure-switch, liquid-level controllers help solve problems of control of extremely viscous or highly corrosive liquids. Since small volume flow is required to actuate a pressure switch, in contrast to the volume flow required to actuate a float-type controller, a buffer fluid in a pressure-switch level controller can be easily utilized (Figure 15-14). The pressure switch may also be isolated by a thin flexible diaphragm (Figure 15-15), since there is no pressure differential across the diaphragm.

The liquid pressure-switch, liquid-level controller can be adapted to sense and control liquid levels in tanks below ground level or where bottom tank connections are not practicable. Air or gas is bubbled slowly through a down pipe (Figure 15-16). Pressure required to force the gas out of the bottom of the down pipe will be proportional to the head of liquid above the discharge point. This indirect pressure measurement can be used to actuate a pressure switch in the same way

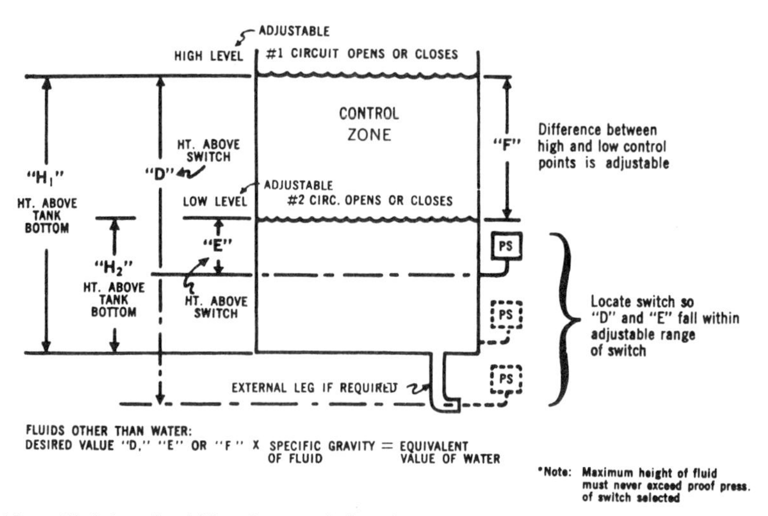

Fig. 15-13. High-low liquid level control showing possible locations of level sensors.

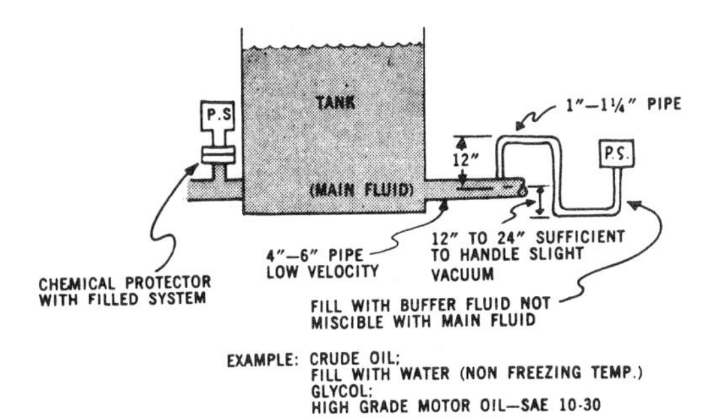

Fig. 15-14. Use of buffer liquid.

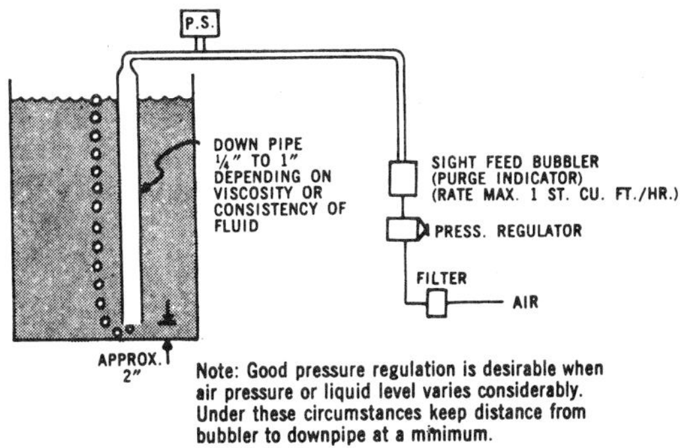

Fig. 15-16. Bubbler system for corrosive fluids.

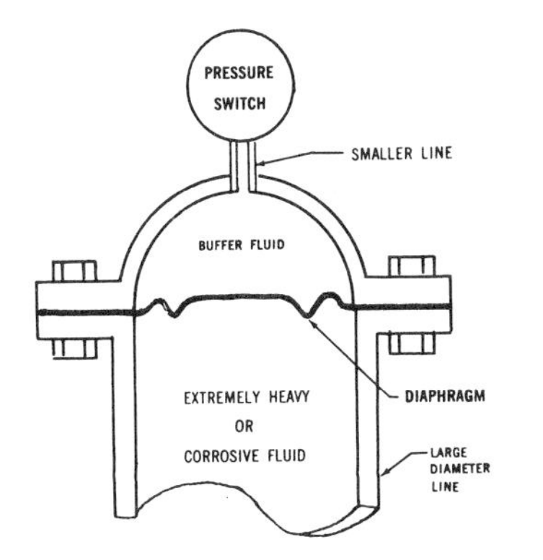

Fig. 15-15. Use of diaphragm and buffer liquid.

as the direct pressure measured. Rate of air or gas flow can and should be kept low, on the order of one cfh so the gas flow will neither be expensive nor harmful to the liquid in the tank. This same type of liquid-level sensor can be made into a portable unit to automatically control filling of tank cars, tank trucks or other types of shipping or storage containers.

For a combination of versatility, accuracy, dependability and economy, the pressure-switch method of sensing and controlling liquid levels exceeds that of any other method. Use of the pressure-switch method can also frequently aid in solution of liquid-level control problems which are difficult to solve by other methods.

XVI Other Device Applications

OTHER APPLICATIONS FOR FLOW SWITCHES

- Air conditioning systems
- Hot water space heating systems
- Hot water supply systems
- Pump systems
- Water cooled equipment
- Blending or additive systems
- Liquid transfer systems
- Fire sprinkler systems
- Water treatment systems
- Duct type heating systems
- Exhaust systems
- Make-up air systems

A. FLOW SWITCH

FIRE SPRINKLER SYSTEM

Waterflow indicator used as a safety control to actuate alarms or signals, and to start or stop mechanical equipment when fire occurs.

Operation of Flow Switch

Installed directly in the piping, a waterflow indicator offers an economical and positive means of detecting the flow of water in the

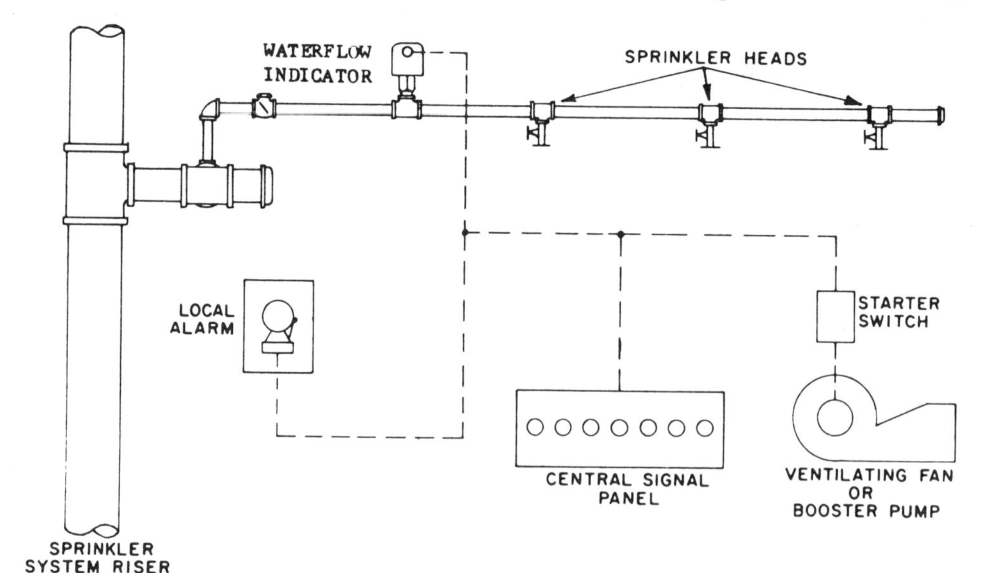

Schematic diagram of system

SOME FLOW SWITCH APPLICATIONS

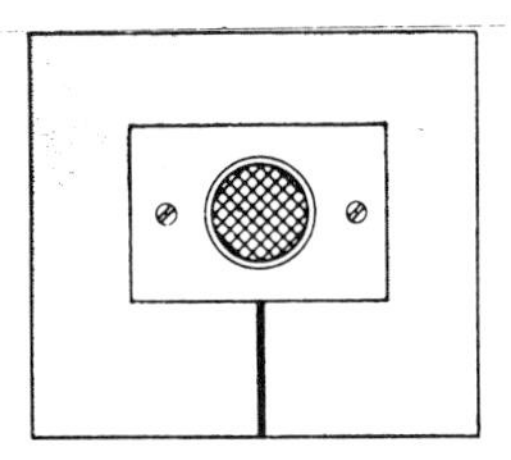

Signal Devices—With increasing automation, knowledge of flow, or no-flow, is important. A flow switch can provide a visual or audible report from any location, either nearby or remote.

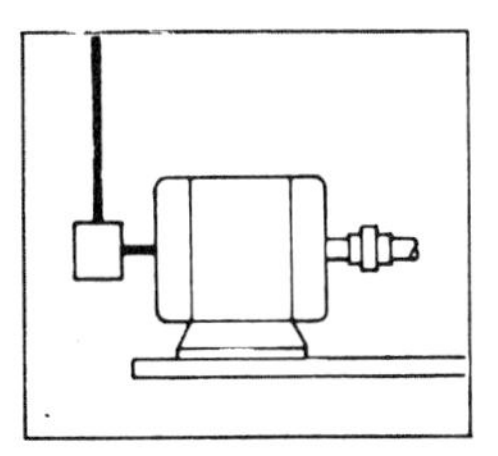

Motors—The operation of pumps, compressors and similar apparatus frequently depends upon fluid flow in a pipeline. A flow switch provides a dependable way to control such motors or other prime movers.

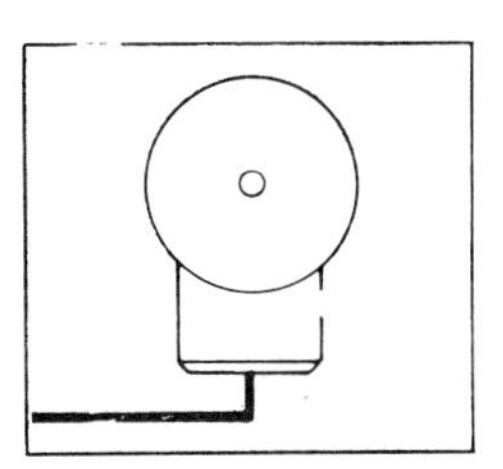

Alarms—Where flow failure is critical, or where a flow occurs in an emergency line—as in a fire sprinkler system—a flow switch can warn of trouble and pinpoint its exact location.

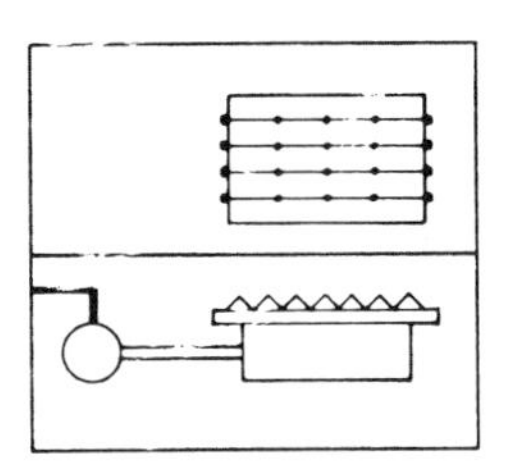

Heating Units—The widespread use of water heaters and duct heaters ranges from commercial processing to personal comfort. A flow switch in pipe or duct can start the heating unit to speed recovery, or stop it if flow fails.

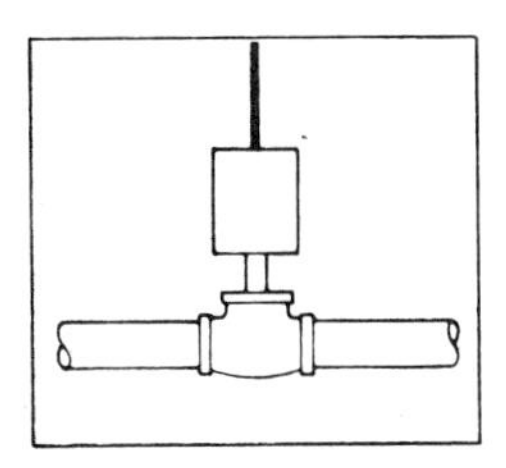

Metering Devices—Many liquids are improved or altered between their source and point of use—chlorination of domestic water, for example. A flow switch can start and stop additive equipment as flow dictates.

distribution or branch piping of a fire sprinkler system, or partial system.

Connected to a local alarm, as shown above, it will alert anyone in the vicinity that the sprinklers are operating, and that a hazardous fire condition may exist. Connected to a central signal panel it will immediately indicate where the sprinkler system is functioning. This quick detection and notification pinpoints the location of the fire, speeds up the ability to extinguish it, assists in safe evacuation of the building, and can minimize the amount of water damage.

Further, when flow occurs the waterflow indicator can make or break the electrical circuit to various types of mechanical equipment. For example, if there is a ventilating fan in the fire area, the waterflow indicator has the ability to stop the fan, thus reducing the draft and impeding acceleration of the fire.

For another example, in areas of relatively low water supply pressure the waterflow indicator can be connected to a booster pump. Whenever the sprinkler system operates, the waterflow indicator quickly starts the pump, thus increasing the supply pressure and providing maximum deluge capacity. (See schematic below.)

DUCT TYPE HEATING SYSTEM

Air flow switch used as a safety shut off switch should air flow fail in duct heating system.

Operation of Flow Switch

The practice of using direct fired heating units or electrical heating elements positioned within a duct system, offers some advantages in certain types of installations.

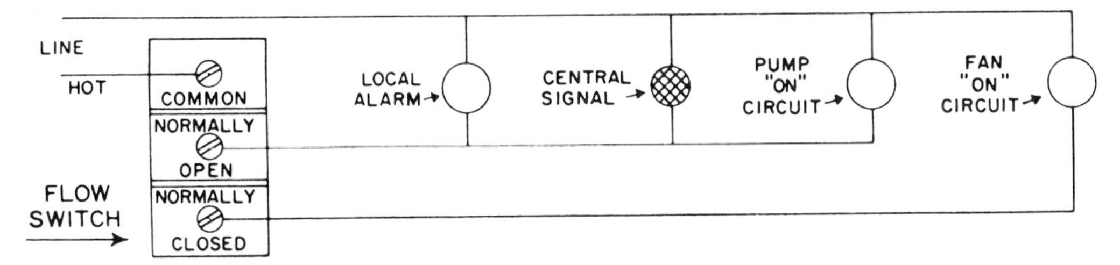

Schematic diagram of system

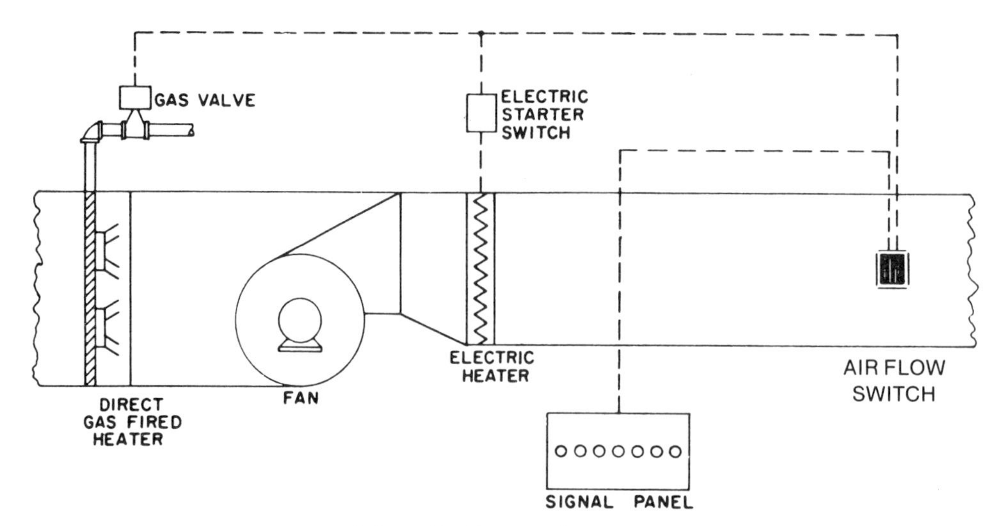

Schematic diagram of system

For example, direct fired gas burners are incorporated in some types of roof mounted air make-up systems. These are usually used in commercial or industrial installations where a large volume of make-up air is required, and heating the air is necessary to reduce draft possibilities.

Electrical heating elements, commonly described as strip heaters, are used as primary heating sources where the power is more readily available than other heating means. They are also used as supplemental heaters where add-on load conditions have occurred, or where they increase the capability of other heating units such as heat pumps, etc.

Because of the temperatures generated by these confined heating units, it is essential that the air flow through them be maintained at a relatively high velocity. Should the air flow fail, and the heating units remain on, an immediate overheating condition would exist. This could cause possible system damage within the duct, could cause the tripping of fusible link fire dampers, or could overheat the external area surrounding the duct.

The air flow switch, as shown above, is installed downstream from the heating unit to verify air flow in the duct. Should the air flow cease because of an inoperative fan, closed inlet damper, or some other cause, the air flow switch would immediately sense this lack of air flow velocity and shut off the electric heating element or close the gas valve.

In addition, an air flow "On" signal could be wired to a central control panel or monitoring point. (See schematic below.)

PUMP SYSTEM

Flow switch used to actuate a booster pump in a system where water supply pressure is inadequate.

Operation of Flow Switch

In many localities the available water supply pressures are continually on the low

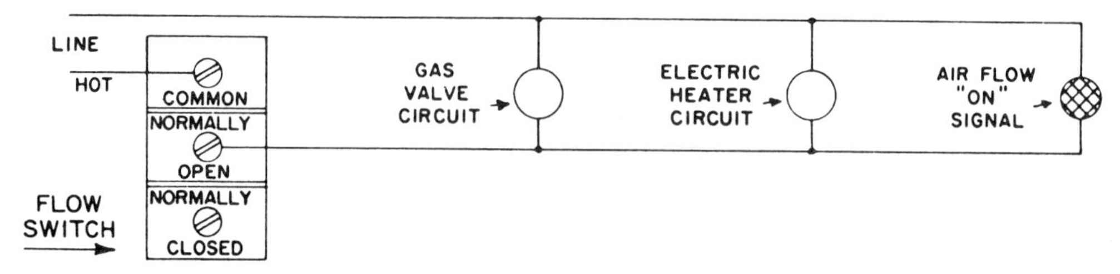

Schematic diagram of system

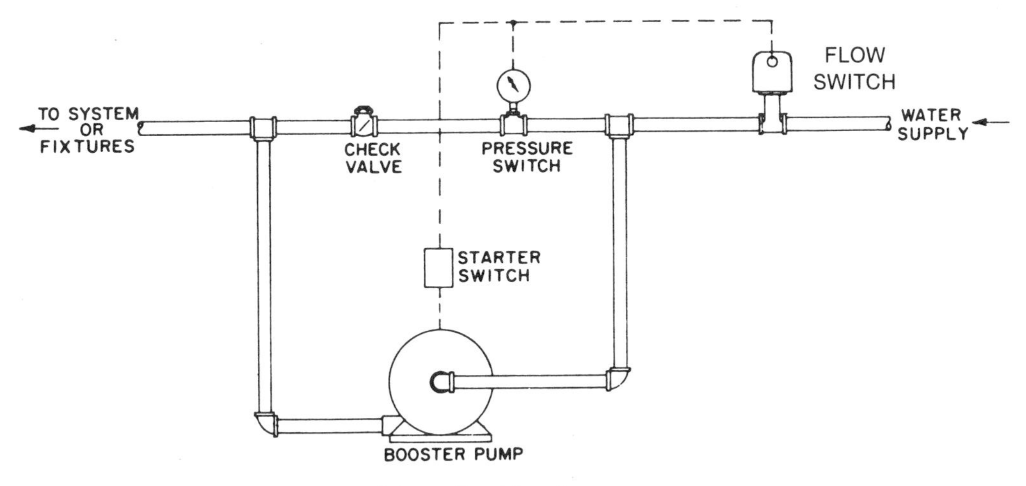

Schematic wiring diagram

side, or are intermittently low during periods of greatest water consumption. In these areas it is necessary to employ some means of increasing the pressure to meet the requirements of modern water usage. One popular and inexpensive method of accomplishing this objective is the use of a tankless pressure booster system as shown above.

In this system a booster pump is connected into the water supply piping, with a check valve in the by-pass arrangement. A pressure switch and a flow switch are also installed in the water supply piping. The pressure switch is adjusted to make contact when the water supply drops below the lowest allowable pressure which will take care of system requirements.

As long as the water supply pressure is sufficient to meet the system requirements, the contacts in the pressure switch will be open and prevent the flow switch from actuating the booster pump. If the water supply pressure drops below the setting of the pressure switch, the switch contacts will close, and any usage of water will cause the flow switch to start the booster pump. The pump will continue to operate until either the draw of water has ceased or pressure has been restored.

In this system *both* low pressure and water flow must exist before the booster pump operates. In any area in which the supply pressure is *always* below system requirements, the pressure switch can be eliminated, and the pump can be actuated by the flow switch whenever any draw of water occurs. See schematic below.

LIQUID TRANSFER SYSTEM

Flow switch used to control the operation of liquid transfer from one location to another.

Operation of Flow Switch

The movement of liquid from one location to another frequently requires close monitoring. A typical example is illustrated in the above system where liquid is transferred from one storage area to another tank or receiver used for process or further storage.

In this application the transfer of liquid often takes considerable time for completion;

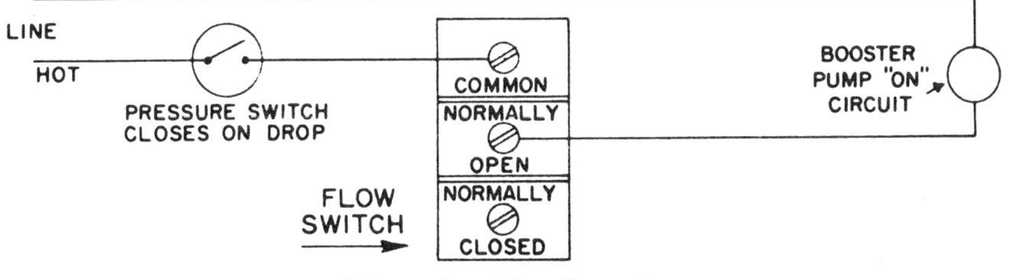

Schematic wiring diagram

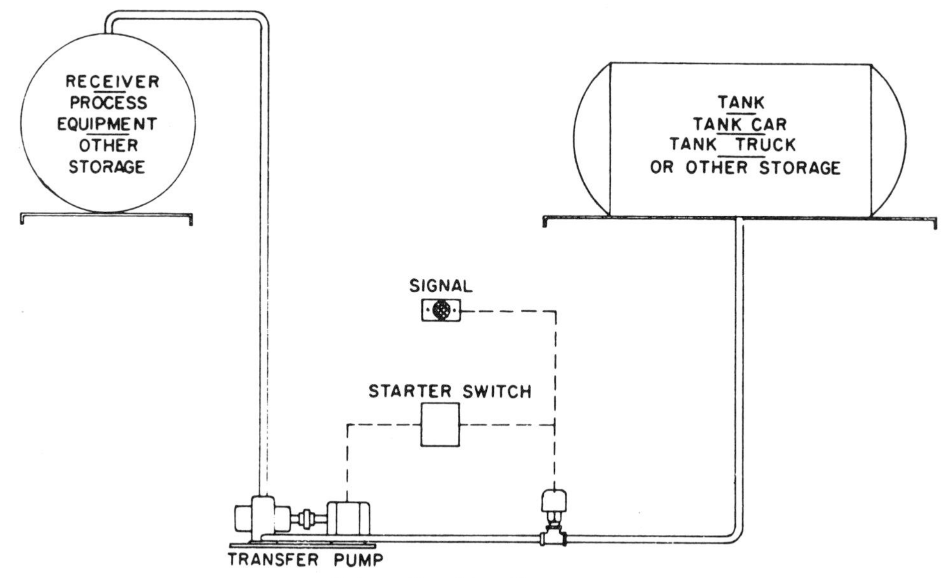

Schematic wiring diagram

therefore it is advantageous to provide automatic control of the operation. With a flow switch installed in the system, once the transfer process has begun, the flow switch takes over and automatically controls the operation until it is finished.

This type of control provides several advantages:

- Eliminates stand-by manpower.
- Safeguards pump against running dry.
- Provides automatic operation to stop pump.
- Provides switch contact to actuate visual signal.

The flow switch is installed in the pipe line between the supply tank and transfer pump. After the pump is started manually, the flow switch—wired through an electrical interlock—takes over the job of keeping the pump operating until the transfer is complete.

B. FLOAT OPERATED SWITCHES

In this section we shall show two of the basic liquid level switches.

Figure 16-1 shows a liquid level switch that operates mechanically by the buoyancy of the float which is connected to a mechanical link that operates a switch when a given fluid level has been reached.

Figure 16-2 is another type of liquid level control switch designed to shut down machinery or turn on warning devices when the liquid supply fails or recedes to a predetermined level. The switch is used for remote indication of low liquid levels vital to protecting bearings against breakdown due to insufficient lubrication. It can be wired to flash a warning light, sound a howler, shut down a machine, or signal a computer.

Simplicity makes this switch reliable in operation. A float carries a ring magnet inside, and moves up and down on a central pipe that encloses a magnetic reed switch. When the liquid recedes, the float is lowered, and the magnet actuates the reed switch at a predetermined level.

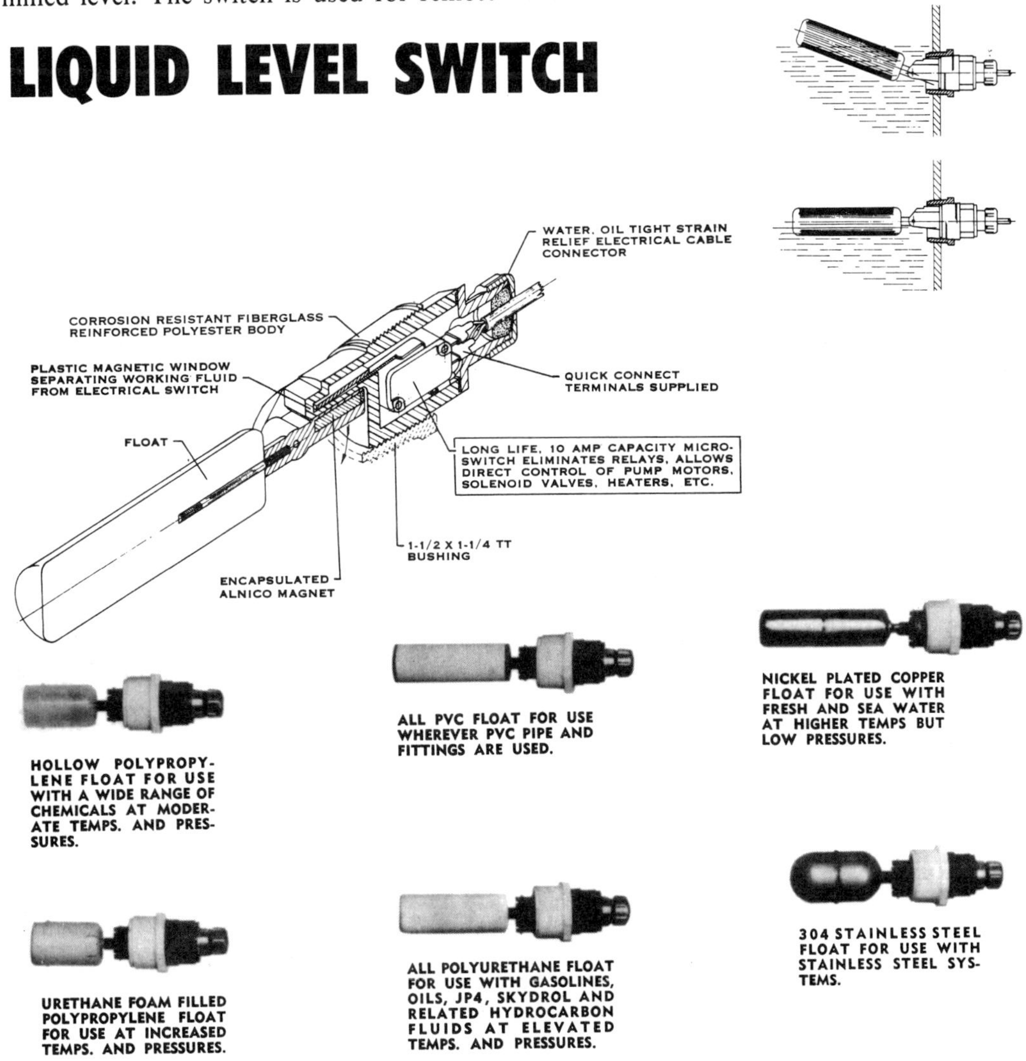

LIQUID LEVEL SWITCH

WATER, OIL TIGHT STRAIN RELIEF ELECTRICAL CABLE CONNECTOR

CORROSION RESISTANT FIBERGLASS REINFORCED POLYESTER BODY

PLASTIC MAGNETIC WINDOW SEPARATING WORKING FLUID FROM ELECTRICAL SWITCH

QUICK CONNECT TERMINALS SUPPLIED

FLOAT

LONG LIFE, 10 AMP CAPACITY MICRO-SWITCH ELIMINATES RELAYS, ALLOWS DIRECT CONTROL OF PUMP MOTORS, SOLENOID VALVES, HEATERS, ETC.

1-1/2 X 1-1/4 TT BUSHING

ENCAPSULATED ALNICO MAGNET

NICKEL PLATED COPPER FLOAT FOR USE WITH FRESH AND SEA WATER AT HIGHER TEMPS BUT LOW PRESSURES.

ALL PVC FLOAT FOR USE WHEREVER PVC PIPE AND FITTINGS ARE USED.

HOLLOW POLYPROPYLENE FLOAT FOR USE WITH A WIDE RANGE OF CHEMICALS AT MODERATE TEMPS. AND PRESSURES.

304 STAINLESS STEEL FLOAT FOR USE WITH STAINLESS STEEL SYSTEMS.

ALL POLYURETHANE FLOAT FOR USE WITH GASOLINES, OILS, JP4, SKYDROL AND RELATED HYDROCARBON FLUIDS AT ELEVATED TEMPS. AND PRESSURES.

URETHANE FOAM FILLED POLYPROPYLENE FLOAT FOR USE AT INCREASED TEMPS. AND PRESSURES.

Fig. 16-1. Liquid level switch. *(Courtesy of Harwil Company, Santa Monica, California.)*

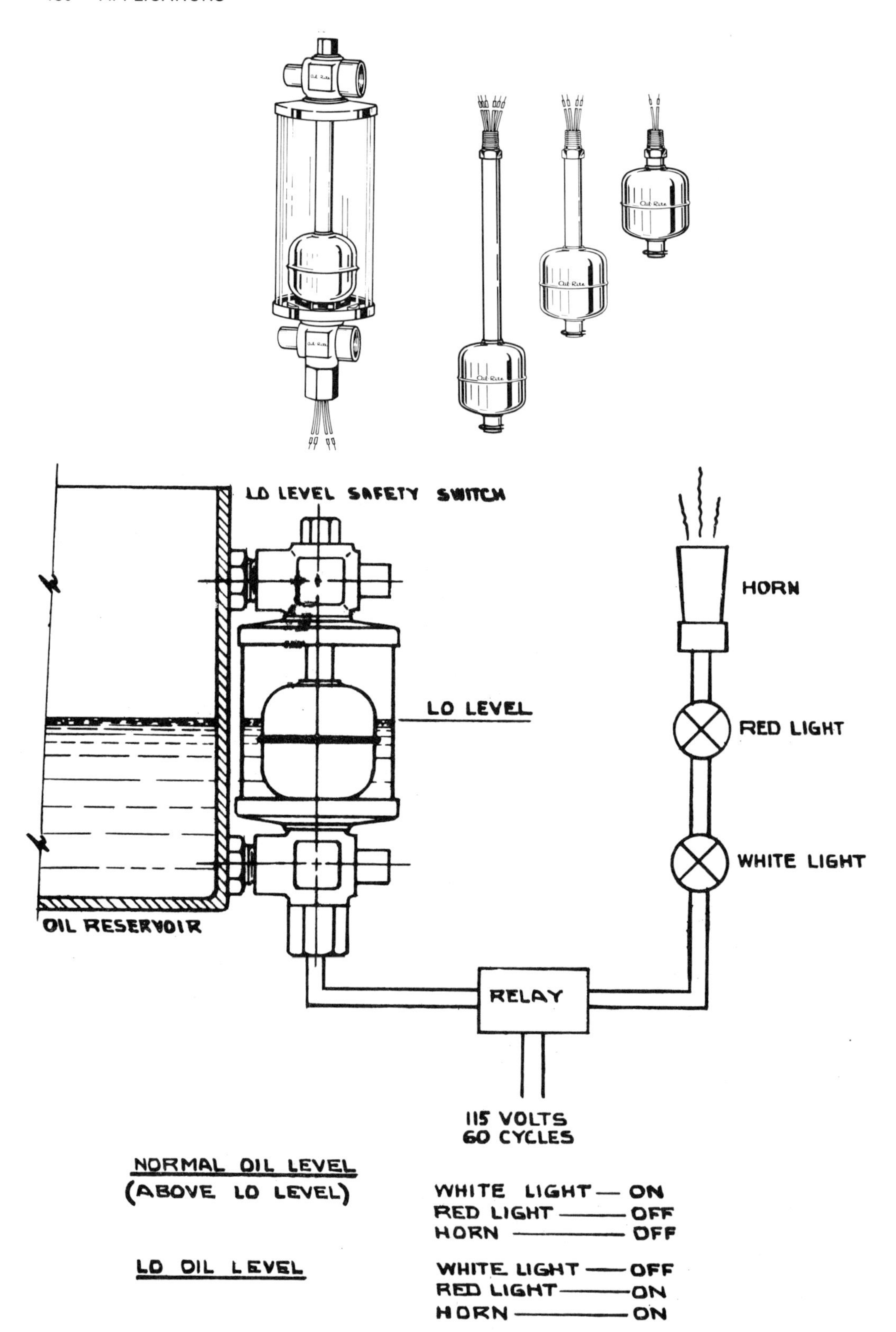

Fig. 16-2. Lo level safety switch. *(Courtesy of Oil-Rite Corporation, Manitowoc, Wisconsin.)*

C. PRESSURE CONTROL†

Here, for the novice, is an introduction and an incentive toward further study, for the seasoned instrument engineer, a stimulant to further simplify descriptions of pressure instrumentation, and for other engineers, a help toward overcoming the uncomprehending acceptance of instrumentation so prevalent in the industry.

Pressure measurement and control serves three purposes: the protection of equipment, the protection of personnel, and the production of on-spec product.

The average applications engineer is not often asked to design pressure instruments. Many instrument companies have perfected instruments that are proven, reliable, relatively inexpensive, and often available in quantity as off-the-shelf items. Thus the instrument engineer's function becomes one of choosing.

This means that the instrument applications engineer must be concerned with fundamentals of pressure measurement and control. These are most easily understood in terms of functional parts, i.e., of pressure-sensitive elements, transmitters and controllers.

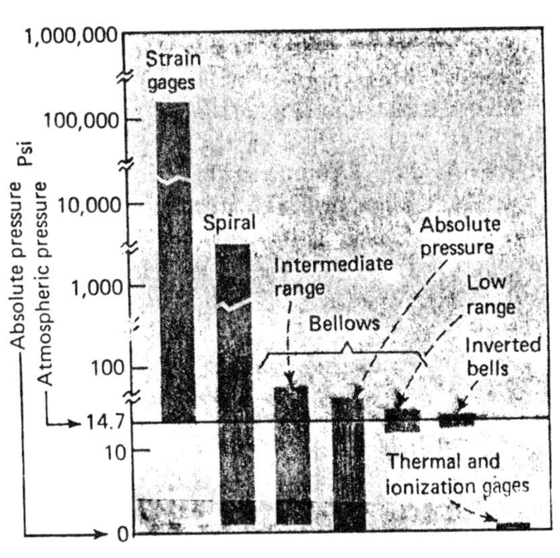

Fig. 1. Normal pressure for elements and gages.

PRESSURE-SENSITIVE ELEMENTS

Pressure instruments are usually categorized according to their pressure-sensitive element and range (Figure 1). However, all of the elements now used were initiated by the development of the bourdon tube.

Eugene Bourdon, a French engineer who died in 1884, invented a pressure-sensitive element that has since carried his name and has served as the basis for a steady accretion of additional developments. He bent a piece of good quality tubing into the shape of a horseshoe, sealed one end, and connected the other end to a source of fluid pressure.

When pressure was applied, the tube would flex, as if trying to straighten out. Since the elastic limit of the metal was not exceeded, the tube would return to its original form when the pressure was released; and the amount of flexure was proportional to the increase or decrease in pressure. If the open end of the tube were fixed, and the closed end linked to a pointer, it could be made to indicate the pressure.

The stage was thus set for pressure measurement. Different wall thicknesses and various alloys could be used to make bourdon tubes for various pressure ranges. Clockwork-type fittings, such as a pinion gear, a quadrant, a pointer, a dial, and a suitable hairspring to take up slack from the gears, could be used to provide precision. And a set of calibration attachments completes the parts necessary to a pressure gage (Figure 2).

The bourdon tube has been a very important milestone toward control of process pressures previously considered unsafe. Before a pressure can be safely controlled, it must be measured, and this gage provided the necessary reliability for such measurement.

However, the bourdon tube provides only a limited distance of travel at low pressures. This led to development of the *helix,* an

†By J. B. Ryan, Fluor Corp. Reprinted by special permission from *Chemical Engineering,* Feb. 3, 1975, by McGraw-Hill, Inc., New York, N.Y. 10020.

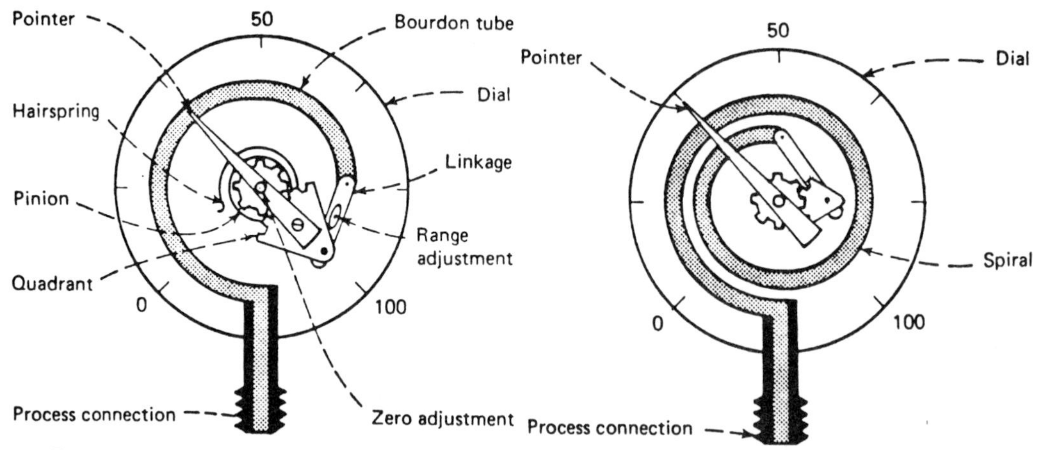

Fig. 2. Bourdon tube type pressure gage.

Fig. 4. Spiral tube type pressure gage.

extended bourdon tube in the form of a helical coil, which gives additional movement of the closed travelling end (Figure 3). Such additional movement is used to power transmitters, recorders, or controllers. Even at high pressures, it gives protection against overpressure and less hysteresis.

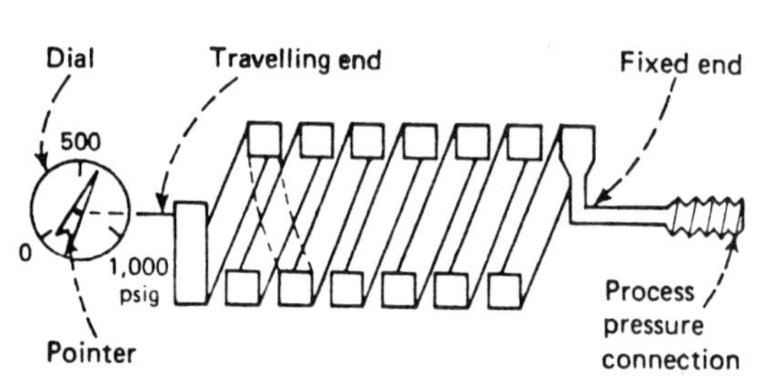

Fig. 3. Helical tube type pressure gage.

Usually the helical element can handle pressures ranging about 10:1 and be capable of precise measurement over continuous fluctuations. Accuracy is normally ±1% of the calibrated span. Such elements are available in various materials provided by various manufacturers.

The *spiral* element permits more compact housings than the helix, but retains extra travel not available with the bourdon tube. Thus, it does not require linkages to amplify the movement of the indicator (Figure 4).

The same principle of elastic metal deformed under pressure has been used for the *bellows* type of element (Figure 5), which is

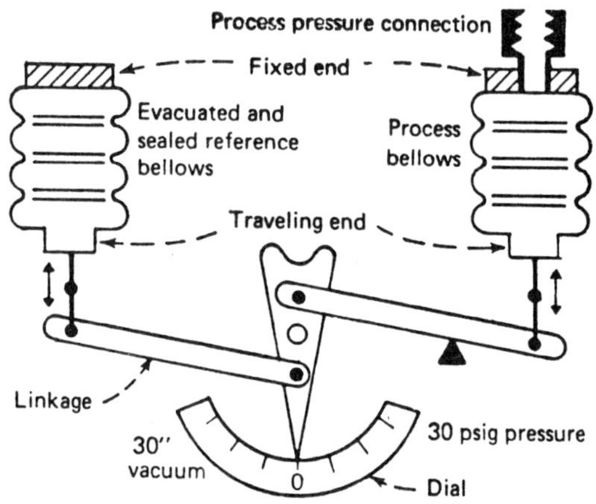

Fig. 5. Absolute-pressure-bellows indicator.

suited to lower pressures. Originally beryllium was used because it is quite rugged and elastic. Other materials such as stainless steel, nickel, and brass are now used extensively. In time, a bellows element tends to become work-hardened and needs replacement. Thus bellows-type instruments require more maintenance than other pressure devices. Normally, a return spring is integral with a bellows assembly (Figure 10).

THE TRANSMITTER

Although mechanical linkages from element to needle can translate pressure to a reading in a gage mounted at the equipment, they cannot transmit that information on to a control room or a remote location. Nor are they powerful enough to move a valve to regulate the pressure. Some sort of system is needed to transport the measured signals and to convert them to strong forces.

Two such systems have evolved—pneumatic and electronic. The pneumatic system usually employs air at 3–15 psig pressure. The electronic system usually employs 4–20 milliamps of direct current.

Pneumatic transmitters translate the position of a pressure element into air pressure that varies in strict proportion to the element position, so that this instrument air can be piped to other components as a measured signal.

Within the transmitter (Figure 6) the main flow of instrument air passes through a pneumatic pilot valve that regulates the signal pressure, while a sidestream passes through a restriction orifice to a nozzle. A connection transmits the pressure in this nozzle to one side of the diaphragm actuator for the pilot valve, and a flapper over the nozzle outlet lets air escape according to the position of the primary element.

Since air flow to the nozzle is restricted, a change in element and flapper positions causes a simultaneous change in nozzle-air pressure. This in turn changes the diaphragm pressure and the pilot valve position to put the signal air at a different pressure level. The signal air is stabilized at this new pressure by means of a feedback bellows that acts to oppose the continued action of the flapper. And the instrument can be calibrated at this feedback balance point (Figure 6).

The system in Figure 6 is known as a force-balance-type transmitter. There are also motion-balance transmitters. In either type, the process pressure is balanced against the spring action of the pressure-sensitive element, whether it be bellows, bourdon tube, spiral, helix, etc. A spiral or helical element would be used where accuracy could be a problem because of loss of motion.

Calibration adjustments for rangeability relate the pressure-sensitive element to the fulcrum of the force beam, for a corresponding relationship between the flapper and the air output of the transmitter. This latter relation depends on the coordination between the feedback bellows and the pilot diaphragm. An increase in output air pressure is felt by the feedback bellows which forces the flapper back to its throttling position.

Normally, the output air from the transmitter is fed to a controller, which in turn actuates a final control component, such as

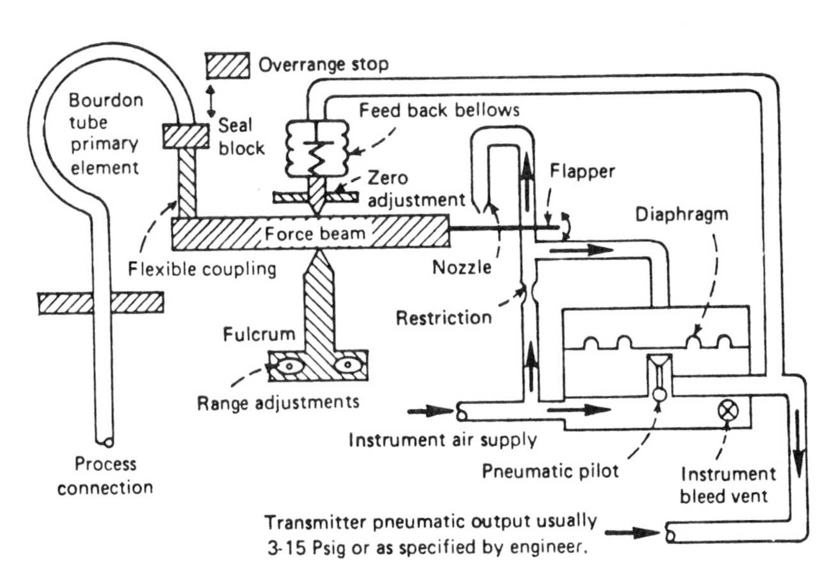

Fig. 6. Bourdon tube primary element activating a force-balance type of pneumatic transmitter.

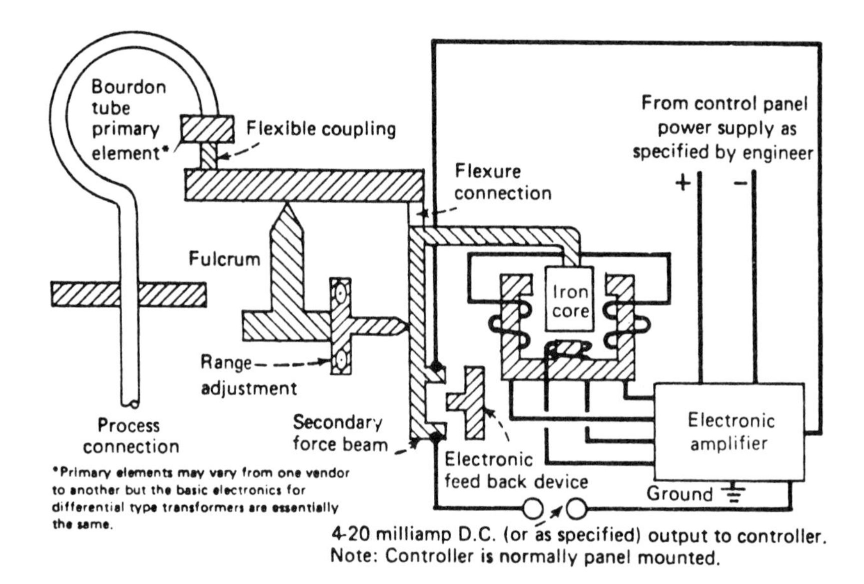

Fig. 7. Force-balance type electronic transmitter using a different type of transformer.

a valve. The normal calibrated range of these components would be 3 psig for 0% pressure and 15 psig for 100% pressure, so that midrange would be 9 psig. Similarly, electronic transmitters (Figure 7) and controllers normally use an input of 4 milliamps (mA) and an output of 20 mA, so that 4 mA is 0, 12 mA is 50% and 20 mA is 100% of the range.

Similar relationships exist for instruments having ranges of 6–30 psig or 10–50 mA.

Differential-pressure elements, normally in force-balance-type transmitters, are used in addition to the direct-acting elements to measure pressure, as well as flow and liquid level. They sense a difference in pressure between two sources acting on opposite sides of a diaphragm (Figure 8). Great care should be exercised in specifying them. They are available in exotic metals, as well as most of the common materials compatible with process fluids and conditions.

Because of their many uses, differential-pressure elements command a unique position in pressure instrumentation. Within the transmitter, a primary force-balance beam passes through a seal and fulcrum integral in the diaphragm housing to either a pneumatic or electronic system. Movement of the diaphragm is thus transmitted to the secondary beam, and the changing output from the pilot is the measured variable.

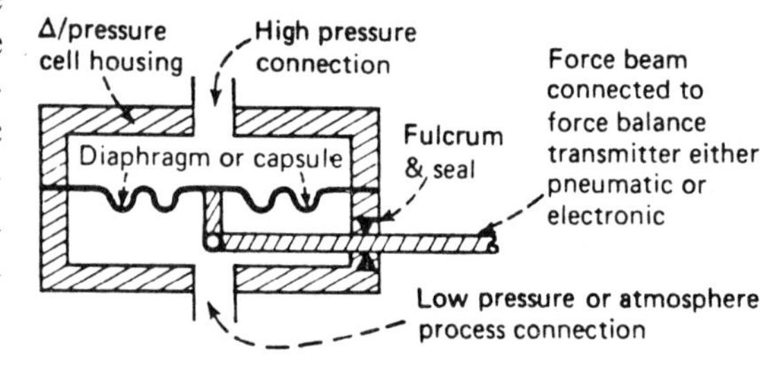

Fig. 8. Differential pressure type element.

Still other pressure-sensing and transmitting devices (i.e., slack diaphragm detectors, pressure repeaters and boosters, absolute pressure sensors, opposed bellows type detectors, fused-quartz helical pressure sensors, inverted bell manometer type sensors, strain gages, U-tube manometers, inclined manometers, etc.) are available and described in detail by their manufacturers' literature; but the foregoing principles of operation are generally applicable.

All such devices do not have equal electronic/pneumatic possibilities. The strain gage, for example, is available in an electronic version only (Figure 9). It employs a wire resister, which is stretched like a violin string,

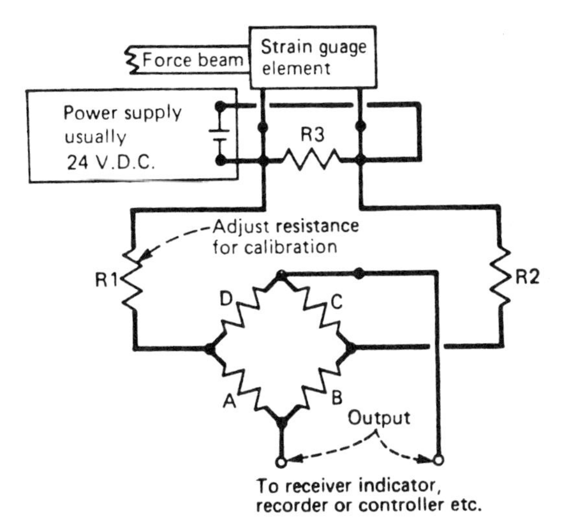

Fig. 9. Strain gage with electronic circuit.

thereby varying the electrical resistance of the wire according to the pressure. Its readout is translated through an electronic bridge circuit into a signal which can do the work (i.e., measure, indicate, record, or control). Such an instrument is expensive and should be applied only where very high pressures warrant its use (see Figure 1). Nevertheless, it is very accurate if properly applied.

Also, the repeatability of the bellows can be extended, when fitted with a positive return spring (Figure 10). In an electronic transmitter (Figure 10), this element can be used when pivoted across a fulcrum, so that a change in pressure changes the position of a noble-metal contactor in a variable potentiometer. The resistance signal is fed to a wheatstone bridge

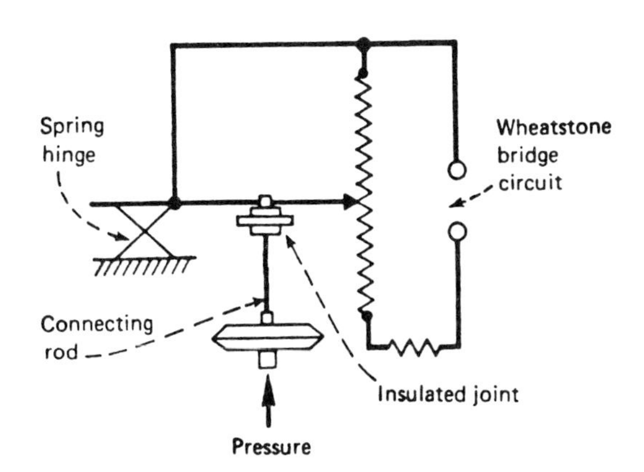

Fig. 10. Resistance type of transmitter.

circuit. Readout devices can take the form of an indicator, a recorder, or a controller. Accuracy is about ±2%.

Where the process warrants higher accuracy, a capacitance-type pressure transmitter with a differential-pressure element should be used (Figure 11). Such devices are expensive and very sensitive to the dielectric properties of the process, so that they should be limited to dielectric processes and to dry noncondensible gases. However, they are accurate to ±0.2% of their range, and with the proper diaphragms, can handle pressures from 3 in. of water to 5,000 psig.

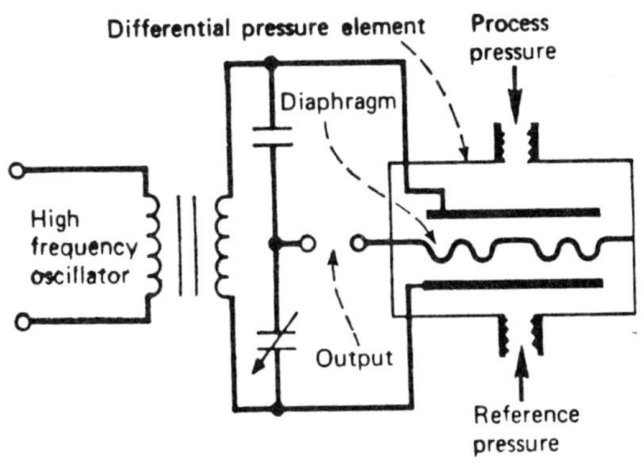

Fig. 11. Twin capacitance plate type transmitter.

As with other electric or electronic instruments, the developers have taken a single basic principle with predictable characteristics and developed it. One side of a diaphragm is exposed to the process pressure, and the other to a known reference pressure. Any change in process pressure moves the diaphragm; and in Figure 11 the capacitance is measured by the wheatstone bridge circuit. The output is linear and temperature compensation is available. Such units can be used to measure pressure or vacuum.

The following points are recommended for choosing pressure components:

1. *Transmitters:* For suppressed-range control, transmitters should normally be nonindicating force-balance type, piped in parallel with an indicating pressure gage. For full-range control,

or for indicating or recording service only, transmitters may be motion balance, indicating type, or nonindicating type, piped in parallel with an indicating pressure gage. Transmitter output should be 3-15 psig where pneumatic instruments are used and 4-20 or 10-50 mA direct current where electronic instruments are used.

2. *Controllers:* For applications where transmission to remote receivers is not required, controllers should be nonindicating, pilot-operated type, with the pressure-sensitive element and controller mounted in a weatherproof case suitable for installation by the valve or remotely. Direct-connected pressure gages should be piped in parallel with local indication. An adjustable proportional band of at least 2-20% and convenient control point setting should be provided. Output should be 3-15 or 6-30 psig as is normally required for the final control element.
3. *Elements:* Except where the process requires special material, process-measuring elements should be AISI Type 316 stainless steel or "Ni-Span C." Bronze or steel may be used for utility services.
4. Differential pressure instruments should be liquid-filled manometers, bellows, or force-balance type, according to requirements.
5. Control modes should normally be of adjustable proportional control with automatic reset. Local-mounted pilot-type controllers may be limited-proportional type. Derivative response should be provided only when required by process conditions.

PRESSURE INSTRUMENTS FOR TEMPERATURE

Some pressure instruments are used to determine other properties, particularly temperature. The sensing elements of temperature instruments are usually modifications based on the expansion and contraction of a liquid or gas with temperature, and the system is essentially a pressure gage fitted with a readout device. The gage is connected by a small-bore capillary tube to the temperature-sensing device, filled with a suitable fluid, and sealed. If the gage is a bellows, for example, the change in temperature would act like a change in the pressure to expand or contract the bellows; and the movement due to fluid in a large bulb could be used to actuate a direct-connected control valve. Such self-actuated temperature control valves are in widespread use in heating systems for storage tanks and other rough applications, where a few degrees, plus or minus, is not acute.

PRESSURE CONTROLLERS

Before selecting a pressure controller from the wide variety available, the engineer must know the purpose and preferred mode of operation for the entire control circuit, including transmitter, controller and valve. Normally the transmitter output signals the controller, which in turn positions the valve.

A pressure controller should include an adjustment for proportional control in addition to other calibrational adjustments, so as to obtain proportional action over a wider or narrower range according to the needs of the process. Since the advent of electronic controllers, this proportional action is sometimes called "gain." Another desirable feature, called "rate" or "derivative" action is used to smooth out undesirable pressure peaks that cannot be corrected by proportional action only.

These adjustments are critical to the operation of a control circuit and should not be tampered with by unqualified operating personnel. Such tampering is responsible for a large proportion of the instrument troubles experienced during the startup of new plants.

The operating principles of controllers resemble those of transmitters, whether pneumatic or electronic. Whereas the transmitter receives input data in a bourdon tube or similar sensing element, the controller receives the transmitted signal in a bellows or electronic bridge, which has an adjustable setpoint. When the transmitted signal differs from this setpoint, the controller generates an output signal to the valve for corrective action, which is supposed to return the transmitter signal to the setpoint.

The choice between pneumatic or electronic control should depend largely on the plant's management, who will know the qualifications of operating and maintenance personnel. Sometimes plant managers convert an existing unit from pneumatic to electronic control, because technicians have gained exposure to electronic circuitry from work in television and space programs. By themselves, the components would rarely command precedence one over the other; and when the choice is made by an engineering contractor, it is based largely on economics, availability, and the area where a new plant is to be located.

The following points might be helpful in choosing between the two:

Pneumatic	Electronic
Less expensive	More expensive
Small plants	Large plants
Short transmission	Long transmission
Simple process	Complex process
Slow response	Fast response
Greater dependability	More maintenance
No computer capability without transducer	Adapted to computers
Intrinsically safe	Added cost for safe components
Analog signals not available with digital	Digital signals; also available with analog

INSTALLATION

The engineer should also be aware of accessories, such as snubbers and blowout backs, as well as good practice regarding such items as dirty services and instrument position.

Snubbers: Pressure snubbers on reciprocating pump discharges and other pressure-cycling streams tend to smooth out the peaks and thus give longer life to an instrument. They should always be considered for pressure transmitters and direct-connected pressure controllers.

The wear on the quadrant of a pressure gage, for example, usually occurs in the center portion, and the pinion gear also wears. If such wear is left unchecked, the hairspring can no longer take up slack movement, with resulting loss of accuracy. In recent years, several manufacturers have filled the cavity in the pressure-gage housing with a nonfreezing liquid to absorb the shock of cycling pressures as a means of smoothing out peaks.

Blowout backs: These are provided on some models of pressure gages, so that a rupture directs the process fluid out the back end of the gage, and not into the face of a bystander in front of the unit. However, if the proper metal and ranges are specified for the pressure-sensing device, chances of the element rupturing are considerably lessened.

Dirty or corrosive services: The chemistry of the process should always be evaluated before specifying any pressure devices, so as to choose the proper type of element for the service. Suppose, for instance, that the pressure from a 1,200 psig crude-feed supply line is let down to 50 psig for a distillation unit, with the crude containing particles of sand and traces of water, but not enough to warrant a strainer or filter on the upstream side of the transmitter. A measuring element other than a bourdon tube, spiral or helix should be chosen, since such elements present a collection chamber for foreign matter.

Water or other foreign matter can cause corrosion, thereby limiting the choice of metal to stainless steel or Monel. Diaphragms have been developed by various instrument vendors for such service. These work reasonably well but are limited to the ranges shown in Figure 1. Because of the increased surface area of a diaphragm, relative to that of bourdon-tube type elements, the latter requires less maintenance and offers longer life.

Position: A pressure instrument applied to piping and vessels should be mounted in an upright position, so that particles carried in the process stream will drop out of the sensing element before they become lodged there. Also, water can act like particles, when it is carried in an immiscible lighter liquid such as hydrocarbons. In time such impurities can cause an unnecessary shutdown of the unit. Regardless of the instrument selected, periodic maintenance for such dirty service is necessary, and perhaps a redundant pressure control circuit is warranted.

Usually the housing of the capsule diaphragm is complete with a blowdown valve, and the controller in that particular loop will have been placed on manual control before the maintenance operation is undertaken.

CALIBRATION

All pressure devices require proper calibration, and a number of standard instruments have been developed for this. The first and most essential is the simple glass-tube manometer. No instrument shop would be complete without one. Generally known as "U" tubes because of the shape of their glass tubes, these instruments are partially filled with water or mercury, and one end fitted to a variable pressure source. The pressure can be read in the difference of liquid levels on a scale located between the legs of the U tube.

The component to be tested is usually connected in parallel with the manometer, and a reference pressure applied to give the component's 0%, 50% and 100% points, with the readings checked on the manometer. Other points can also be checked, especially the known operating point.

Inclined manometers are used for calibrating instruments such as draft gages, where the U-tube is limited by its sensitivity. Still other limitations relate to the length or size of glass manometer that can be built. If the desired calibration range lies outside that of the standard water-filled manometer, a mercury-filled unit might be used. But if the range lies still outside that of such heavier solutions, the most logical standard calibration device would be a "dead weight" tester. No well-equipped instrument shop should be without a dead weight tester.

This device functions by measuring hydraulic pressure of oil in a chamber with dead weights loaded onto a vertical piston that rests down against the pressurized liquid. A connection puts the instrument to be calibrated against the same liquid pressure. In use, the instrument is connected, weights are loaded, and a hand pump used to bring liquid from a built-in reservoir into the chamber until the piston can just lift the weights. Then these are checked against the indication on the instrument.

Dead weight testers are certified by a governmental bureau of standards and must not be altered in any way. Their care and maintenance is of paramount importance. Before using such sensitive calibration equipment, the instrument to be tested should be thoroughly cleaned, especially if it has been exposed to corrosive chemicals or dirty substances.

Although normally a first task for new technicians, calibrating pressure gages should not be taken lightly as a menial task; every time an instrument technician or mechanic says a given instrument is calibrated, that person has taken responsibility for the safety of fellow workers and the safe operation of a process.

In conclusion, it should be noted that safety is a primary function of pressure control. Consequently, a few extra dollars are always justified for a more reliable installation. It follows that only properly qualified and well-trained technical people should be used in its application.

REFERENCES

1. Moore, J. F., others, Process Control and Guide to Process Instrument Elements, *Chem. Eng.,* June 2, 1969, pp. 94–164, also *Chem. Eng.* reprint no. 68.
2. Chalfin, Sanford, Specifying Control Valves, *Chem. Eng.,* Oct. 14, 1974, pp. 105–114.
3. Harvey, Glenn F., ed. *Standards and Practices for Instrumentation,* 3rd ed., Instrument Soc. of America, 400 Stanwix Street, Pittsburgh, Pennsylvania 15222.
4. Liptak, Bela G., ed., *Instrument Engineers' Handbook,* Vol. I, *Process Measurement,* Chilton Book Co., New York.

Appendix A Glossary of Terminology

Absolute Pressure The total pressure measured from an absolute vacuum. It is the sum of the gage pressure and the prevailing atmospheric pressure.

Absolute Pressure Switch A pressure switch designed to actuate at a specified absolute pressure. The pressure sensor is referenced to a near-perfect vacuum.

Absolute Temperature Temperature measured from absolute zero.

Absolute Viscosity The resistance of a fluid to relative motion of its parts. Also called **dynamic viscosity.**

Absolute Zero The zero point on the Kelvin temperature scale. It is the temperature at which a system undergoes a reversible isothermal process without transfer of heat.

Acceleration The time rate of change of velocity.

Acceleration Error The maximum difference, at any measured value within the specified range, between output readings taken with and without the application of specified constant acceleration along specified axes.* *Note:* See **Transverse Sensitivity** when applied to acceleration transducer.

Accuracy The ability of a pressure switch to operate at its set points repetitively. It is normally specified as a percentage of full scale.

Actuation Point The exact point at which the electrical circuit of a pressure switch will operate on either increasing or decreasing pressure.

Actuator The device such as bellows, bourdon tube, diaphragm, piston, etc., that senses the applied pressure and actuates the electrical switch.

Actuator Spring See **Loading Spring.**

Actuator Stem A rod extending from the pressure sensor that presses on the actuator button of the electrical switch.

Adjustable Differential The capability of a pressure switch to allow adjustment of the differential over a given range.

Adjustable Range The pressure range of a pressure switch within which the actuation point may be adjusted.

Adjusting Screw A screw used to regulate or control the compression setting of a pressure switch. See **Loading Spring.**

Adjusting Screw Cap A cover for the adjusting screw used to prevent accidental changes in setting.

Adjustment Range See **Adustable Range.**

Air The naturally occurring mixture of nitrogen, oxygen, carbon dioxide, and other gases.

Air, Standard Air at a temperature of 68°F, pressure of 14.70 psia, and 36% relative humidity. Some industries consider the temperature of standard air to be 60°F.

Airtight The property of being impermeable to air. It generally refers to a seal that is tight enough to prevent leakage when tested with air.

Altitude The vertical distance above a stated reference level.* *Note:* Unless otherwise specified, this reference is mean sea level.

Ambient Conditions The conditions (pressure, temperature, etc.) of the medium surrounding the case of the transducer.*

Ambient Pressure The pressure of the medium

immediately surrounding a pressure switch. It is not necessarily atmospheric pressure.

Ambient Pressure Error The maximum change in output, at any measured value within the specified range, when the ambient pressure is changed between specified values.*

Ambient Temperature The temperature in the immediate vicinity of the pressure switch. Usually, it refers to room temperature.

Amplifying With integral output amplifier.*

Analog Output Transducer output which is a continuous function of the measurand, except as modified by the resolution of the transducer.*

Aneroid See **Capsule.**

Aneroid Barometer An instrument used for measuring atmospheric pressure.

A.N.S.I. American National Standards Institute.

A.S.M.E. American Society of Mechanical Engineers.

A.S.T.M. American Society for Testing Materials.

Atmospheric Pressure The pressure exerted by the atmosphere. At sea level it is taken as 14.7 psia.

Attitude The relative orientation of a vehicle or object represented by its angles of inclination to three orthogonal reference axes.

Attitude Error The error due to the orientation of the transducer relative to the direction in which gravity acts upon the transducer (see **Acceleration Error**).

Auxiliary Actuator A mechanism, sold separately, to provide basic switches with easier means of operation and adjustment and adapt switches to different operating motions by supplying supplemental overtravel.**

Barometer A device used to measure prevailing atmospheric pressure. It is usually of the aneroid or mercury column type.

Basic Switch A self-contained switching unit. It can be used alone, gang-mounted, built into assemblies, or enclosed in metal housings.**

Bellows A flexible, thin-walled, circumferentially corrugated cylinder. It can expand or contract axially under a change in pressure.

Best Straight Line A line midway between the two parallel straight lines closest together and enclosing all output vs. measurand values on a calibration curve.*

Bifurcated Contact A movable contact that is forked to provide two contact mating surfaces in parallel, for more reliable contact.**

Bondable Designed to be permanently mounted to a surface by means of adhesives.*

Bonded Permanently attached over the length and width of the active element.*

Bounce See **Contact Bounce.**

Bourdon Tube A tube closed at one end that will move under change in internal pressure. Usually, the tube is coiled and will try to straighten out as pressure increases.

Boyle's Law The relation between pressure and volume of a fixed mass of gas for an isothermal process. The product of pressure and volume is constant if the temperature is held constant.

Break An opening or interruption of an electrical circuit.

Break-Before-Make The breaking of double throw contacts where the moving contact interrupts one circuit before establishing the other.

Break Distance The minimum open gap distance between stationary and movable contacts.**

Breakdown Voltage Rating (S) The dc or sinusoidal ac voltage that can be applied across specified insulated portions of a transducer without causing arcing or conduction above a specified current value across the insulating material.* *Note:* Time duration of application, ambient conditions, and ac frequency must be specified.

Bridge Resistance (See **Input Impedance** and **Output Impedance**.)*

British Thermal Unit The amount of heat necessary to raise the temperature of one pound of water from 63° to 64°F.

Buoyancy The tendency of an object to float in a fluid. It depends on the densities of the object and the fluid.

Burst Pressure The maximum pressure that may be applied to a pressure switch without causing leakage or rupture.

Burst Pressure Rating (S) The pressure that may be applied to the sensing element or the case (as specified) of a transducer without rupture of either the sensing element or the transducer case as specified.* *Note:* (1) Minimum number of applications and time duration of each application must be specified; (2) in the case of transducers intended to measure a property of a pressurized fluid, burst pressure is applied to the portion subjected to the fluid.

Calibration A test during which known values of the measurand are applied to the transducer and corresponding output readings are recorded under specified conditions.*

Calibration Curve A graphical representation of the calibration record.*

Calibration Cycle The application of known values of the measurand, and recording of corresponding output readings, over the full (or specified portion of the) range of a transducer in an ascending and descending direction.*

Calibration Range The pressure range over which the switch can be adjusted to actuate.

Calibration Record A record (e.g., table or graph) of the measured relationship of the transducer output to the applied measurand over the transducer range.* *Note:* Calibration records may contain additional calculated points so identified.

Calibration Simulation Provisions Electrical connections or circuitry, contained within a transducer, designed to permit the calibration of the associated measuring system by causing output changes of known magnitude without varying the applied measurand.*

Calibration Traceability The relation of a transducer calibration, through a specified step-by-step process, to an instrument or group of instruments calibrated by the National Bureau of Standards.* *Note:* The estimated error incurred in each step must be known.

Calibration Uncertainty The maximum calculated error in the output values, shown in a calibration record, due to causes not attributable to the transducer.*

Capacitive Converting a change of measurand into a change of capacitance.*

Capillary A very thin tube, generally with an internal diameter in the range of a few thousandths of an inch.

Capsule A pressure-sensing element consisting of two metallic diaphragms joined around their peripheries.*

Case Pressure (See **Burst Pressure Rating**, **Proof Pressure**, or **Reference Pressure**.)*

Center of Seismic Mass The point within an acceleration transducer where acceleration forces are considered to be summed.*

Characteristics This term is used by Micro Switch in a restricted sense and refers only to switch operating characteristics such as pretravel, operating force, etc.**

Charles' Law The volume of a fixed mass of gas varies directly with absolute temperature if the pressure of the gas is held constant.

Chatter Prolonged undesirable opening and closing of electronic contacts.

Circuit The contact arrangement with switch actuator and contacts in their normal position.**

Clearance Air space, usually 1/16″ minimum, between live metal parts of opposite polarity or to ground.

Closed Ends Ground End finish of compression springs as illustrated on compression spring specification chart. Also called **squared and ground ends** or **ends squared and ground**.

Closed Length See **Solid Height**.

Coefficient of Linear Expansion The change in unit length of a material per degree rise in temperature.

Coils Per Inch See **Pitch**.

Cold Flow The change of dimension or distortion caused by sustained application of a force.

Collapse Pressure The difference between external and internal pressures that causes structural failure.

Compensation Provision of a supplemental device, circuit, or special materials to counteract known sources of error.*

Condensate A liquid that has condensed from the vapor state.

Conduction Error The error in a temperature transducer due to heat conduction between the sensing element and the mounting of the transducer.*

Conformance (See **Accuracy** and **Error Band**.)*

Connector A device used to join a conductor to a component part.

Contact Bounce Rapid opening and closing of electrical contacts due to their elasticity.

Contact Gap The air space between mating contacts when contacts are open.

Contact Resistance The electrical resistance of closed contacts.

Contaminant Material that reduces the purity of another material.

Continuous Rating The rating applicable to specified operation for a specified uninterrupted length of time.

Convection The transfer of heat through the associated transfer of mass.

Corona Discharge of electricity that appears on the surface of the conductor when the potential gradient exceeds a certain value.

Corrosion Deterioration of materials due to chemical and/or electrical action.

Creep A change in output occurring over a specific time period while the measurand and all environmental conditions are held constant.*

Creepage The distance over the surface of an insulator between live metal parts of opposite polarity or to ground.

Critical damping This term is defined under **Damping.***

Critical Set Point The set point of the unit that is held to the closest tolerance. It can be either the actuation or the deactuation point.

Cross-Axis Acceleration See **Transverse Acceleration.***

Cross Sensitivity, Cross-Axis Sensitivity See **Transverse Sensitivity.***

Damper A device used to dampen or reduce the amplitude of a pressure wave.

Damping The energy-dissipating characteristic that, together with natural frequency, determines the limit of frequency response and the response-time characteristics of a transducer.* *Note 1:* In response to a step change of measurand, an underdamped (periodic) system oscillates about its final steady value before coming to rest at that value; an overdamped (aperiodic) system comes to rest without overshoot; and a critically damped system is at the point of change between the underdamped and overdamped conditions. *Note 2:* **Viscous Damping** uses the viscosity of fluids (liquids or gases) to effect damping. *Note 3:* **Magnetic Damping** uses the current induced in electrical conductors by changes in magnetic flux to effect damping.

Damping Ratio The ratio of the actual damping to the damping required for critical damping.*

Deactuate Pressure The pressure at which transfer of contacts occurs on decreasing pressure.

Deadband The difference between the actuation point and the deactuation point.

Dead Break An unreliable contact made near the trip point, at low contact pressure. The circuit is interrupted, but the switch does not "snap over."

Dead Volume The total volume of the pressure port cavity of a transducer with room barometric pressure applied.*

Deflection Rate The change in deflection (output) caused by a unit change in pressure.

Density Mass per unit volume.

Detector See **Transducer.***

Detent A catch or holding device.

Diaphragm A device, usually metallic, used to convert pressure into a linear force.

Diaphragm, Rolling See **Rolling Diaphragm.**

Dielectric Strength The maximum potential gradient that an insulating material can withstand without rupture.

Differential See **Deadband.**

Differential Pressure The difference between the reference pressure and the sense pressure.

Differential Pressure Switch A pressure switch with high and low pressure ports. Actuation occurs when the pressure to the high port exceeds pressure to the low port by a predetermined value.

Differential Value See **Deadband.**

Digital Output Transducer output that represents the magnitude of the measurand in the form of a series of discrete quantities coded in a system of notation.* *Note:* Distinguished from analog output.

Directivity The solid angle, or the angle in a specified plane, over which sound or radiant energy incident on a transducer is measured within specified tolerances in a specified band of measurand frequencies.*

Discrete Increment Providing an output that represents the magnitude of the measurand in the form of discrete or quantized values.*

Displacement The change in position of a body or point with respect to a reference point.* *Note:* Position is the spatial location of a body or point with respect to a reference point.

Distortion See **Harmonic Content.***

Dithering The application of intermittent or oscillatory forces just sufficient to minimize static friction within the transducer.*

Double Amplitude The peak-to-peak value.*

Double Break A contact arrangement in which the moving switch element bridges across two fixed contacts so that the circuit is broken in two places simultaneously.

Double Throw A switch that alternately completes a circuit at each of its two extreme positions.

Double Pole-Double Throw A switching element with six electrical terminals. It is equivalent to two single pole-double throw switches combined for simultaneous operation.

Drift A sustained change in value of a variable.

Dual Output Providing two separate and noninteracting outputs that are functions of the applied measurand.*

Durometer Hardness The hardness value of an elastomer as measured on a durometer scale.

Dynamic Characteristics Those characteristics of a transducer that relate to its response to variations of the measurand with time.*

Effective Pressure Area The area of a unit that responds to a pressure change.

Elastic Limit The maximum stress to which a material may be subjected without causing a permanent set.

Electrical Switching Element Opens or closes an electrical circuit as a result of the movement and force it receives from the pressure-sensing element.

Electromagnetic Converting a change of measurand into an output induced in a conductor by a change in magnetic flux, in the absence of excitation.*

Enclosed Switch A basic switch unit (contact block) enclosed in a durable metal housing. The enclosure protects the switching unit, provides a mounting means and fitting for the conduit connection.**

End Device, End Instrument See **Transducer.***

End Points The outputs at the specified upper and lower limits of the range.* *Note:* (S) Unless otherwise specified, end points are averaged during any one calibration.

End-Point Line The straight line between the end points.*

Ends Squared and Ground See **Closed Ends Ground**.

Endurance Limit The maximum stress at which any given material will operate indefinitely without failure.

Environmental Conditions Specified external conditions (shock, vibration, temperature, etc.) to which a transducer may be exposed during shipping, storage, handling, and operation.*

Environmental Conditions, Operating-Environmental Condtions Conditions during exposure to which a transducer must perform in some specified manner.*

Environment-Proof Switch A switch that is completely sealed to ensure constant operating characteristics. Sealing normally includes an "O" ring on an actuator shaft and fused glass-to-metal terminal seals or complete potting and an elastomer plunger-case seal.**

Error The algebraic difference between the indicated value and the true value of the measurand.* *Note 1:* (S) It is usually expressed in percent of the full scale output, sometimes expressed in percent of the output reading of the transducer. *Note 2:* (S) A theoretical value may be specified as true value.

Error Band The band of maximum deviations of output values from a specified reference line or curve due to those causes attributable to the transducer.* *Note 1:* (S) The band of allowable deviations is usually expressed as "± ____ percent of full scale output," whereas in test and calibration reports the band of maximum actual deviations is expressed as "+ ____ percent, − ____ percent of full scale output." *Note 2:* (S) The error band should be specified as applicable over at least two calibration cycles, so as to include repeatability, and verified accordingly.

Error Curve A graphical representation of errors obtained from a specified number of calibration cycles.*

Excitation The external electrical voltage and/or current applied to a transducer for its proper operation.* *Note 1:* In the sense of a physical quantity to be measured by a transducer, use the measurand. *Note 2:* (S) Usually expressed as range(s) of voltage and/or current values. *Note 3:* Also see **Maximum Excitation**.

Explosion-Proof Switch A UL-listed switch capable of withstanding an internal explosion of a specified gas without igniting surrounding gases.**

Field of View The solid angle, or the angle in a specified plane, over which radiant energy incident on a transducer is measured within specified tolerances.*

Fixed Differential A term that refers to the difference between the actuation point and the reactuation point of a pressure switch. It further signifies that this differential is a fixed function of the pressure switch and not adjustable.

Flow Rate The time rate of motion of a fluid, usually contained in a pipe or duct, expressed as fluid quantity per unit time.*

Flow Switch An instrument that makes or breaks an electrical circuit at a predetermined flow rate.

Frequency Number of cycles of operation per minute.

Frequency-Modulated Output An output in the form of frequency deviations from a center frequency, where the deviation is a function of the applied measurand.*

Frequency Output An output in the form of frequency that varies as a function of the applied measurand (e.g., angular speed and flow rate).*

Frequency, Natural The frequency of free (not forced) oscillations of the sensing element of a fully assembled transducer.* *Note 1:* It is also defined as the frequency of a sinusoidally applied measurand at which the transducer output lags the measurand by 90°. *Note 2:* (S) Applicable at room temperature unless otherwise specified. *Note 3:* Also see **Frequency, Resonant** and **Frequency, Ringing,** which are considered of more practical value than natural frequency.

Frequency, Resonant The measurand frequency at which a transducer responds with maximum output amplitude.* *Note 1:* (S) When major amplitude peaks occur at more than one frequency, the lowest of these frequencies is the resonant frequency. *Note 2:* (S) A peak is considered major when it has an amplitude at least 1.3 times the amplitude of the frequency to which the specified frequency response is referred. *Note 3:* For subsidiary resonance peaks see **Resonances**.

Frequency, Ringing The frequency of the oscillatory transient occurring in the transducer output as a result of a step change in the measurand.*

Frequency Response The change with frequency of the output/measurand amplitude ratio (and of the phase difference between output and measurand), for a sinusoidally varying measurand applied to a transducer within a stated range of measurand frequencies.* *Note 1:* (S) It is usually specified as "within ± ___ percent (or ± ___ db) from ___ to ___ Hertz." *Note 2:* (S) Frequency response should be referred to a frequency within the specified measurand frequency range and to a specific measurand value.

Frequency Response, Calculated The frequency response of a transducer calculated from its transient response, its mechanical properties, or its geometry, and so identified.*

Friction See **Friction Error.***

Friction Error The maximum change in output, at any measurand value within the specified range, before and after minimizing friction within the transducer by dithering.*

Friction-Free Error Band The error band applicable at room conditions and with frictions within the transducer minimized by dithering.*

Full Scale See **Range.***

Full-Scale Output The algebraic difference between the end points.* *Note:* (S) Sometimes expressed as "± (half the algebraic difference)"; e.g., "± 2.5 volts."

Gage Factor A measure of the ratio of the relative change of resistance to the relative change in length of a resistive strain transducer (strain gage).*

Gage Pressure Pressure above or below (plus or minus) atmospheric pressure. (Zero psig equals 14.7 psia at standard sea level.) Gage pressure is also defined as the difference between atmospheric pressure and a variable pressure.

Gage Pressure (and Vacuum) Switches The pressure-sending element in both pressure and vacuum configurations is ported to the pressure being monitored. Monitored pressure is opposed by atmospheric pressure.

Gradient See **Rate per Inch**.

Gyro (A contraction of gyroscope.) A transducer that makes use of a self-contained spatial directional reference.*

Harden To cause steel to become harder by heating it to above the critical temperature and quenching it in oil or another cooling medium. Also called **quenching**.

Harmonic Content The distortion in a transducer's sinusoidal output, in the form of harmonics other than the fundamental component.* *Note:* (S) It is usually expressed as a percentage of rms output.

Heat Flux The quantity of thermal energy transferred to a unit area per unit time.*

Hermetically Sealed Switch A switch completely sealed to provide constant operating characteristics. All junctures made with metal-to-metal or glass-to-metal fusion.**

Humidity, Absolute The mass of water vapor present in a unit volume of air or other fluid.*

Humidity, Relative The ratio of the water vapor pressure actually present to the water vapor pressure required for saturation at a given temperature, expressed in percent.*

Hysteresis The difference between up-scale and down-scale deflections of a unit at the same pressure, expressed as a percentage of full range deflection.*

Inaccuracy See **Error**.*

Inductive Converting a change of measurand into a change of the self-inductance of a single coil.*

Input See **Excitation** or **Measurand**.*

Input Impedance The impedance (presented to the excitation source) measured across the excitation terminals of a transducer.*

Instability See **Stability**.*

Insulation Resistance (S) The resistance measured between specified insulated portions of a transducer when a specified dc voltage is applied at room conditions unless otherwise stated.*

Integrating Providing an output that is a time integral function of the measurand.*

Intermittent Rating The rating applicable to specified operation over a specified number of time intervals of specified duration; the length of time between these time intervals must also be specified.*

Internal Pressure See **Burst Pressure, Proof Pressure**, or **Reference Pressure**.*

Ionizing Converting a change of measurand into a change in ionization current, such as through a gas between two electrodes.*

Jerk The time rate of change of acceleration. Expressed in feet/s^3, cm/s^3/g/s.*

Lead See **Pitch**.

Leakage Rate The maximum rate at which a fluid is permitted or determined to leak through a seal.* *Note:* (S) The type of fluid, the differential pressure across the seal, the direction of leakage, and the location of the seal must be specified.

Least-Squares Line The straight line for which the sum of the squares of the residuals (deviations) is minimized.*

Life, Cycling (S) The specified minimum number of full range excursions or specified partial range excursions over which a transducer will operate as specified without changing its performance beyond specified tolerances.*

Life, Operating (S) The specified minimum length of time over which the specified continuous and intermittent rating of a transducer applies without change in transducer performance beyond specified tolerances.*

Life, Storage (S) The specified minimum length of time over which a transducer can be exposed to specified storage conditions without changing its performance beyond specified tolerances.*

Light An electromagnetic radiation whose wavelength is between approximately 10^{-2} and 10^{-6} cm.* *Note:* By strict definition only visible radiation (4×10^{-5} to 7×10^{-5} cm) can be considered as light.

Linearity The closeness of a calibration curve to a specified straight line.* *Note:* (S) Linearity is expressed as the maximum deviation of any calibration point on a specified straight line, during any one calibration cycle. It is expressed as "within ± ____ percent of full scale output."

Linearity, End Point Linearity referred to the end point line.*

Linearity, Independent Linearity referred to the best straight line.*

Linearity, Least Squares Linearity referred to the least-squares line.*

Linearity, Terminal Linearity referred to the terminal line.*

Linearity, Theoretical Slope Linearity referred to the theoretical slope.*

Line Pressure See **Reference Pressure**.*

Load See **Load Impedance**.*

Load Factor See **Rate per Inch**.

Load Impedance The impedance presented to the output terminals of a transducer by the associated external circuitry.*

Loading Error An error due to the effect of the load impedance on the transducer output.* *Note:* In the case of force transducers the term loading has been applied to application of force.

Loading Spring A pressure switch loading spring is generally that device which compensates and/or adjusts the tripping point that is used in bellows, piston, and/or diaphragm type sensing devices. (See Figure 1-4.)

Magnetic Blow-Out Switch Contains a small permanent magnet that provides a means of switching high dc loads. The magnet deflects the arc to quench it.**

Maintained Contact Switch Device designed for applications requiring sustained contact after the plunger has been released, but with provision for resetting.**

Make To close or establish an electrical circuit.**

Manual Reset With reference to a pressure switch, this term indicates that when the actuation point is reached on either increasing or decreasing pressures, the pressure switch automatically operates the electrical circuit. As the pressure returns beyond the reactuation point, the switch does not automatically reactuate, but it may be manually reactuated.

Maximum (Minimum) Ambient Temperature The value of the highest (lowest) ambient temperature that a transducer can be exposed to, with or without excitation applied, without being damaged or subsequently showing a performance degradation beyond specified tolerances.*

Maximum Excitation (S) The maximum value of excitation voltage or current that can be applied to the transducer at room conditions without causing damage or performance degradation beyond specified tolerances.*

Maximum (Minimum) Fluid Temperature (S) The value of the highest (lowest) measured-fluid temperature that a transducer can be exposed to, with or without excitation applied, without being damaged or subsequently showing a performance degradation beyond specified tolerances.* *Note:* (S) When a maximum or minimum fluid temperature is not separately specified, it is intended to be the same as any specified maximum or minimum ambient temperature.

Maximum Sustained Pressure (Proof Pressure) Maximum sustained pressure is the highest pressure including surges to which a pressure switch may be subjected without damage.

Mean Output Curve The curve through the mean values of output during any one calibration cycle or a different specified number of calibration cycles.*

Measurand A physical quantity, property, or condition that is measured.* *Note:* The term measurand is preferred to "input," "parameter to be measured," "physical phenomenon," "stimulus," or "variable."

Measured Fluid The fluid that comes in contact with the sensing element.* *Note:* The chemical and/or physical properties of this fluid may be specified to ensure proper transducer operation.

Modulus of Rigidity A coefficient that expresses the stiffness of a material mathematically, in thousands of pounds per square inch. It is independent of hardness.

Momentary Switch A switch with contacts that return from operated condition to normal condition when actuating force is removed. Unless otherwise stated, all switches in this handbook are momentary.**

Mounting Dimensions All dimensions on the mounting dimension drawings furnished by vendor.**

Mounting Error The error resulting from mechanical deformation of the transducer caused by mounting the transducer and making all measurand and electrical connections.*

Natural Frequency See **Frequency, Natural**. See also **Frequency, Resonant**.*

Noncritical Set Point The noncritical set point is the least important setting, and the tolerances are not as close as for the critical set point. It can be either the actuation or the deactuation point.

Nonlinearity The difference between the actual deflection curve of a unit and a straight line drawn between the upper and lower range terminal values of the deflection, expressed as a percentage of full range deflection.

Nonoperating Conditions The external condition (shock, vibration, temperature, etc.) to which a transducer or pressure switch may be exposed during shipping, storage or handling.

Nonrepeatability See **Repeatability**.*

Normally Closed Switch A switch in which the contacts are normally closed. A pressure change opens the contacts.

Normally Closed Switching Element Current flows through the switching element until the switch is actuated. In a normally closed switch a plunger pin is held down by a snap action leaf spring, and force must be applied to the plunger pin to open the circuit.

Normally Open Switch A switch in which the contacts are normally open. A pressure change closes the contacts.

Normally Open Switching Element No current can flow through the switching element until the switch is actuated. In a normally open switch a plunger pin is held down by a snap action leaf spring, and force must be applied to the plunger pin to close the circuit.

Nuclear Radiation The emission of charged and uncharged particles and of electromagnetic radiation from atomic nuclei.*

Null A condition, such as of balance, that results in a minimum absolute value of output.*

Null Switch A floating contact switch with a zone of no electrical current flow for operation of reversible motors.

Open Ends Ground As illustrated on compression spring specification chart. Also called **plain ends ground**.

Open Ends Not Ground As illustrated on compression spring specification chart. Also called **plain ends**.

Operating Conditions See **Environmental Conditions.***

Operating Differential The difference between the actuate pressure and the deactuate pressure. The operating differential is expressed in psi but is a function of the calibration range of the switch. For example, a switch with a calibration range of 0–15 psi would have a smaller operating differential than one with a 0–100 psi calibration range.

Output The electrical quantity, produced by a transducer, that is a function of the applied measurand.*

Output Impedance The impedance across the output terminals of a transducer presented by the transducer to the associated external circuitry.*

Output Noise The rms, peak, or peak-to-peak (as specified) ac component of a transducer's dc output in the absence of measurand variations.* *Note:* (S) Unless otherwise specified, output impedance is measured at room conditions and with the excitation terminals open/circuited, except that nominal excitation and measurand between 80 and 100 percent-of-span is applied when the transducer contains integral active output-conditioning circuitry.

Output Regulation The change in output due to a change in excitation.* *Note:* (S) Unless otherwise specified, output regulation is measured at room conditions and with the measurand applied at its upper range limit.

Overload The maximum magnitude of the measurand that can be applied to a transducer without causing a change in performance beyond specified tolerance.*

Overrange See **Overload.***

Overshoot The amount of output measured beyond the final steady output value, in response to a step change in the measurand.* *Note:* (S) Expressed in percent of the equivalent step change in output.

Parameter (To be measured.) See **Measurand.***

Peak-to-Peak See **Double Amplitude.***

Photoconductive Converting a change of measurand into a change in resistance or conductivity of a semiconductor material by a change in the amount of illumination incident upon the material.*

Photovoltaic Converting a change of measurand into a change in the voltage generated when a junction between certain dissimilar materials is illuminated.*

Physical Input See **Measurand.***

Pickup See **Transducer.***

Piezoelectric Converting a change of measurand into a change in the electrostatic charge or voltage generated by certain materials when mechanically stressed.*

Pitch The distance from center to center of adjacent coils in a spring when it is free of load. Also called **lead** and sometimes expressed as **coils per inch**.

Plain Ends See **Open Ends Not Ground**.

Plain Ends Ground See **Open Ends Not Ground**.

Potentiometeric Converting a change of measurand into a voltage-ratio change by a change in the postition of a movable contact on a resistance element across which excitation is applied.*

Power Input See **Excitation.***

Precision See **Repeatability and Stability.***

Precision Snap-Acting Switch An electromechanical switch having predetermined and accurately controlled characteristics, and having a spring-loaded quick make and break contact action.**

Press See **Remove Set**.

Pressure Pressure is defined as the force exerted over a surface divided by the area of that surface.

Pressure, Absolute The difference between zero pressure (a perfect vacuum) and some known pressure. It may be arrived at by adding barometric pressure to gage pressure.

Pressure, Ambient The pressure (usually but not necessarily atmospheric) surrounding a pressure switch.

Pressure, Atmospheric The actual weight of the earth's atmosphere at a given locale and altitude. Atmospheric pressure at sea level is approximately 14.7 psi or 30 inches of mercury or 408 inches of water.

Pressure, Barometric Actual weight of the earth's atmosphere at a given locale and altitude as measured by a barometer, usually expressed in measurement of a column of mercury. Standard barometer reading at sea level is 29.92 inches of mercury at 32°F.

Pressure, Differential The difference between a reference pressure and a variable pressure.

Pressure, Gage Gage pressure uses atmospheric pressure as a reference, and therefore will vary according to the barometer reading.

Pressure, Proof Proof pressure (normally 1½ times system pressure) is the maximum pressure that may be applied to any switch without causing permanent degradation.

Pressure, Ratio The ratio between two pressures. Actuation point of a ratio switch is based on the ratio of two pressures.

Pressure, Reference A specified pressure on which the actuation point is based; may be absolute, or controlled pressure.

Pressure-Sensing Element The pressure-sensing element is the component part that moves as pressure increases or decreases and, in the process, actuates or deactuates an electrical switch at a predetermined point. Most common types of sensing elements are the bourdon type, piston, diaphragm, and bellows. CCS Dual-Snap switches employ a unique Belleville negative-rate spring device to establish set points.

Pressure Sensitivity The smallest pressure change to which the unit will exhibit a measurable response, expressed in percent of rated pressure range.

Pressure, Static The pressure maintained by a head of stationary fluid.

Pressure, Surge A transient pressure varying in amplitude, frequency, and duration. It is difficult to measure without complicated electronic instrumentation. If the duration and/or amplitude of the pressure surge is large, a pressure-sensing element rated for the maximum surge pressure should be used. There are surge-dampening devices and techniques that can reduce the amplitude of pressure surges.

Pressure Switch An instrument that upon the increase or decrease of a pressure or vacuum, opens or closes one or more electrical switching elements at a predetermined actuation point (setting).

Pressure, System The nominal pressure level at which a system will operate including work load.

Pressure, Variable A changing pressure, normally the one that actuates the pressure switch.

Pressure, Working The pressure of the system in which the pressure switch is used.

Primary Element, Primary Detector See **Sensing Element.***

psia Absolute pressure in lb/sq in.

psid Differential pressure in lb/sq in.

psig Gage pressure in lb/sq in.

psi Vac Gage Vacuum gage pressure in lb/sq in. as measured from 0 psig.

Pulse Switch Provides a single pulse of current for each cycle of operation.**

Quick Acting See **Snap Action**.

Range The span of differential pressures within which the pressure-sensing element can be set to actuate an electrical switching element.

Range of Stress The difference between operating stress at maximum load and stress at minimum load.

Rated Pressure The total pressure that the pressure-sensing element of the differential pressure switch can withstand without damage (burst pressure less a reasonable safety factor).

Rate per Inch The load in pounds required to

deflect a spring a distance of one inch. Also called **scale, gradient,** and **load factor**.

Reactuation Point (Reset Point, Deactuation Point) After the presssure in a pressure switch has reached the actuation point and operated the electrical switch, it must return to a point called the reactuation point before the electrical switch can return to its original position. This reactuation point may be an adjustable point, depending upon the design of the pressure switch.

Recovery Time The time interval, after a specified event (e.g., overload, excitation transients, output shortcircuiting), after which a transducer again performs within its specified tolerances.*

Reference Pressure The pressure to which the measured or controlled pressure is being compared. In a gage or vacuum switch the reference pressure is atmospheric pressure, in a differential unit it is the pressure plumbed to the "lo" side, and in an absolute switch it is a near-perfect vacuum.

Reference Pressure Error The error resulting from variations of a differential-pressure transducer's reference pressure within the applicable reference pressure range.* *Note:* (S) It is usually specified as the maximum change in output, at any measurand value within the specified range, when the reference pressure is changed from ambient pressure to the upper limit of the specified reference pressure range.

Reference Pressure Range (S) The range of reference pressures that can be applied without changing the differential-pressure transducer's performance beyond specified tolerances for reference pressure error. When no such error is specified, none is allowed.*

Reference-Pressure Sensitivity Switch The sensitivity shift resulting from variations of a differential-pressure transducer's reference pressure within specified limits.*

Reference-Pressure Zero Shift The change in the zero-measurand output of a differential-pressure transducer resulting from variations of reference pressure (applied simultaneously to both pressure ports) within its specific limits.*

Reluctive Converting a change of measurand into an ac voltage change by a change in the reluctance path between two or more coils or separated portions of one coil when ac excitation is applied to the coil(s).* *Note:* Included among reluctive transducers are those employing differential-transformer, inductance-bridge, and synchro elements.

Remove Initial Set See **Remove Set**. The word "initial" is added because ordinary springs may take some further set if held under prolonged loading at fairly high stress, after the first removal of set.

Remove Set The process of closing solid a compression spring that has been coiled longer than the desired finished length, in order to increase the elastic limit. Also called **remove initial set, press,** and **set**.

Repeatability (Accuracy) Repeatability is the maximum allowable set-point deviation of a single pressure switch under one given set of environmental and operational conditions.

Repetitive Accuracy The ability of a pressure switch to operate repetitively at its set point under consistent conditions.

Reproducibility See **Repeatability**.*

Resistive Converting a change of measurand into a change of resistance.*

Resolution The magnitude of output step changes as the measurand is continuously varied over the range.* *Note 1:* This term is related primarily to potentiometric transducers. *Note 2:* (S) Resolution is best specified as average and maximum resolution; it is usually expressed in percent of full scale output. *Note 3:* It is used in the sense of the smallest detectable change in measurand use threshold.

Resolution, Average (S) The reciprocal of the total number of output steps over the range, multiplied by 100 and expressed in percent voltage ratio (for a potentiometric transducer) or in percent of full scale output.*

Resolution, Maximum (S) The magnitude of the largest of all output steps over the range, expressed as percent voltage ratio (for a potentiometric transducer) or in percent of full scale output.*

Resonances Amplified vibrations of transducer components, within narrow frequency bands, observable in the output, as vibration is applied along specified transducer axes.*

Resonant Frequency See **Frequency, Resonant**.*

Response Time The length of time required for the output of a transducer to rise to a specified percentage of its final value as a result of a step change of measurand.* *Note 1:* (S) To indicate this percentage it can be worded so as to precede the main term, e.g., "98% response time: ____ milliseconds, max." *Note 2:* Also see **Time Constant** and **Rise Time**.

Ringing Period The period of time during which the amplitude of output oscillations, excited by a step change in the measurand, exceeds the steady-state output value.* *Note:* (S) Unless otherwise specified, the ringing period is considered terminated when the output oscillations no longer exceed 10% of the subsequent steady state output value.

Rise Time The length of time for the output of a transducer to rise from a small specified percentage of its final value to a large specified percentage of its final value as a result of a step change of the measurand.* *Note 1:* (S) Unless otherwise specified, these percentages are assumed to be 10 and 90% of the final value. *Note 2:* Also see **Time Constant**.

Rolling Diaphragm The rolling diaphragm is a long-stroke, deep-convolution, constant-area diaphragm that is free-positioning with complete relaxation at any point in its stroke.

Room Conditions Ambient environmental conditions, under which transducers must commonly operate, which have been established as follows: (a) temperature: 25 ± 10 °C (77 ± 18 °F); (b) (relative humidity: 90% or less; (c) barometric pressure: 26 to 32 inches of mercury. *Note:* Tolerances closer than these are frequently specified for transducer calibration and test environments.*

Scale See **Rate per Inch**.

Self-Generating Providing an output signal without applied excitation. Examples are piezoelectric, electromagnetic, and thermoelectric transducers.*

Self-Heating Internal heating resulting from electrical energy dissipated within the transducer.*

Semiconductor Materials, used for sensing elements or transduction elements, whose resistivity falls between that of conductors and insulators (e.g., germanium, silicon, etc.). Examples of useful phenomena associated with these materials are: Hall effect, temperature coefficient of resistance, photoresistivity, photovoltaic effect, piezoresistance, etc.*

Sensing Element That part of the transducer which responds directly to the measurand.* *Note:* This term is preferred to "primary element," "primary detector," or "primary detecting element."

Sensitivity The ratio of the change in transducer output to a change in the value of the measurand.* *Note:* In the sense of the smallest detectable change in measurand use threshold.

Sensitivity Shift A change in the slope of the calibration curve due to a change in sensitivity.*

Sensor See **Transducer**.*

Servo A shortening of servomechanism. A transducer type in which the output of the transduction element is amplified and fed back so as to balance the forces applied to the sensing element or its displacements. The output is a function of the feedback signal.*

Set See **Remove Set**.

Set Point (Actuation Point) The exact pressure at which the switch will actuate an electrical circuit.

Single Pole Double Throw (SPDT) Switching Element A SPDT switching element has one normally open, one normally closed, and one common terminal. Three terminals mean that the switch can be wired with the circuit either normally open (N/O) or normally closed (N/C).

Snap Action The action of a pair of contacts that open and close quickly enough to immediately extinguish any arc that may form, and which snap closed with sufficient pressure to firmly establish an electrical circuit.

Solid Height Length of a compression spring when under sufficient load to bring all coils into contact with adjacent coils. Also called **solid length** or **closed height**.

Solid Length See **Solid Height**.

Sound Pressure The total instantaneous pressure at a given point in the presence of a sound wave, minus the static pressure of that point.*

Source Impedance The impedance of the excitation supply presented to the excitation terminals of the transducer.*

Span The algebraic difference between the limits of the range.*

Specific Pressure Any stated pressure with limits established by specification.

Speed of Response See **Response Time** and **Time Constant**.*

Spring Index Ratio of mean diameter to wire diameter, or D/d.

Spring Rate (Force Constant) The energy per unit deflection of a diaphragm, capsule, or element available to actuate a device. A perfectly linear unit will have a constant spring rate, expressed in lb/in.

Squared and Ground Ends See **Closed Ends Ground**.

Stability The ability of a transducer to retain its performance characteristics for a relatively long period of time.* *Note:* (S) Unless otherwise stated, stability is the ability of a transducer to reproduce output readings obtained during its original calibration, at room conditions, for a specified period of time; it is then typically expressed as "within ____ percent of full scale output for a period of ____ months."

Static Calibration A calibration performed under room conditions and in the absence of any vibration, shock, or acceleration (unless one of these is the measurand).*

Static Pressure That pressure produced by a stationary fluid.

Stimulus See **Measurand.***

Strain The deformation per unit length produced in a solid as a result of stress.*

Strain Error The error resulting from a strain imposed on a surface to which the transducer is mounted.* *Note 1:* This term is not intended to relate to strain transducers (strain gages). *Note 2:* Also see **Mounting Error**.

Strain-Gage To convert a change of measurand into a change in the emf generated by a temperature difference between the junctions of two selected dissimilar materials.*

Strain Relieve To subject springs to a low temperature heat treatment after coiling to remove coiling strains. Also called **stress relieve** and **blue**.

Stress When a spring at rest supports a load, there is an internal force resisting further deflection under the load. To maintain static equilibrium under load, this force must be equal to, and opposite in direction to, the external load. This internal force is stress, and is expressed in thousands of pounds per square inch of sectional area.

Stress Relieve See **Strain Relieve**.

Switch Assembly Unit ASCO uses this term to describe that portion of a pressure switch which incorporates the adjusting mechanism that reacts to a pressure transducer or sensor to cause the

Tapping See **Dithering.***

Temper To heat a hardened steel to some temperature that is below the critical one, and will produce the hardness desired. Also called **drawing**.

Temperature Error The maximum change in output, at any measurand value within the specified range, when the transducer temperature is changed from room temperature to specified temperature extremes.*

Temperature Error Band The error band applicable over stated environmental temperature limits.*

Temperature Gradient Error The transient deviation in output of a transducer at a given measurand value when the ambient temperature or the measured fluid temperature changes at a specified rate between specified magnitudes.*

Temperature Range, Compensated See **Temperature Range, Operating.***

Temperature Range, Fluid The range of temperature of the measured fluid, when it is not the ambient fluid, within which operation of the transducer is intended.* *Note 1:* (S) Within this range of fluid temperature all tolerances specified for temperature error, temperature error band, temperature gradient error, thermal zero shift, and thermal sensitivity shift are applicable. *Note 2:* (S) When a fluid temperature range is not separately specified, it is intended to be the same as the operating temperature range.

Temperature Range, Operating The range of ambient temperatures, given by their extremes, within which the transducer is intended to operate; (S) within this range of ambient temperature all tolerances specified for temperature error, temperature error band, temperature gradient error, thermal zero shift, and thermal sensitivity shift are applicable.*

Temperature Switch An instrument designed to convert temperature into motion in order to actuate an electrical switch—thereby making or breaking an electrical circuit.

Terminal Enclosure A housing that fits over switch terminals to protect against electrical shock and accidental shorting, and facilitate wiring.**

Terminal Line A theoretical slope for which the theoretical end points are 0 and 100% of both measurand and output.*

Theoretical Curve The specified relationship (table, graph, or equation) of the transducer output to the applied measurand over the range.*

Theoretical End Points The specified points

between which the theoretical curve is established and to which no end point tolerances apply.* *Note:* The points can be other than 0 and 100% of both measurand and output.

Theoretical Slope The straight line between the theoretical end points.*

Thermal Coefficient of Resistance The relative change in resistance of a conductor or semiconductor per unit change in temperature over a stated range or temperature.* *Note:* (S) Expressed in ohms per ohm per degree F or C.

Thermal Compensation See **Compensation.***

Thermal Sensitivity Shift (S) The sensitivity shift due to changes of the ambient temperature from room temperature to the specified limits of the operating temperature range.*

Thermal Zero Shift (S) The zero shift due to changes of the ambient temperature from room temperature to the specified limits of the operating temperature range.*

Thermoelectric Converting a change of measurand into a change in the emf generated by a temperature difference between the junctions of two selected dissimilar materials.*

Threshold The smallest change in the measurand that will result in a measurable change in transducer output.* *Note:* When the threshold is influenced by the measurand values, these values must be specified.

Time Constant The length of time required for the output of a transducer to rise to 63% of its final value as a result of a step change of measurand.*

Tolerance Tolerance is the normal variation in production pressure switches of the same specifications. It affects the actuation value and the reactuation point and not the accuracy of the set point. For example: Three switches are set to actuate at 100 psi. They may all respond to this pressure with an accuracy of plus or minus 1/2%. However, one switch may reactuate at 94 psi, another at 95 psi, and the third at 96 psi.

Torque Error See **Mounting Error.***

Total Error Band See **Error Band.***

Transducer A device that provides a usable output in response to a specified measurand.* *Note:* The term transducer is usually preferred to "sensor" and "detector" and to such terms as "flowmeter," "accelerometer," and tachometer"; it is always preferred to "pickup," "gage" (when not equipped with a dial-indicator), "transmitter" (which has an entirely different meaning in telemetry technology), "cell," and "end instrument."

Transducer Assembly Unit (Sensor) ASCO uses the term transducer assembly unit to describe that portion of a pressure switch to which a pressure is connected and thereby converted to another form of energy, which, in this case, is movement and force.

Transduction Element The electrical portion of a transducer in which the output originates.*

Transient Response The response of a transducer to a step change in measurand.* *Note:* (S) Transient response, as such, is not shown in a specification except as a general heading, but is defined by such characteristics as time constant, response time, ringing period, etc.

Transverse Acceleration An acceleration perpendicular to the sensitive axis of the transducer.*

Transverse Response See **Transverse Sensitivity.***

Transverse Sensitivity The sensitivity of a transducer to transverse acceleration or other transverse measurand.* *Note:* (S) It is specified as maximum transverse sensitivity when a specified value of measurand is applied along the transverse plane in any direction, and is usually expressed as percent of the sensitivty of the transducer in its sensitive axis.

Turbine A bladed rotor that turns at a speed nominally proportional to the volume rate of flow.*

Two-Circuit Switch In one position, moving contacts complete one circuit; in the other position, contacts complete another separate circuit.**

Two-Stage (Dual) ASCO uses the term two-stage to describe a pressure switch that encompasses two electrical switches, each independently adjustable with its own adjusting nut. This arrangement is equivalent to two fixed differential pressure switches.

Ultrasonic Using frequencies above the audio-frequency range, i.e., above 20 kHz.*

Unbonded Stretched and unsupported between ends (usually refers to strain-sensitive wire).*

Variable See **Measurand.***

Variable Pressure A fluctuating pressure having characteristics of sufficient magnitude to operate a pressure-actuated switch. This pressure is usually the one that is sensed by a pressure switch.

Vibration Error The maximum change in output, at any measurand value within the specified range, when vibration levels of specified amplitude and range of frequencies are applied to the transducer along specified axes.*

Vibration Sensitivity See **Vibration Error**.*

Voltage Ratio For potentiometric transducers, the ratio of output voltage to excitation voltage, usually expressed in percent.*

Warm-up Period The period of time, starting with the application of excitation to the transducer, required to assure that the transducer will perform within all specified tolerances.*

Wind To form wire into a helix by wrapping it around an arbor.

Working Range Working range is the pressure range a switch may see under normal working conditions. This is normally the adjustable range.

Zero-Measurand Output The output of a transducer, under room conditions unless otherwise specified, with nominal excitation and zero measurand applied.*

Zero Shift A change in the zero-measurand output over a specified period of time and at room conditions.* *Note:* This error is characterized by a parallel displacement of the entire calibration curve.

REFERENCES

*These are definitions obtained from the Instrument Society of America Standard, 400 Stanwix Street, Pittsburgh, Pennsylvania 15222. (ISA S37.1, 1969).

**These are definitions obtained from Micro Switch Company, Freeport, Illinois 61032, A Division of Honeywell.

Appendix B Engineering Standards and Specifications

1. REFERENCE STANDARDS (UL, CSA, NEMA)
2. LIST OF MILITARY SPECIFICATIONS
3. NFPA (PRESSURE SWITCH PRESSURE RATING)

1. REFERENCE STANDARDS (UL, CSA, NEMA)*

UL (Underwriter's Laboratories, Inc.) is chartered as a nonprofit organization without capital stock under the laws of the state of Delaware, to establish, maintain, and operate laboratories for the examination and testing of devices, systems, and materials. Founded in 1894, the enterprise is operated for service, not for profit. CSA (Canadian Standards Association) is a nonprofit, nongovernmental association established as a national standardization body for Canada. Included in its scope of activities is the responsibility of investigating and approving products and materials in the interest of safety. NEMA (National Electrical Manufacturer*s Association) prepares standards that define a product, process, or procedure with reference to one or more of the following: nomenclature, composition, construction, dimensions, tolerances, safety, operating characteristics, performance, quality, electrical rating, testing, and the service for which designed. The reference standards herein reflect the latest data in the NEMA Standards Publication on Enclosures for Industrial Controls and Systems. (CAUTION: Enclosures are based, in general on the broad definitions outlined in NEMA Standards. Therefore, it will be necessary to ascertain that a particular enclosure will adequately meet any unusual conditions that might exist in intended applications.)

*Courtesy of Micro Switch.

Type 1—General Purpose—Indoor. Type 1 enclosures are intended for use indoors, primarily to prevent accidental contact of personnel with the enclosed equipment, in areas where unusual service conditions do not exist.

Type 3—Dusttight, Raintight, and Sleet (Ice)-Resistant—Outdoor. Type 3 enclosures are intended for use outdoors to protect the enclosed equipment against wind-blown dust and water. They are not sleet (ice)-proof.

Type 3R—Rain-proof and Sleet (Ice)-Resistant —Outdoor. Type 3R enclosures are intended for use outdoors to protect the enclosed equipment against rain. They are not dust-, snow-, or sleet (ice)-proof.

Type 3S—Dusttight, Raintight, and Sleet (Ice)-Proof—Outdoor. Type 3S enclosures are intended for use outdoors to protect the enclosed equipment against wind-blown dust and water and to provide for its operation when the enclosure is covered by external ice or sleet. These enclosures do not protect the enclosed equipment against malfunction resulting from internal icing.

Type 4 and 4X—Watertight and Dusttight—Indoors. Type 4 enclosures are intended for use indoors to protect the enclosed equipment against splashing water, seepage of water, falling or hose-directed water, and severe external condensation. Type 4X is the same as Type 4 except 4X includes corrosion resistance.

Type 7—Hazardous Locations—Class 1. These enclosures meet the application requirements of the National Electrical Code for Class 1 hazardous locations, Groups B, C, and D. Group B atmospheres contain butadiene, ethylene, oxide, hydrogen (or gases or vapors equivalent in hazard to hydrogen, such as manufactured gas), or propylene oxide. Group C atmospheres contain acetaldehyde, cyclopropane, diethyl ether, ethylene, isoprene, or unsymmetrical dimethyl hydrazine (UDMH). Group D atmospheres contain acetone, acrylonitrile, alcohol, ammonia, benzine, benzol, butane, ethylene dichloride, gasoline, hexane, lacquer solvent vapors, naphtha, natural gas, propane, propylene, styrene, vinyl acetate, vinyl chloride, or xylenes.

Type 9—Hazardous Locations—Class II. These enclosures are designed to meet the application requirements of the National Electrical Code for Class II hazardous locations, Groups E, F, and G. Group E atmospheres contain metal dust, including aluminum, magnesium, and their commercial alloys and other metals having similar hazardous characteristics. Group F atmospheres contain carbon black, coal, or coke dust. Group G atmospheres contain flour, starch, or grain dust.

Type 13—Oiltight and Dusttight—Indoor. Type 13 enclosures are intended for use indoors primarily to house pilot devices such as limit switches, foot switches, pushbuttons, selector switches, pilot lights, etc., and to protect these devices against lint and dust, seepage, external condensation, and spraying of water, oil, or coolant.

2. LIST OF MILITARY SPECIFICATIONS

NUMBER	TITLE OF SPECIFICATION/STANDARD
MIL-S-9395/48	Switch, Pressure, Absolute Type 1, 10 Amperes
MIL-S-8933	Switch, Pressure, Aircraft, Absolute, Types 1 and 2, Class 1
MIL-S-8936	Switch, Pressure, Aircraft, Approximate Gage, Types 1 and 2, Class 4
MIL-S-8934	Switch, Pressure, Aircraft, Differential, Types 1 and 2, Class 2
MIL-S-8935	Switch, Pressure, Aircraft, Gage, Types 1 and 2, Class 3
MIL-S-8932	Switch, Pressure, Aircraft, General Specification for
MS-25276	Switch, Pressure, Bulkhead Mounted (Type 2)
MIL-S-9395/4B	Switch, Pressure, Differential
MIL-S-9395/7A	Switch, Pressure, Differential (Type 1ii, Class 2) Spst, 4 Amperes
MIL-S-9395/32A	Switch, Pressure, Differential (Type 1ii), 4 Amperes
MIL-S-9395/28A	Switch, Pressure, Differential (Type 1ii), 5 Amperes
MS-25275	Switch, Pressure, Engine Mounted (Type 1)
MIL-S-45869	Switch, Pressure, for T4016 (T310) Test Set
MIL-S-9395/9A	Switch, Pressure, Gage (Type 1i, Class 1) Dpdt, 10 Amperes
MIL-S-9395/20B	Switch, Pressure, Gage (Type 1i, Class 1) Dpdt, 8 Amperes
MIL-S-9395/8B	Switch, Pressure, Gage (Type 1i, Class 1) Spdt, 10 Amperes
MIL-S-9395/25A	Switch, Pressure, Gage (Type 1i, Class 1) Spdt, 3 Amperes
MIL-S-9395/12A	Switch, Pressure, Gage (Type 1i, Class 1) Spdt, 3 Amperes
MIL-S-9395/2C	Switch, Pressure, Gage (Type 1i, Class 1) Spdt, 4 Amperes
MIL-S-9395/19A	Switch, Pressure, Gage (Type 1i, Class 1) Spdt, 5 Amperes
MIL-S-9395/22A	Switch, Pressure, Gage (Type 1i, Class 1) Spst, 3 Amperes
MIL-S-9395/13A	Switch, Pressure, Gage (Type 1i, Class 1) Spst, 3 Amperes
MIL-S-9395/27C	Switch, Pressure, Gage (Type 1i), 3 Amperes
MIL-S-9395/31(1)	Switch, Pressure, Gage (Type 1i), 5 Amperes
MIL-S-60999	Switch, Opener, Cockpit Setter (Tfd)
MIL-S12211C(2)	Switch, Pressure
MIL-S-9395/3B	Switch, Pressure
MIL-S-9395/1B	Switch, Pressure, Gage (Type 1i) Spdt, 3 Amperes
MIL-S-9395/5B	Switch, Pressure, Gage (Type 1i) Spst, 5 Amperes
MIL-S-9395/10C	Switch, Pressure (Type 1i)
MS-27152B	Switch, Pressure-warning, Low Air Pressure, 60 psi, Spst, Waterproof
MS-75062A	Switch, Pressure-spotlight, Vehicular, Air Brake System, 24 Volt, Waterproof
MS-75063B	Switch, Pressure-stoplight, Vehicular, Hydraulic Brake System, 24 Volt, Waterproof
MS-90530D	Switch, Pressure-warning, Spst, Waterproof, 24 Volt, dc
MIL-S-9395/33	Switch, Pressure, Differential (Type 1ii), 5 Amperes
MIL-S-9395/40	Switch, Pressure, Gage (Type 1i), Low Level to 5 Amperes
MIL-S-9395/30C(1)	Switch, Pressure, Gage (Type 1i), 5 Amperes
MIL-S-9395E, SUPP 1	Switch, Pressure, Gage (Type 1i), 5 Amperes
MIL-S-9395/41A	Switch, Pressure, Gage (Type 1i), Low Level to 1 Ampere
MIL-S-9395/39A	Switch, Pressure, Gage (Type 1i), Low Level to 5 Amperes
MIL-S-9395/38A	Switch, Pressure, Gage (Type 1i), Low Level to 5 Amperes
MIL-S-9395/15A	Switch, Pressure, Absolute (Type 1, Class 1) Spdt, 3 Amperes
MIL-S-9395/18A	Switch, Pressure, Absolute (Type 1, Class 1) Spdt, 5 Amperes
MIL-S-9395/16A	Switch, Pressure, Absolute (Type 1, Class 1) Spdt, 6 Amperes
MIL-S-9395/21A	Switch, Pressure, Absolute (Type 1, Class 1) Spdt, 7.5 Amperes
MIL-S-9395/26B	Switch, Pressure, Absolute (Type 1), 3 Amperes
MIL-S-9395/23A	Switch, Pressure, Absolute (Type 1, Class 1) Spdt, 6 Amperes

3. PRESSURE SWITCH PRESSURE RATING SUPPLEMENT NO. 3 TO NFPA RECOMMENDED STANDARD FOR VERIFYING THE FATIGUE AND STATIC PRESSURE RATINGS OF THE PRESSURE CONTAINING ENVELOPE OF A METAL FLUID POWER COMPONENT

Foreword

(This foreword is not part of Pressure Switch Pressure Rating Supplement No. 3 to NFPA Recommended Standard for Verifying the Fatigue and Static Pressure Ratings of the Pressure Containing Envelope of a Metal Fluid Power Component, NFPA/T2.6.1 S3 [T3.29.2]-1976.)

Early in 1974, the approval of NFPA/T2.6.1-1974 established a group of common requirements intended to provide an industry-wide philosophy and basic standard, providing a rationale for judging a component's ability as a pressure containing envelope. Although the specific applicability of NFPA/T2.6.1-1974 is limited, it immediately established a uniform base for subsequent, more specific proposed NFPA recommended standards for individual components.

The Pressure Switch Pressure Rating Project Group held its first meeting on 22 May 1973. The TSP was approved at the Technical Board meeting on 4 October 1973. On 20 November 1973 the Project Group met to discuss the development of the first draft, which was completed on 3 January 1974. Draft No. 1 was circulated among the Project Group members on 9 January 1974. Their comments were forwarded to Chairman Blickley, who prepared the Final Working Draft and submitted it to NFPA Headquarters on 6 September 1974. Headquarters' Technical Staff prepared the General Review Draft on 16 December 1974.

Comments resulting from General Review were resolved on 22 July 1975, and the document received approval to ballot from the Technical Board on 21 August 1975. The Ballot Draft was prepared on 28 August 1975. The one negative ballot received on this standard was resolved on 30 January 1976. One affirmative comment offering some suggested changes was also received; it was discussed at the Pressure Switch Section meeting on 6 November 1975, but none of the suggestions was incorporated in the document.

On 4 February 1976, the Technical Board unanimously voted to recommend that this

*Courtesy of the National Fluid Power Association.

document be approved as an NFPA Recommended Standard. The NFPA Board of Directors granted approval to NFPA/T2.6.1 S3 (T3.29.2)-1976 on 25 February 1976.

Contents

Pressure Switch Pressure Rating Supplement No. 3 to NFPA Recommended Standard for Verifying the Fatigue and Static Pressure Ratings of the Pressure Containing Envelope of a Metal Fluid Power Component

HEADQUARTERS NOTE—The Project Group which developed this Supplement intended it for use only with the basic pressure rating document, NFPA/T2.6.1-1974. This Supplement provides a list of additions, deletions and changes to the basic document which are necessary for the establishment of pressure ratings for metal pressure switches. All clauses which have been modified are so indicated in this Supplement. Read all other clauses as they appear in the basic pressure rating document, NFPA/T2.6.1-1974.

Since revisions of the basic pressure rating document will apply to this Supplement, the reader is cautioned to always use the most recent edition of the basic document. When reading this Supplement with the basic document, *substitute the phrase "pressure switch(es)" whenever the term "component(s)" appears in the basic document.*

INTRODUCTION

CHANGE the first three paragraphs to read:

In fluid power systems, power is transmitted and controlled through a fluid (liquid or gas) under pressure within an enclosed circuit.

A pressure switch is a device that measures positive or negative pressures (with respect to atmosphere) or a differential pressure with one or more pressure sensing elements which produce a motion and/or force which is utilized to operate a switching element. The switching element is generally an electrical switch or contact(s), but may also be a pneumatic or hydraulic element such as a valve.

The functions of a pressure switch in a fluid power system are many and varied, but they generally fall into one or more of the following three categories:

1. Alarm. In this case the pressure switch initiates a visual or audible alarm to

inform an operator of a certain condition of pressure.

2. Control. In this case the pressure switch starts or stops a pump, or opens or closes a valve.
3. Safety. In this case the pressure switch initiates an action that protects the system from a failure mode of pressure.

A basic requirement of fluid power pressure switches is that they should be capable of completely containing the pressurized fluid.

The basic pressure rating document, NFPA/T2.6.1-1974 established a group of common requirements intended to provide an industry-wide philosophy and basic standard, providing a rationale for judging a component's ability as a pressure containing envelope. Although the specific applicability of NFPA/T2.6.1-1974 is limited, it immediately established a uniform base for subsequent, more specific NFPA Proposed Recommended Standards for individual fluid power components. This Recommended Standard implements NFPA/T2.6.1-1974 and specifically applies to pressure switches.

Ratings verified in accordance with this standard do not replace pressure ratings based on considerations such as performance, overrange, leakage and heat rejection. Instead, these new ratings are intended to supplement those that are provided by present practice and may be numerically different from prior ratings.

No attempt is made in this standard to deal with the electrical, pneumatic or hydraulic switching elements of pressure switches. The ratings for pneumatic and hydraulic switching elements (such as valves) are dealt with in other NFPA documents. The ratings for electrical switching elements (such as snap-action switches) are covered thoroughly by other industry standards.

The system designer should be aware that the cyclic tests in this standard do not include the switching element (either under load or with no load) and the life of the switching element may be quite different from the test cycles performed under this standard.

1. SCOPE

   ADD the following clause:

   1.1.5

   Made of the above listed materials (clause 1.1.4) but may contain seals or interfaces of elastomeric, organic or other nonmetallic elements. Such elements may be part of the pressure containing envelope, but are not verified by these procedures.

   CHANGE clause 1.2 to read:

   1.2

   These verification procedures do not include the functioning of a pressure switch in its ability to maintain a set point when either RFP or RSP is applied. It is expected that a future standard will cover the methods for determining Rated Overrange Pressure or that static pressure above which there can be expected a shift in, or malfunction of, the set point.

2. PURPOSE

   NO CHANGE from the basic document.

3. TERMS AND DEFINITIONS

   CHANGE clause 3.2 to read:

   3.2

   Rated Static Pressure. A pressure that a component pressure containing envelope is represented to sustain for only one cycle without failure.

   CHANGE clause 3.9 to read:

   3.9

   Pressure Containing Envelope Failure. A structural fracture, crack or any seal leakage caused by deformation or permanent deformation resulting from an RFP verification test; or a structural fracture, crack or any seal leakage caused by deformation or permanent deformation which interfered with the functioning of the pressure containing envelope of the pressure switch resulting from an RSP verification test.

   CHANGE clause 3.10 to read:

   3.10

   Rated Overrange Pressure. The pressure to which a pressure switch can be subjected for extended time without change in operating characteristics, shift in set point, or damage to the device.

4. UNITS OF MEASUREMENT
NO CHANGE from the basic document.
5. LETTER SYMBOLS
NO CHANGE from the basic document.
6. OUTLINE OF PROCEDURES FOR RFP VERIFICATION BY TEST
NO CHANGE from the basic document.
7. RFP VERIFICATION BY SIMILARITY
CHANGE the entire Section to read:
Verification of the rated static pressure ratings of the pressure containing envelope of a pressure switch by similarity is allowed when the similar pressure switch differs from the tested pressure switch only in the size of the external port openings, provided those ports are of the same type and smaller than in the tested pressure switch.
8. OUTLINE OF PROCEDURES FOR RSP VERIFICATION BY TEST
NO CHANGE from the basic document.
9. PREPARATIONS FOR TESTING
CHANGE clause 9.3 to read:
9.3
Bleed the entrapped air from the circuit and from the pressure containing envelope being tested.
CHANGE clause 9.8 to read:
9.8
Perform cyclic and static tests on separate pressure containing envelopes.
ADD the following clause:
9.12
If the sensing element is separable from the total assembly, and is not normally restrained in either force or motion by an external means, it may be removed from the total assembly for cyclic testing.
10. TEST EQUIPMENT
CHANGE clause 10.1.3 to read:
10.1.3
Mount the pressure measuring instrument directly into the pressure containing envelope, through a pressurized port which is not being used to supply the test fluid.
CHANGE clause 10.1.4 to read:
10.1.4
When no such port (clause 10.1.3) is available, minimize restrictions between the instrument and the pressure containing envelope by mounting the pressure measuring instrument on the pressure supply line.
11. TEST CONDITIONS ACCURACY
NO CHANGE from the basic document.
16. CRITERIA FOR RSP VERIFICATION
CHANGE clause 16.4 to read:
16.4
Consider any leakage at threaded inlet ports due to distortion, a failure.
ADD the following clause:
16.5
Consider any permanent deformation which interferes in any way with the proper functioning of the pressure containing envelope, but which is consistent with clause 1.2 of the Scope, a failure.
17. DATA PRESENTATION
CHANGE clause 17.2.1 to read:
17.2.1
All physical values obtained in Sections 6 and 8 as selected in clause 13.1.2.
CHANGE clause 17.2.2 to read:
17.2.2
Modifications to the pressure switch as permitted by clauses 9.11, 9.12, 12.1.1, 12.1.4, 14.1, 14.2 and 14.3.
ADD the following clause:
17.2.3
Accuracies and ranges of pressure gages used to verify test pressures.
18. SUMMARY OF DESIGNATED INFORMATION
NO CHANGE from the basic document.
19. JUSTIFICATION STATEMENT
CHANGE the entire Section to read:
This recommended standard verification procedure is based upon the combined expert experiences of those who have participated in its preparation and review, and in the preparation and review of the basic document, NFPA/T2.6.1-1974, and its tutorial reference.
Additional justification data will be obtained and compiled as the procedures set forth in this document become more

widely used. Recommendations for the further improvement of this recommended standard are always welcome. Forward them directly to NFPA at the following address:

National Fluid Power Association, Inc.
3333 North Mayfair Road
Milwaukee, Wisconsin 53222

20. TEST/PRODUCTION SIMILARITY
NO CHANGE from the basic document.
21. IDENTIFICATION STATEMENT
CHANGE the quoted statement to read: "Method of verifying the rated fatigue and rated static pressures of the pressure containing envelope of a pressure switch conforms to NFPA Recommended Standard, NFPA/T2.6.1 S3 (T3.29.2)-1976, category ____*."
22. KEY WORDS
The following Key Words, useful in indexes and in information retrieval systems, are suggested for this recommended standard:
fluid power
pressure, cyclic test
pressure, rated fatigue
pressure, rated static
pressure, static test
pressure rating, by similarity
pressure rating, by test
pressure rating, pressure switch
pressure switch, fluid power

This is a supplement to a basic standard related to all metal fluid power components. Please refer to the latest standard available from NFPA.

PROJECT GROUP MEMBERS WHO DEVELOPED THIS STANDARD

Blickley, George—SOR, Inc.
Project Chairman and Section Chairman (1975 to present)
Brown, James—DeLaval Turbine, Inc.
Section Chairman (1972 to 1975)
Mueller, John—The Weatherhead Co.
Technical Auditor
Luecke, John R.—National Fluid Power Association, Director of National Technical Services
Lyons, J.—Chemetron Corporation
Van Skike, O.—DeLaval Turbine, Inc.

REFERENCES

1. American National Standard Glossary of Terms for Fluid Power, ANSI/B93.2-1971, and Supplements thereto. (ISO/TC 131/SC 1 [USA-2]3).
2. SI units and recommendations for the use of their multiples and of certain other units, ISO 1000-1973.
3. National Fluid Power Association Recommended Standard Method for Verifying the Fatigue and Static Pressure Ratings of the Pressure Containing Envelope of a Metal Fluid Power Component, NFPA/T2.6.1-1974.

BACKGROUND REFERENCE

1. Proposed National Fluid Power Association Recommended Standard Glossary of Terms for Pressure Switches, NFPA/T3.29.1-19xx.

Appendix C
Specifications and Tests

SPECIFICATIONS AND TESTS OF POTENTIOMETRIC PRESSURE TRANSDUCERS

Instrument Society of America

1. PURPOSE

This Standard establishes the following for potentiometric pressure transducers:

1.1 Uniform minimum specifications for design and performance characteristics.

1.2 Uniform acceptance and qualification test methods, including calibration techniques.

1.3 Uniform presentation of minimum test data.

1.4 A drawing symbol for use in electrical schematics.

2. SCOPE

2.1 This Standard covers potentiometric pressure transducers, primarily those used in measuring systems.

2.2 Included among the specific versions of potentiometric pressure transducers to which this Standard is applicable are the following:

Absolute Pressure Transducers
Differential Pressure Transducers
Gage Pressure Transducers

2.3 Technology used in this document is defined in ISA Standard 37.1. Additional terms considered applicable to potentiometric pressure transducers are defined in section 4.3 of this document. An asterisk appears before those terms defined in S37.1. The terms defined in section 4.3 appear in italics.

3. DRAWING SYMBOL

3.1 The drawing symbol for a potentiometric pressure transducer is a square with an added equilateral triangle, the base of which is the left side of the square. The letter "P" in the triangle designates "pressure" and the subscripts denote the second modifier (the illustration shows an absolute pressure transducer, as symbolized by "P_A").

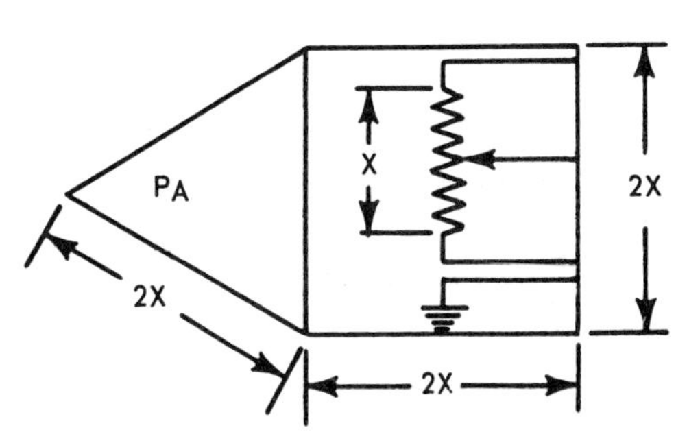

*Reproduced by permission of Instrument Society of America.

Subscripts

A = Absolute
D = Differential
G = Gage
S = Sealed Reference Differential (see 4.3)

3.2 The potentiometer is symbolized by a variable resistance of length x. The lines from it and to ground represent the electrical leads or terminations.

4. SPECIFICATION CHARACTERISTICS

4.1 Design Characteristics

4.1.1 Required Mechanical Design Characteristics

The following mechanical design characteristics shall be listed.

4.1.1.1 TYPE OF PRESSURE SENSED

* Absolute Pressure
* Differential Pressure, Unidirectional
* Differential Pressure, Bidirectional
Differential Pressure, Sealed Reference
* Gage Pressure

Note: At present, no provision is made by the SI system of units for abbreviations following the pressure units to indicate the type of pressure, as is done in the U.S. customary system of units, e.g., psia for absolute pressure in psi. In the interim it is recommended that for the SI system, the type of pressure be indicated in this manner: " . . . An absolute pressure of ____ Pa." " . . . A differential pressure range of ____ kPa," etc.

Note: For differential pressure transducers, the allowable range of reference pressures shall be listed e.g., "0 to 1MPa" or "0 to 100 psi."

4.1.1.2 * Measured Fluids

The fluids in contact with pressure port(s) shall be listed, e.g., nitric acid, liquid oxygen. Requirements for and limitations on the *isolating element* (if used) shall be listed.

4.1.1.3 Configuration and Dimensions

The outline drawing shall show the configuration with dimensions in millimeters (inches). Unless pressure and electrical connections are specified (Reference 4.1.1.5, 4.1.3.4), the outline shall include limiting maximum dimensions for these connections.

4.1.1.4 Mountings and Mounting Dimensions

Unless the pressure connection serves as a mounting, the

outline drawing shall indicate the method of mounting with hole size, centers, and other pertinent dimensions in millimeters (inches), including thread specifications for threaded holes, if used.

4.1.1.5 Pressure Connection

The pressure connection(s) shall be indicated on the outline drawing. For threaded fittings, specify: Applicable Military or Industry standards or nominal size, number of threads per millimeter (threads per inch), thread series, and thread class. For hose tube fitting, specify tube size.

4.1.1.6 Mounting Effects

The maximum mounting force or torque shall be specified if it will tend to affect transducer performance (Reference 4.2.28).

4.1.1.7 Mass

The mass of the transducer shall be specified to grams (ounces).

4.1.1.8 Case Sealing

If case sealing is necessary, the mechanism and materials used for sealing should be described. The same requirement applies to the electrical connector. The resistance of the sealing materials to cleaning solvents and commonly used measured fluids should be stated.

4.1.1.8 Identification

The following characteristics shall be permanently inscribed on the outside of the transducer case or on a suitable nameplate permanently attached to the case.

Nomenclature of transducer (acc. to ISA-S37.1, Section 3).
Name of Manufacturer, (Part number to reflect one controlled configuration), and Serial Number.
* Range in Pa(psi) and designation of type of pressure (see 4.1.1.1.). *Maximum* * *excitation.*
* Transduction element resistance (*Potentiometric Element*).
Identification of * Measured and Reference Ports (for differential pressure transducers).
* Reference Pressure Range (for differential pressure transducers).
Identification of Electrical Connections.

Inscription of the following characteristics is optional:

Customer Specification or Part Number or both.
Type of Electrical Connector and Mating Connector (if applicable).
Operating Temperature Range.
* Proof Pressure.

Note: Identificaton of Pressure Ports may be abbreviated MEAS & REF.

4.1.2 Supplemental Mechanical Design Characteristics

Listing of the following mechanical design characteristics is optional:

4.1.2.1 Case Material

Where applicable, state the surrounding environmental condition requirement or compatibility.

4.1.2.2 Sensing Element

The sensing element type shall be specified, e.g.,

* Diaphragm (flat or corrugated)
* Capsule
* Bellows
* Bourdon Tube-"C" or "U" shaped, spiral, helical, or twisted.

Note: Where used, an isolating element with transfer fluid shall be detailed as to composition.

4.1.2.3 Damping Fluid

Where used, the type, composition, temperature characteristics and compatibility with transducer components shall be specified.

4.1.2.4 Number of *Potentiometric Elements or Taps*

Where more than one potentiometric transducer element or a tapped element is required, they shall be specified.

4.1.2.5 Dead Volume

The dead volume shall be given in cubic millimeters (cubic inches). For differential pressure transducers the volume of both cavities should be listed.

4.1.2.6 Volume Change Due to Full Scale Pressure

The change in volume of the sensing element due to application of full scale pressure shall be given in cubic millimeters (cubic inches).

4.1.2.7 Materials in Contact with the Measured Fluid

The materials in contact with the measured fluid shall be listed.

Note: For differential pressure transducers, materials in both ports must be considered.

4.1.2.8 Gage Vent (Port)

In gage pressure transducers where the transduction element is exposed to ambient atmosphere, the allowable types and concentrations of atmospheric contaminant shall be specified.

4.1.2.9 Maximum and Minimum Temperatures

The maximum and minimum temperatures of fluids or environments which can be applied to the transducer and which will not cause permanent calibration shift shall be listed.

Note: Exposure time shall be specified, if relevant.

4.1.3 Required Electrical Design Characteristics

The following electrical design characteristics shall be listed. All are applicable at *Room Conditions.

4.1.3.1 *Excitation

Expressed as " ______volts (milliamperes) DC" or " ______ volts (milliamperes) AC rms at ___ hertz."

4.1.3.2 Maximum Excitation

Expressed as " _______ volts (milliamperes) DC" or " ______ volts (milliamperes) AC rms at ___ hertz."

4.1.3.3 * Transduction Element Resistance

Expressed as " ______ ± ______ohms."

4.1.3.4 Electrical Connections

Whether the electrical termination is by means of a connector or a cable, the pin designations or wire color code shall conform to the following transducer wiring standard.

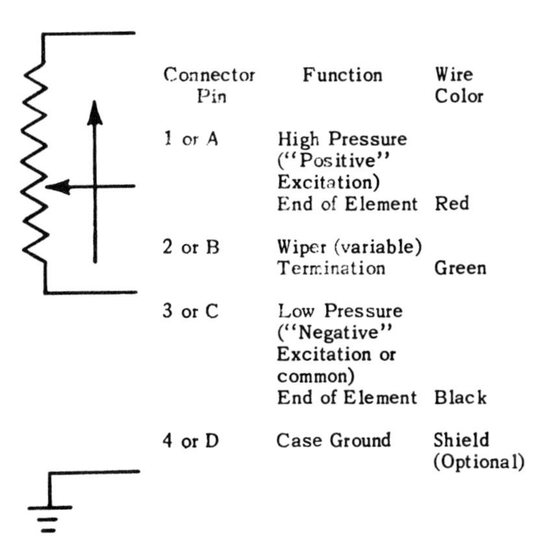

Connector Pin	Function	Wire Color
1 or A	High Pressure ("Positive" Excitation) End of Element	Red
2 or B	Wiper (variable) Termination	Green
3 or C	Low Pressure ("Negative" Excitation or common) End of Element	Black
4 or D	Case Ground	Shield (Optional)

Notes:

1. For differential pressure transducers, the arrow indicates pressure at the measurand port greater than pressure at the reference port.

2. The transduction element(s) shall be arranged with increasing "positive" voltage output as caused by increasing resistance between the wiper and low pressure (common) end of winding.

4.1.3.5 *Insulation Resistance

Expressed as " _____ megohms at _______ volts DC between all transduction element terminals in parallel and the transducer case at a temperature of ______ °C and 90% relative humidity."

4.1.3.6 *Breakdown Voltage Rating

Expressed as "Capable of withstanding volts ac-rms at _____ hertz, at a temperature of ______°C and 90% relative humidity for _____ minutes."

4.1.3.7 *Load Impedance

Expressed as " ______ ohms" (See 4.2.29).

Note: Although load impedance (the impedance presented to the output terminals of the transducer by the associated external circuitry) is not a transducer but a system characteristic, it should be specified in order to define loading error. A single, close-tolerance value of load impedance shall be specified for use during all tests where not otherwise noted.

4.2 The performance characteristics of the potentiometric pressure transducers should be tabulated in the order shown. Unless otherwise specified, they apply at Room Conditions as defined in ISA S37.1. Characteristics are usually referred to the ouput and expressed as "% VR," i.e. "percent *Voltage Ratio".

4.2.1 * **Range**

Expressed as "______ to _____ Pa (psi)" or "+ _____ Pa (psid)."

Note: Equivalent pressure units in the SI system are expressed in Pascals.

1 psi = 6894.8 Pa
10KPa = 1.4504 psi

4.2.2 * **End Points**

Expressed as " _______ % ± _______% and _______% ± _______ % VR."

Note: "% VR" is "percent *Voltage Ratio."

End Points shall be omitted where adequately defined using Error Band specifications.

4.2.3 * **Full Scale Output**

Expressed as " _____ % ± _____ % VR."

Note: Full scale output shall be omitted where adequately defined using End Points or Error Band specifications.

4.2.4 * **Linearity**

Expressed as " ______ linearity shall be within ± ______% VR."

Note: The linearity modifier shall be one of those defined in ISA S37.1; namely end point, independent, least squares, terminal or theoretical slope.

4.2.5 * **Hysteresis**

Expressed as " ______ % VR."

Alternately, 4.2.4 and 4.2.5 may be combined as 4.2.6.

4.2.6 Combined *Hysteresis and *Linearity

Expressed as "combined hysteresis and linearity within ± ____% VR."

Note: The linearity modifier shall be stated.

4.2.7 ***Friction Error**

Expressed as "within ____% VR."

4.2.8 ***Repeatability**

Expressed as "within ____% VR over a period of ___ hours."

Alternately, 4.2.2, 4.2.3, 4.2.4, 4.2.5, 4.2.7, and 4.2.8 may be combined as 4.2.9.

4.2.9 ***Static Error Band**

Expressed as "± ____% VR as referred to _____(curve)_____."

Note: The curve shall be stated as:

1) "End Point Line"—a straight line between end points.

2) "Best Straight Line"—a line midway between the two parallel lines closest together and enclosing all output versus measurand values.

3) "Least Square Line"—a straight line for which the sum of the squares of the residuals is minimized.

4) "Terminal Line"—a straight line between 0 and 100% of both measurand and output.

5) "Theoretical Slope"—a straight line between the theoretical end points.

6) Other curves shall be defined if specified, e.g., mean-output curve.

4.2.10 ***Friction-Free Error Band**

Expressed as "± ____% VR as referred to _____(curve)_____."

Note: The reference curve shall be specified (See 4.2.9).

4.2.11 ***Resolution (See also 4.3)**

Expressed as "*average resolution* within ____% VR and *maximum resolution* within ____% VR."

4.2.12 Reference-Pressure Error

Expressed as "change in end points within ± ____% VR for a reference pressure change of ____ Pa (psi) over a reference pressure range of ___ to ____ Pa (psi)." Alternately expressed as "operation at reference pressures from ____ Pa (psi) to ____ Pa (psi) shall not cause output readings which exceed the specified error band."

Instrument Society of America

4.2.13 *Frequency Response

Expressed as "within ± ___% from zero to ___hertz."

Note: Frequency response should be referred to response at a frequency within the specified frequency range, preferably zero, and to a specific static pressure. Mounting conditions and measured fluid should be specified, as should length and inside diameter of attached tubing.

Alternately, 4.2.13 may be replaced by 4.2.14, 4.2.15, and 4.2.16.

4.2.14 *Resonant Frequency

Expressed in "hertz" or "kilohertz."

Note: If a number of acoustic or mechanical, resonant frequencies exist, the lowest shall be listed and so identified.

4.2.15 *Damping Ratio

Expressed as " ___% of critical damping." Or, as " (ratio) of critical damping."

Note: For other than a single-degree-of-freedom system, the ringing period shall be stated.

4.2.16 *Ringing Period*

Expressed as " ___ milliseconds."

For transducers with relatively high damping and little overshoot, either 4.2.13 or 4.2.14, and 4.2.16 may be replaced by 4.2.17 and 4.2.18.

4.2.17 *Time Constant

Expressed as " ___milliseconds (microseconds) for step change in measurand."

4.2.18 *Overshoot

Expressed as "maximum of ___% VR, settling within ___ cycles to ___% full range, at a frequency of ___ hertz."

4.2.19 *Proof Pressure

Expressed as " ___ Pa (psi) for ___ minutes" (will not cause changes in end points or in transducer performance characteristics beyond specified tolerances), "with output reading taken within ___ minutes after pressure removal."

4.2.20 *Burst Pressure Rating

Expressed as " ___ Pa (psi) applied ___ times for a period of ___ minutes each."

Note: Applicability to sensing element or case, or both, shall be stated.

4.2.21 *Operating Temperature Range*

Expressed as "from ___ °C to ___ °C."

4.2.22 *Temperature Error

Expressed as "within ___ % VR per ___ °C." Or "within ___% VR over the *operating temperature range*."

Alternately 4.2.22 may be specified by 4.2.23.

4.2.23 *Temperature Error Band

Expressed as "within ± ___% VR from the reference curve established for the Static Error Band and over the *operating temperature range*."

4.2.24 *Acceleration Error

Expressed as "within ___% VR per g_n along ___axis at steady acceleration levels to ___g_n."

Note: The error shall be listed either for each of the three axis or for the axis with the largest error, i.e., most sensitive axis.

Alternately 4.2.24 may be replaced by 4.2.25.

4.2.25 *Acceleration Error Band

Expressed as "within ± ___% VR from the reference curve established for the Static Error Band for steady accelerations up to ___g_n along ___axis."

Note: The error band shall be listed either for each of the three axes or for the axis with the largest error, i.e., most sensitive axis.

4.2.26 *Vibration Error

Expressed as "within ± ___% VR along ___axis over the specified vibration program." Signal "dropout" or discontinuities shall be noted.

Note: The error shall be listed either for each of the three axis or for the axis with the largest error, i.e., most sensitive axis; and the program shall be detailed, possibly by a graph.

Alternately 4.2.26 may be replaced by 4.2.27.

4.2.27 *Vibration Error Band

Expressed as "within ± ___% VR along ___axis from the reference curve established for the Static Error Band over the specified vibration program."

Note: The error band shall be listed either for each of the three axes or for the axis with the largest error, i.e., most sensitive axis; and the program shall be detailed, possibly by a graph.

4.2.28 *Mounting Error

Expressed as "within ± ___% VR" or "within the Static Error Band."

4.2.29 *Loading Error

Expressed as "within ± ___% VR" or "within the Static Error Band."

4.2.30 *Cycling Life

Expressed as " ____ cycles at one fourth the designated maximum operating frequency of the transducer."

4.2.31 Other Environmental Conditions

Other pertinent environmental conditions which shall not change transducer performance beyond specified limits should be listed, examples are:

*Ambient Pressure
Shock-Triaxial
High Level Acoustic Excitation
Humidity
Salt Atmosphere
Nuclear Radiation
Magnetic Fields
Sand and Dust
Total Immersion (and in what medium)
Solar (or other) Heat Radiation
Temperature Shock

4.2.32 *Storage Life

Expressed as " ___ months (years) without changing performance characteristics beyond their specified tolerances."

Note: Environmental storage conditions shall be described in detail. Where performance characteristics require additional tolerances over the storage life they shall be specified.

4.3 Additional Terminology

Average Resolution

The reciprocal of the total number of output steps over the unit range multiplied by 100 and expressed in % VR.

Damping Fluid

A fluid used to damp the single-degree-of-freedom spring/mass system, usually surrounding the reference side (transduction element side) of the sensing element.

Isolating Element

A movable membrane, usually of metal, that physically separates the measured fluid from the sensing element. Usually this membrance is considerably more flexible than the sensing element and is coupled to the sensing element using a *transfer fluid*. Its purpose is to provide material compatibility with the measured fluid while maintaining the performance integrity of the sensing element.

Maximum Excitation

The maximum allowable voltage (current) applied to the potentiometric element at Room Conditions while maintaining all other performance characteristics within their limits. (Note: The excitation value is particularly associated with temperature.)

Mounting Effects

The effects (errors) introduced into transducer performance during installation caused by fastening of the unit or its mounting hardware or by irregularities of the surface on (or to) which the transducer is mounted.

Operating Temperature Range

The range in extremes of ambient temperature within which the transducer must perform to the requirements of the "Temperature Error" or "Temperature Error Band." (See paragraphs 4.2.22 and 4.2.23, respectively.)

Potentiometric Element

The resistive part of the transduction element upon which the wiper (movable contact) slides and across which excitation is applied. It may be constructed of a continuous resistance or of small diameter wire wound on a form (mandrel).

Pressure Connection (Pressure Port)

The opening and surrounding surface of a transducer used for measured fluid access to transducer sensing element (or isolating element). This can be a standard industrial or military fitting configuration, a tube hose fitting or a hole (orifice) in a base plate. For differential pressure transducers there are two pressure connections: the measurand port and the reference.

Reference Pressure Error

The maximum change in output at specified measurand values due to a specified change in the Reference Pressure applied at both ports simultaneously.

Ringing Period

The period of time during which the amplitude of measurand step-function-excited oscillations exceed 10% of the step amplitude.

Sealed Reference Differential Pressure Transducer

A transducer which measures the pressures difference between an unknown pressure and the pressure of a gas in an integral sealed reference chamber.

Tap

A connection to a potentiometric element along its length, frequently at the element's center for use in providing bidirectional output.

Transfer Fluid

A degassed liquid used between an isolating element and a sensing element to provide hydraulic coupling of the pressure between both elements.

Worst Resolutions

The magnitude of the largest of all output steps over the unit range expresed as a percentage of VR.

4.4 Tabulated Characteristics versus Test Requirements

This table is intended for use as quick reference for design and performance characteristics and tests of their proper verification as contained in this Standard.

				Verified During	
Characteristic	**Paragraph**	**Basic**	**Supple-mental**	**Individual Acceptance Test**	**Qualifi-cation Test**
Type of Pressure Sensed	4.1.1.1	x		No Test	
Measured Fluids	4.1.1.2	x			Special Test
Configuration, Dimensions, Mounting Pressure Connection	4.1.1.3 through 4.1.1.5	x		5.2.1	
Mounting Effects	4.1.1.6	x			6.6
Weight	4.1.1.7	x			6.3
Case Sealing	4.1.1.8	x			5.2.1
Identification	4.1.1.9	x		5.2.1	
Case Material	4.1.2.1		x		5.2.1
Sensing Element	4.1.2.2.		x		5.2.1
Damping Fluid	4.1.2.3		x		5.2.1
Number of Potentiometric Elements or Taps	4.1.2.4		x	5.2.2 through 5.2.6	
Dead Volume	4.1.2.5		x		6.4
Volume Change Due to Full Scale Pressure	4.1.2.6		x		6.5
Materials in Contact with Measured Fluid	4.1.2.7		x		Special Test
Gage Vent	4.1.2.8		x		Special Test
Maximum Temperature	4.1.2.9		x		Special Test
Excitation	4.1.3.1	x		5.2.6	
Maximum Excitation	4.1.3.2	x			Special Test
Transduction Element Resistance	4.1.3.3	x		5.2.2	
Electrical Connections	4.1.3.4	x		5.2.2	

Characteristic	Paragraph	Basic	Supple-mental	Verified During Individual Acceptance Test	Verified During Qualification Test
Insulation Resistance	4.1.3.5	x		5.2.3	
Breakdown Voltage Rating	4.1.3.6	x		5.2.4	
Load Impedance	4.1.3.7	x		5.2.5 (partially)	6.7
Range	4.2.1	x		5.2.6	
End Points	4.2.2	x		5.2.6	
Full Scale Output	4.2.3	x		5.2.6	
Linearity	4.2.4	x		5.2.6	
Hysteresis	4.2.5	x		5.2.6	
Combined Hysteresis and Linearity	4.2.6	x		5.2.6	
Friction Error	4.2.7	x		5.2.6	
Repeatability	4.2.8	x		5.2.6	
Static Error Band	4.2.9	x		5.2.6	
Friction-Free Error Band	4.2.10	x		5.2.6	
Resolution	4.2.11	x			6.2
Reference-Pressure Error	4.2.12	x		5.2.7	
Frequency Response	4.2.13				6.8
Resonant Frequency	4.2.14				6.8
Damping Ratio	4.2.15				6.8
Ringing Period	4.2.16				6.8
Time Constant	4.2.17	x			6.8
Overshoot	4.2.18	x			6.8
Proof Pressure	4.2.19	x		5.2.8	
Burst Pressure Rating	4.2.20	x			6.14
Operating Temperature Range	4.2.21	x			6.9
Temperature Error	4.2.22	x			6.9
Temperaure Error Band	4.2.23	x			6.9
Acceleration Error	4.2.24	x			6.10
Acceleration Error Band	4.2.25	x			6.10
Vibration Error	4.2.26	x			6.11
Vibration Error Band	4.2.27	x			6.11
Mounting Error	4.2.28				6.6
Loading Error	4.2.29				6.7
Cycling Life	4.2.30	x			6.13
Other Environmental Conditions	4.2.31	x			6.12
Storage Life	4.2.32				Special Test (accelerated)

5. INDIVIDUAL ACCEPTANCE TESTS AND CALIBRATIONS

5.1 Basic Equipment Necessary to Perform Individual Acceptance Tests and Calibrations of Potentiometric Pressure Transducers.

The basic equipment for acceptance tests and calibrations consists of a source of pressure, a source of electrical excitation for the potentiometer, and a device which measures the electrical output of the transducer directly or as a ratio to excitation input (VR). The errors or uncertainties of the measuring system comprising these three components should be less than one-fifth of the permissible tolerance of the transducer performance characteristic under evaluations. The traceability to national standards for this measuring system shall be well known.

5.1.1 Pressure Source

A pressure medium similar to the one which the transducer is intended to measure should be used for testing. The accuracy of the pressure source should be at least five times greater than the permissible tolerance of the transducer performance characteristic under evaluation. The range of the pressure source and monitoring equipment should be selected to provide the necessary pressure, and accuracy, respectively, to 125% of the full scale of the transducer.

The pressure source may be either continuously variable over the range of the instrument, or may give discrete steps as long as the steps can be programmed in such a manner that the transition from one pressure to the next during calibration is accomplished without eliminating an existing hysteresis (or friction) error in the transducer by overshoot or fluctuation.

Note: By "similar" is meant a fluid with similar properties, bearing in mind safety and availability, i.e., H_2, N_2, O_2, silicone oils, and the like.

EXAMPLES OF PRESSURE SOURCES/MONITORING EQUIPMENT MERCURY MANOMETER

(Pressure Indicating Device)

Typical Ranges

100 KPa (about 30 in. Hg) . . .
Accuracy ± 0.02% Full Scale
200 KPa (about 60 in. Hg) . . .
Accuracy ± 0.02% Full Scale
340 KPa (about 100 in Hg) . . .
Accuracy ± 0.01% Full Scale

AIR PISTON (Pressure Source)

Typical Ranges

About 2 to 10 KPa (0.3 to 1.5 psi) . . .
Accuracy ± 0.15% of Reading
About 10 to 350 KPa (1.5 to 50 psi) . . .
Accuracy ± 0.015% of Reading
About 100 KPa to 1 MPa (15 to 150 psi) . . .
Accuracy ± 0.025% of Reading
About 100 KPa to 3.5 MPa (1 to 500 psi) . . .
Accuracy ± 0.025% of Reading

PRECISION DIAL GAGE (Pressure Indicating Device)

Typical Ranges

0 to 30 KPa (about 0 to 120 Hz0) . . .
Accuracy ± 0.1% Full Scale
0 to 100 KPa (about 0 to 30 in. Hg)
Accuracy ± 0.1% Full Scale
0 to 100 KPa (about 0 to 100 psi) . . .
Accuracy ± 0.1% Full Scale
0 to 700 MPa (about 0 to 10 000 psi) . . .
Accuracy ± 0.1% Full Scale

Note: Presssure indicating devices generally require a supply of dry gas, e.g., dehumified air, or nitrogen, or helium, required for reasons of safety.

OIL PISTON GAGE (Pressure Source)

Typical Ranges

About 40 KPa to 30 MPa (6 to 4000 psi) . . .
Accuracy ± 0.01% of Reading
About 400 KPa to 300 MPa (60 to 40 000 psi) . . .
Accuracy ± 0.01% of Reading
About 14 MPa to 700 MPa . .
Error in Piston Area Less Than ± 0.009%
About 30 MPa to 1400 MPa . . (400 to 200 000 psi)
Error in Piston Area Less Than 0.012%

Note: The accuracies cited may be greater than needed for the calibration of many potentiometric pressure transducers, but may be required for the calibration of other types of pressure sensing instruments. Economic considerations suggest acquisition of minimum number of pressure sources/monitors to meet calibration needs of majority of transducers in a given installation.

5.1.2 Stable Excitation Source of accurately known amplitude (unless VR is being measured).

Commonly used sources are chemical batteries such as dry cells and storage batteries or line-powered, electronically regulated, power supplies. (A current limiting device shall be inserted in series with the transducer to preclude accidental damage of the potentiometric element.)

5.1.3 Electronic Indicating Instrument

Examples of suitable devices are:

Manually Balanced Ratiometer
Achievable Accuracy
1 part in 10 000

Self Balancing Ratiometer
Achievable Accuracy
1 part in 10 000

Digital Electronic Voltmeter/Ratiometer
Achievable Accuracy

± 0.01% of Reading + 1 digit (4 digits display)
± 0.005% of Reading + 1 digit (5 digits display)

Note: The input impedance of the readout instrument shall be sufficiently high to produce negligible loading error. Suggested value is 100 times the resistance of the transduction element.

5.2 Calibration and Test Procedures

Results obtained during the calibration and testing shall be recorded on a data sheet similar to the sample date sheet in Section 7, Figure 7.1 (7.7 for static error and calibration) of this standard. Calibration and testing shall be performed under Room Conditions as defined in ISA-S37.1 unless otherwise specified.

Notes:

1. The definitive paragraph under Performance Characteristics (Section 4) of this document is listed beside each of the parameters for which the test results are to be compared.

2. If more than one potentiometric element is used in the transducer, the performance of every element shall be recorded on its own form.

5.2.1 The transducer shall be inspected visually for mechanical defects, poor finish, and other applicable mechanical characteristics of 4.1.1.

Configuration and Dimensions	4.1.1.3
Mounting and Mounting Dimensions	4.1.1.4
Pressure Connection	4.1.1.5
Identification	4.1.1.9

By use of special equipment, or by formal verification of production methods and materials used, the following can be additionally determined:

Case Sealing	4.1.1.8
Case Material	4.1.2.1
Sensing Element	4.1.2.2
Damping Fluid	4.1.2.3

5.2.2 A precision resistance measuring device shall be used to measure:

Transduction Element Resistance	4.1.3.3
and can be used to verify:	
Number of Potentiometric Elements or Taps	4.1.2.4
Electrical Connections	4.1.3.4

5.2.3 Measure the insulation resistance between all transduction element terminals (or leads) connected in parallel and the case (and ground pin) of the transducer with a megohmmeter device, using a potential of 50 volts unless otherwise specified.

Insulation Resistance	4.1.3.5

5.2.4 Verify the Breakdown Voltage Rating, using sinusoidal ac voltage test with all transduction element terminals (or leads) paralleled, and tested to case and ground pin.

Breakdown Voltage Rating (ac-rms)	4.1.3.5

5.2.5 The transducer shall be connected to the pressure source and secured as recommended for its use. The appropriate excitation source and indicating instruments shall be properly connected to the transducer and turned on. Adequate warm-up time for indicating instruments shall be allowed before tests are conducted. The pressure source, connecting tubing, and transducer system shall pass a leak test to assure absence of calibration errors. Electrical connections shall be checked for correctness of hook-up including the appropriate load impedance (See 4.1.3.7).

5.2.6 Two or more complete calibration cycles shall be run consecutively. A minimum of eleven data points shall be obtained including both ascending and descending directions. Excitation amplitude shall be monitored as required unless VR is measured.

In order to verify performance between the discrete levels and to assure absence of noise, a full-scale X-Y plot shall be obtained, preferably inscribed diagonally across the test record form, by applying increasing then decreasing pressure to the transducer, and simultaneously to a reference transducer having continuous resolution and suitable linearity, each connected to one axis input of the plotter.

From the data obtained during these tests, the following characteristics should be determined:

End Points	4.2.2
Full Scale Output	4.2.3
Linearity	4.2.4
Hysteresis	4.2.5
(or Hysteresis and Linearity)	4.2.6
Friction Error	4.2.7
Repeatability	4.2.8

Note: To determine Friction Error or Friction-Free Error Band, at least one calibration cycle shall be run with the transducer dithered (light but sufficient vibration or shock).

5.2.7 For Differential Pressure Transducers, the performance of a three-point (e.g., 10, 50, and 90%) calibration cycle at both the minimum and maximum specified reference pressures shall establish:

Reference Pressure Error	4.1.12

5.2.8 After application of the proof pressure, at least a three-point calibration shall be performed to establish that the performance characteristics of the transducer are still within specifications. The first output reading shall be recorded within the period of time specified for this.

Proof Pressure	4.2.19

Note: For bidirectional differential transducer, proof pressure shall be applied to both ports individually. For reporting purposes these are identified as "positive" and "negative" proof pressures.

6. QUALIFICATION TEST PROCEDURES

Qualification Tests shall be performed as applicable using the test forms of Section 7 as required. Upon completion of all testing the form of Figure 7.6 shall be used to summarize all testing.

6.1 Initial Performance Tests (Figure 7.2)

The tests and procedures of Section 5, Individual Acceptance Tests and Calibrations, shall be run to establish reference performance during increasing (and decreasing) steps of 0, 20, 40, 60, 80, and 100% of range as a minimum (% of span for bidirectional transducers).

6.2 Resolution Test (Figure 7.6)

An X-Y plotter shall be connected so that the transducer output is connected to the X-Axis and a continuous-resolution reference transducer to the Y-Axis input of the plotter. As the pressure to both transducers is slowly increased (simultaneously on both transducers), the number of steps shall be recorded from 0 to 100% of the test transducer's range. The following shall be determined:

Resolution (Average and Worst)	4.2.11

6.3 Weight Test (Figure 7.6)

The transducer shall be weighed on an appropriate balance or scale. The following shall be established:

Weight	4.1.1.7

6.4 Dead Volume Test (Figure 7.6)

The pressure cavity shall be filled (both cavities for a differential transducer) with a measurable, non-corrosive liquid (under a vacuum if necessary), and the contents poured into a graduate. The following shall be established:

Dead Volume	4.1.2.5

6.5 Volume Change Test (Figure 7.6)

A liquid pressure system shall be connected to a transducer, a parallel pressure gage, and a graduated reservoir. (Provisions for isolating the transducer when filled shall be made.) The pressure system shall be evacuated and filled with liquid, the valve to the transducer closed, the pressure increased to 100% of range, the valve opened, and the following shall be determined:

Volume Change due to Full Scale Pressure	4.1.2.6

6.6 Mounting Test (Figure 7.6)

The mounting of the actual installation shall be duplicated as closely as possible following specific instructions and one calibration run performed. The following shall be established:

Mounting Error	4.2.28
Mounting Effect	4.1.1.6

6.7 Loading Test (Figure 7.6)

Approximately 66% of full range pressure shall be applied to the transducer and the total load resistance varied from the highest to the lowest ohmic value allowed. (Note: Take into account the resistance of the ratiometer or other indicating instrument.) The following shall be verified:

Loading Error	4.2.29

The loading error of a potentiometric pressure transducer is variable with wiper position ranging from zero at both extremes to a maximum value at approximately 66% of VR. As a first approximation the percentage error is equal to fifteen times the ratio of the potentiometric element resistance to the loading resistance. Unless otherwise stated, assembly adjustments of the transducer apply to the open circuit conditions at the output terminals.

6.8 Dynamic Response Test (Figure 7.4a or 7.4b as applicable)

The dynamic response characteristics of pressure transducers may be established either with transient-excitaton devices, or with sinusoidal pressure generators.

6.8.1 Transient Excitation Method

A positive step-function of pressure may be generated in gases with a shock-tube or a quick-opening valve. A hydraulic quick-opening valve is used to generate a positive pressure step function in a liquid medium. A burst diaphragm generator produces a negative pressure step in a gas medium. In all cases, the rise time of the generated step function shall be sufficiently short to shock-excite all resonances in the transducer under test. It shall also be one third or less of the anticipated rise time of the transducer under test.

Since the tubing used to mechanically connect the transducer to the test set-up may affect the dynamic characteristics, it is recommended that the shortest possible tubing be installed, or that the tubing used shall duplicate as closely as possible the actual installation, if this condition was specified instead of the characteristics of the transducer alone. Any tubing used shall be described by length, internal diameter, and curvature.

By applying step functions of pressure at Room Conditions within the full scale range of the transducer, and analyzing the electronic or electro-optical recording of the transducer output, the following can be determined.

Frequency Response (amplitude and phase)	4.2.13
Resonant Frequency	4.2.14
Damping Ratio	4.2.15
Ringing Period	4.1.16

Alternately for transducers with relatively high damping and little overshoot, the following can be determined:

Time Constant	4.2.17
Overshoot	4.2.18

6.8.2 Sinusoidal Excitation Method

For frequencies below a few kilohertz, static pressures below 100 MPa (roughly 15 000 psi) and peak dynamic pressures below 10 MPa (roughly 1500 psi), generators that produce sinusoidal pressure wave-forms in either liquids or gases are available. These sinusoidal pressure generators operate either on the pistonphone principle (which is in common use for the absolute calibration of microphones) or by modulating a fluid flow through an orifice.

A sinusoidal pressure waveform of constant amplitude and varying frequency, over a specified frequency range, shall be applied at a specified static pressure. The following shall be determined:

Frequency Response (amplitude and phase)	4.1.13

If within the frequency range covered, the following can be established from the frequency response recording by suitable calculations:

Resonant Frequency	4.2.14
Damping Ratio	4.2.15
Ringing Period	4.2.16
Time Constant	4.2.17
Overshoot	4.2.18

6.9 Temperature Tests

6.9.1 Low Temperature Test (Figure 7.3)

The transducer shall be placed in a suitable temperature chamber. The temperature of the transducer shall be stabilized for one hour at the lower specified operating temperature and two calibration cycles performed, followed by a positive-proof pressure test and a third calibration cycle. The insulating resistance shall be measured and recorded as in 5.2.3. (For differential pressure transducers only, the second calibration cycle shall be performed with maximum specified reference pressure applied, followed immediately by a negative-proof pressure test.)

These tests shall establish the following:

Temperature Error (at low temperature)	4.2.22
or	
Temperature Error Band (at low temperature)	4.2.23

6.9.2 Post Low Temperature Test (at Room Conditions) (Figure 7.3)

The transducer shall be removed from the temperature chamber and permitted to stablize for one hour at room conditions. The tests of 6.9.1 shall be repeated except that the operating temperature shall be room temperature.

6.9.3 High Temperature Test (Figure 7.3)

The tests of 6.9.1 shall be repeated except that the transducer temperature shall be stabilized for one hour at the highest specified operating temperature.

These tests shall establish the following:

Temperature Error (at high temperature)	4.2.22

or
Temperature Error Band (at high temperature) 4.2.23

6.9.4 Post High Temperature Test (at Room Conditions) (Figure 7.3)

The tests of 6.9.2 shall be repeated after stabilization of the transducer at room temperature for one hour.

Note: If required, thermal and post-thermal zero shift and sensitivity shift may also be calculated from the results of these tests.

6.10 Acceleration Test (Figure 7.5)

Acceleration shall be imposed on the transducer in three orthogonal directions by tilting it in the earth's gravitational field or by placing it on a centrifuge. A specific acceleration level shall be applied on specified axes, and the output measured. The following shall be established:

Acceleration Error 4.2.24
or
Acceleration Error Band 4.2.25

6.11 Vibration Test (Figure 7.5)

With specified measurand levels applied, the transducer shall be vibrated along specified axes at specified acceleration amplitudes over the specified frequency range with an electro-magnetic or hydraulic shaker. The transducer output shall be recorded with a high-speed recorder. The following shall be established:

Vibration Error 4.2.26
or
Vibration Error Band 4.2.27

Note: If so specified, the vibration error band can be established as the algebraic sum of maximum vibration errors and the last previously obtained static error band.

6.12 Tests For Other Environmental Conditions (Figure 7.3)

The transducer shall be exposed to other specified environmental conditions. As specified for each condition, one complete calibration cycle shall be performed during or after the test to establish the ability of the transducer to perform satisfactorily.

See Section 4.2.31

6.13 Life Test (Figure 7.3)

After applying the specified number of full range excursions of measurand, or after completion of each of several specified portions of the total number of cycles, at least one complete calibration cycle shall be performed to establish minimum value of:

Cycling Life 4.2.30

6.14 Burst Pressure Test (Figure 7.6)

The transducer shall be connected to a suitable test setup with adequate protection for equipment and personnel. The pressure shall be increased to the specified limit and applied for the specified number of times and durations. The following shall be established :

Burst Pressure Rating 4.2.20

NOTE: If specified, burst pressure is applied to the inside of the case by first puncturing the sensing element.

7. TEST REPORT FORMS

7.1 The test report forms listed below are recommended for use during the testing of Potentiometric Pressure Transducers.

7.2 When using the forms all pertinent information shall be inserted in its proper place. On some forms, blank space has been provided for additional tests. Where the test is prolonged, e.g., Cycling Life, more than one form may be required.

7.3 Individual Acceptance Tests and Calibrations (Figure 7.1). Used during acceptance testing of Section 5.

Initial Performance Tests and Calibrations (Figure 7.2). Used for establishing the reference performance for comparison to other test results.

Environmental Test Record (Figure 7.3). Used for Temperature, Maximum Temperature, Life, and other environmental tests.

Dynamic Response Tests (Figure 7.4),. Used for recording test results of Frequency Response, Resonant Frequency, Damping Ratio, Ringing Period, Time Constant, and Overshoot. (Note: Use 7.4a or 7.4b as applicable.)

Environmental Test Record (Figure 7.5). Used to record Acceleration and Vibration Test results.

Test Summary (Figure 7.6). Used to compile the results of all testing.

Individual Acceptance Test Record (Static Error Band) (Figure 7.7). Used as an alternate for Figure 7.1 when Static Error Band Calibration is specified.

8. BIBLIOGRAPHY

1. Beckwith, T. G., and Buck, N. L.; "Mechanical Measurements"; Addison-Wesley, 1961.

2. Norton, H. N.; "Transducers for Electronic Measuring Systems"; Prentice-Hall, 1969.

3. *ISA Transducer Compendium,* 2nd Edition Part I "Pressure, Flow, and Level"; Plenum Press, New York, 1969.

4. Neubert, H.K.P.; *Instrument Transducers;* Oxford at the Clarendon Press, 1963.

5. Lederer, P.S.; "Methods for Performance Testing of Electromechanical Pressure Transducers"; National Bureau of Standards Technical Note # 411, February, 1967.

6. Cross, J. L.; "Reduction of Data for Piston Gage Pressure Measurements"; National Bureau of Standards Monogram # 65, June, 1963.

7. ISA-S37.1; "Electrical Transducer Nomenclature and Terminology"; 1969.

8. ISA-S37.3; "Specifications and Tests for Strain Gage Pressure Transducers"; 1970.

9. Schweppe et al.; "Methods for the Dynamic Calibration of Pressure Transducers"; National Bureau of Standards Monograph # 67, December, 1963.

10. MIL-E-5272C (ASG); "Environmental Testing, Aircraft Electronic Equipment, General Specification For."

11. MIL-E-5400C (ASG); "Environmental Testing, Aeronautical and Associated Equipment, General Specification For."

12. MIL-STD-810; "Environmental Test Methods for Aerospace and Ground Equipment."

13. PMC 20.1-1973; "Process Measurement and Control Terminology"; August, 1973.

14. ANSI B88.1-1971; "A Guide for the Dynamic Calibration of Pressure Transducers"; ASME, August, 1972.

15. ANSI Z210.1-1973, "Standard Metric Practice Guide"; ASTM (E 380-72E) March, 1973.

Instrument Society of America

VENDOR'S PART NO.	TEST FACILITY	CUSTOMER'S PART NO.
VENDOR	**POTENTIOMETRIC PRESSURE TRANSDUCER**	SERIAL NO.
CUSTOMER	**INDIVIDUAL ACCEPTANCE TEST AND CALIBRATION RECORD**	RANGE _____ TO _____ Pa _____

Visual Inspection:	Electrical Inspection:
Dimensions ☐ Workmanship ☐	Element Resistance _____ ohms Insulation Resistance _____ Megohms at ___ Vdc
Finish ☐ Nameplate ☐ El. Conn. ☐	Breakdown Voltage Rating @ _____ Vac, _____ Hz ☐ Z_L used _____ MΩ

Calibration @ _____ V _____ Excitation Ambient Temperature _____ °C

Pressure (Pa)	Theor. Output (%VR)	(Undithered) First Calib. Cycle Output (%VR)		(Dithered) First Calib. Cycle Output (%VR)		(Undithered) Second Calib. Cycle Output (%VR)		(Dithered) Second Calib. Cycle Output (%VR)			(Undithered) Max. Error (%VR)		(Dithered) Max. Error (%VR)	
		Increase	Decrease	Increase	Decrease	Increase	Decrease	Increase	Decrease		+	−	+	−

* _____ Linearity : + _____ , − _____ %VR (Allowed: ± _____ %VR); * Hysteresis: _____ %VR (All'd _____ %VR))

* Hysteresis and _____ Linearity (Combined): + _____ , − _____ %VR (Allowed: _____ ± _____ %VR)

Friction-free Error Band: + _____ , − _____ %VR (Allowed: ± _____ %VR); Friction Error: _____ %VR (All'd _____ %VR)

Repeatability: _____ %VR (All'd: _____ %VR); * End Points: _____ and _____ %VR (All'd _____ and _____ %VR)

* Full-Scale Output: _____ %VR Allowed _____ ± _____ %VR NOTE: * Values Determined From _____ Calib. Cycle

		P Only: Performance Test @ _____ Pa _____ Ref. Press.				Δ P Only: Perf. Test ___ Minutes After ___ Pa ___ Neg. Proof Press. (Overload Output: _____ _____)				Perf Test _____ Minutes After ___ Pa ___ Pos. Proof Press. (Overload Output: _____ _____)			
Pressure (Pa)	Theor. Output	Output (_____)		Error (_____)		Output (_____)		Error (_____)		Output (_____)		Error (_____)	
				+	−			+	−			+	−
				+	−			+	−			+	−
				+	−			+	−			+	−

Full-Scale X-Y Plot: _____	Static Error Band: + _____ , − _____ %VR Allowed: ± _____ %VR) Error Bands Ref. To _____ Ref. Press. Error: _____ %VR

Equipment Used:	
	Tested By: _____ Date: _____
	Approved By: _____

FIGURE 7.1

VENDOR'S PART NO.	TEST FACILITY	CUSTOMER'S PART NO.
VENDOR	POTENTIOMETRIC PRESSURE TRANSDUCER	SERIAL NO.
REPORT NO.	☐ INITIAL PERFORMANCE TEST	CUSTOMER
TYPE OF TEST	☐ ______ TEST	RANGE ____ TO ____ Pa ____

Visual Inspection:.
Dimensions ☐ Workmanship ☐
Finish ☐ Nameplate ☐ El. Conn. ☐

Electrical Inspection:
Element Resistance ______ ohms ; Insulation Resistance ____ Megohms at ____ Vdc
Breakdown Voltage Rating @ ____ Vac, ____ H_z ☐ Z_L used ____ M Ω

Calibration @ ______ V ____ Excitation Ambient Temperature ____ °C

Pressure (Pa)	Theor. Output (%VR)	(Undithered) First Calib. Cycle Output (%VR)		(Dithered) First Calib. Cycle Output (%VR)		(Undithered) Second Calib. Cycle Output (%VR)		(Dithered) Second Calib. Cycle Output (%VR)			(Undithered) Max. Error (%VR)		(Dithered) Max. Error (%VR)	
		Increase	Decrease	Increase	Decrease	Increase	Decrease	Increase	Decrease		+	−	+	−

* ______ Linearity : + ____ , − ____ %VR (Allowed: ± ____ %VR); * Hysteresis: ____ %VR (All'd : ____ %VR)

* Hysteresis and ______ Linearity (Combined): + ____ , − ____ %VR (Allowed: ± ____ %VR)

Friction-free Error Band: + ____ , − ____ %VR (Allowed): ± ____ %VR); Friction Error: ____ %VR (All'd: ____ %VR)

Repeatability: ____ %VR (All'd: ____ %VR); * End Points: ____ and ____ %VR (All'd : ____ and ____ %VR)

* Full-Scale Output: ____ %VR Allowed: ____ ± ____ %VR NOTE: * Values Determined From ______ Calib. Cycle

Pressure (Pa)	Theor. Output	Δ P Only: Performance Test @. ____ Pa ____ Ref. Press.				Δ P Only: Perf. Test ___ Minutes After ___ Pa ___ Neg. Proof Press. (Overload Output: ____ ____)				Perf Test ____ Minutes After ___ Pa ___ Pos. Proof Press. (Overload Output: ____ ____)			
		Output (____)		Error (____)		Output (____)		Error (____)		Output (____)		Error (____)	
				+	−			+	−			+	−
				+	−			+	−			+	−
				+	−			+	−			+	−

Full-Scale X-Y Plot: ____

Static Error Band: + ____ , − ____ %VR (Allowed: ± ____ %VR)

Error Bands Ref. To ______ Ref. Press. Error: ____ %VR

Equipment Used:	Defects Noted Or Comments:

Tested By: ______ Date: ______ Approved By: ______

FIGURE 7.2

Instrument Society of America

VENDOR'S PART NO.	TEST FACILITY	CUSTOMER'S PART NO.
VENDOR	POTENTIOMETRIC PRESSURE TRANSDUCERS	SERIAL NO.
REPORT NO.	ENVIRONMENTAL TEST RECORD	CUSTOMER
TYPE OF TEST		RANGE ____ TO ____ Pa ____

Tested While: ☐ Undithered ☐ Dithered	☐ Before ☐ During ☐ After	Type of Environment	Level

Pressure	Theoretical	%VR (Run 1)		%VR (Run 2)		Overload	%VR (Run 3)		Maximum Error %VR	
(Pa)	%VR	Increase	Decrease	Increase	Decrease	%VR	Increase	Decrease	+	–
	(Pos. Proof)									
	(Neg. Proof)									

Error Band: +____% – ____ %VR (Referred To ________) Allowed: ± ____ %VR

Proof Pressure ____ % Rated Range for ____ Minutes (POS) Ins. Resistance: ____ Megohms at __ Vdc

For Diff. Press. Transducers Only: Zero Shift: ________ %VR

Neg. Proof Pressure ____ %Rated Range for ____ Minutes Sensitivity Shift: ________ %VR

Ref. Press., Run 2 : ________ Pa

Comments: ________

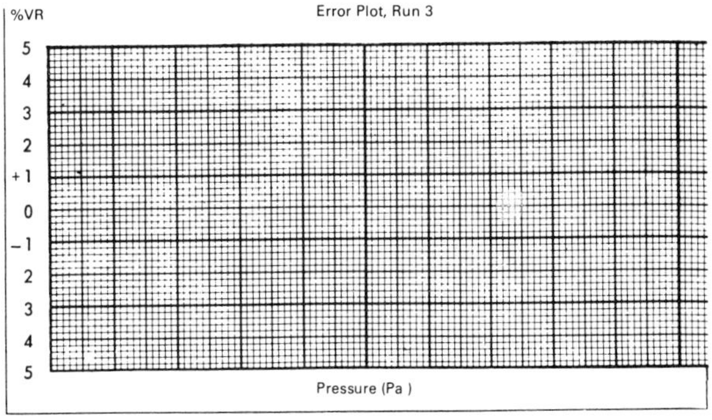

Tested By: ________ Date Test Started: ________ Date Test Finished: ________

Approved By: ________ Approved By: ________

Title: Title:

FIGURE 7.3

VENDOR'S PART NO.	TEST FACILITY	CUSTOMER'S PART NO.
VENDOR	**POTENTIOMETRIC PRESSURE TRANSDUCER**	SERIAL NO.
REPORT NO.	**DYNAMIC RESPONSE TESTS**	CUSTOMER
TYPE OF TEST		RANGE _____ TO _____ Pa _____

1. Ambient Conditions: Temperature: _____ °C, Pressure: _____ Pa; Humidity _____ %
2. Dynamic Response _____
   Excitation Voltage: _____
   Step Function Generator _____ Shock Tube, Dry Air: _____
   Mounting Location: End: _____ Side: _____

Shock No.	Initial Pressures		Shock Velocity	Step Pressure	Pronounced Resonances, Hz			Ringing Period / Time Constant
	Hi	Low						

Frequency Response _____ Hz (Allowed _____ Hz), Damping Ratio _____ (All'd _____)
Ringing Period _____ msec. (All'd _____ msec.),
Time Constant _____ msec. (All'd _____ msec.), Overshoot _____ %VR (All'd _____ %VR)

ATTACH OSCILLOSCOPE PHOTOGRAPHS OF TRANSDUCER RESPONSES

Amplitude Scale — **SHOCK 1** — Time Scale	Amplitude Scale — **SHOCK 3** — Time Scale
Amplitude Scale — **SHOCK 2** — Time Scale	Amplitude Scale — **SHOCK 4** — Time Scale

Tested By: _____ Date Test Started: _____ Date Test Finished: _____

Approved By: _____ Title Approved By: _____ Title

FIGURE 7.4.a

Instrument Society of America

VENDOR'S PART NO.	TEST FACILITY	CUSTOMER'S PART NO.
VENDOR	POTENTIOMETRIC PRESSURE TRANSDUCER	SERIAL NO.
REPORT NO.	DYNAMIC RESPONSE TESTS	CUSTOMER
TYPE OF TEST	(SINUSOIDAL METHOD)	RANGE ____ TO ____ Pa ____

Ambient Conditions: Temperature: ________ °C, Pressure: ______ Pa; ____ Humidity ______ %

Dynamic Response:

Excitation (Volts or ma) ____________ ; Transducer Load ____________ ohms

Sinusoidal Generator ____________ ; Test Fluid ____________

Mounting Configuration ____________

Test Temperature ______ °C Quiescent Static Pressure ________ Pa ______

Sinusoidal pressure ______ Pa peak; Port excited: ____________

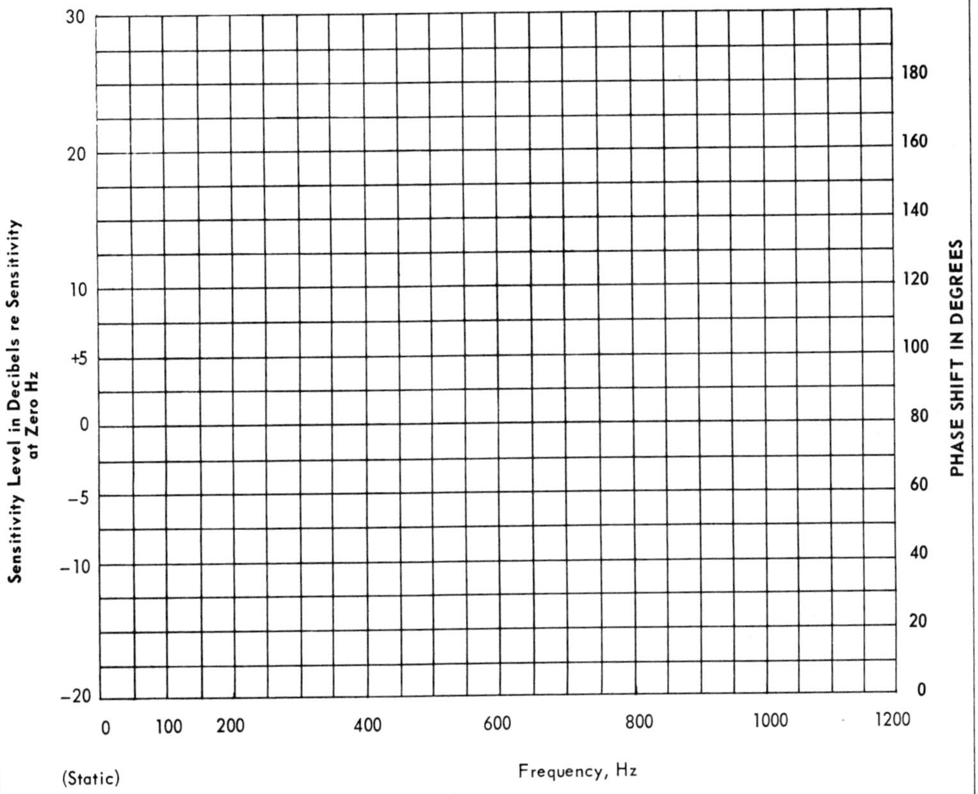

Sensitivity Level in Decibels re Sensitivity at Zero Hz: 30, 20, 10, +5, 0, −5, −10, −20

PHASE SHIFT IN DEGREES: 180, 160, 140, 120, 100, 80, 60, 40, 20, 0

Frequency, Hz: 0 (Static), 100, 200, 400, 600, 800, 1000, 1200

FIGURE 7.4.b

VENDOR'S PART NO.	TEST FACILITY	CUSTOMER'S PART NO.
VENDOR	**POTENTIOMETRIC PRESSURE TRANSDUCER**	SERIAL NO.
REPORT NO.		CUSTOMER
TYPE OF TEST	**ACCELERATION/VIBRATION TEST RECORD**	RANGE ____ TO ____ Pa ____

SKETCH OF TRANSDUCER SHOWING AXIS ORIENTATION:

ACCELERATION TEST

AXIS	+X	−X	+Y	−Y	+Z	−Z
Output Before Accel (%VR)						
Applied Accel. (G)						
Output During Accel. (%VR)						
Accel. Error (%VR)						

COMMENTS

Pressure Level Used: ________ Pa ________

Max. Accel. Error: + ________, − ________ %VR

Pre-Accel.
Static Error Band: + ________, − ________ %VR

Accel. Error Band: + ________, − ________ %VR

(Allowed Accel. Error Band ± ________ %VR)

Tested By: ________ (Technician)

________ (Test Engineer)

Date: ________ Approved By: ________

Witnessed By: ________ (________)

Witnessed By: ________ (________)

VIBRATION TEST

AXIS	X			Y			Z		
Pressure Level Used	Pa			Pa			Pa		
Output Before Vib.	%VR			%VR			%VR		
	Freq. (Hz)	Error Pol.	Error %VR	Freq. (Hz)	Error Pol.	Error %VR	Freq. (Hz)	Error Pol.	Error %VR
Vibration Error									

Max. Vib. Error: + ________, − ________ %VR

Pre-Vib.
Static Error Band: + ________, − ________ %VR

Vib. Error Band: + ________, − ________ %VR

(Allowed Vib. Error Band: ± ________ %VR

Tested By: ________ (Technician)

________ (Test Engineer)

Date: ________ Approved By: ________

Witnessed By: ________ (________)

Witnessed By: ________ (________)

COMMENTS

FIGURE 7.5

Instrument Society of America

VENDOR'S PART NO.	TEST FACILITY	CUSTOMER'S PART NO.
VENDOR		SERIAL NO.
REPORT NO.	**TRANSDUCER TEST REPORT**	CUSTOMER
TYPE OF TEST	**POTENTIOMETRIC PRESSURE TRANSDUCER**	RANGE ______ TO ______ Pa ____

SUMMARY OF RESULTS:

Test	Tested Per Proced. No. or Test Waived Per	Par. No.	Pass	Fail: Error	Fail: Electr.	Fail: Mechan.	Fail: See Comments	☐ Error ☐ Error Band: + %VR	– %VR
Initial P.T. (Performance Test)									
Resolution								Avg.: __ %VR Max.: __ %VR	
Weight								________	________
Dead Volume								________ Cu. ____	
Vol. Change over Press. Range								________ Cu. ____	
Mounting									
Loading Max. Z_L / Min. Z_L									
Frequency Response								Flat (+__%): __ To __ Hz	
Response Time								______ msec., ____ %Ovs.	
Low Temp. ______ °C									
P.T. After Low Temp.									
High Temp. + ______ °C									
Add'l. Temp. ______ °C									
P.T. After High Temp.									
______ g_n Vibration									
P.T. After ______ g_n Vibr.									
Acceleration									
P.T. After Accel.									
Life									
Burst Pressure									

Tested By: ______________________ Date Test Started: ____________ Date Test Finished: ________

Approved By: ______________________ Approved By: ______________________

Title Title

FIGURE 7.6

VENDOR'S PART NO.	TEST FACILITY	CUSTOMER'S PART NO.
VENDOR		CUSTOMER
PURCHASE ORDER NO.	**POTENTIOMETRIC PRESSURE TRANSDUCER**	SERIAL NO.
	INDIVIDUAL ACCEPTANCE TEST AND CALIBRATION RECORD (Static Error Band Calibration)	RANGE ____ TO ________ Pa

Visual Inspection: Dimensions ☐ Threads ☐ Finish ☐ Nameplate ☐ Receptacle (or other Electrical Conn.) ☐

Electrical Inspection: Element Resistance ________ ohms Insulation Resistance ____ Megohms at ________ Vdc

Breakdown Voltage Rating • ______ Vac, ______ Hz. ☐ Electrical Connection ☐

Calibration (Undithered) • ________V ____ Excitation Z_L used Megohms

Pressure (Pa)	Theor. Output (%VR)	First Calib. Cycle (Output (%VR)		Second Calib. Cycle Output (%VR)		Max. Error (%VR)		
		Increase	Decrease	Increase	Decrease	+	−	

All Error Bands Ref. To ____________________

		P Only: Performance Test ±• ________ Pa ___ Ref. Press.				Perf. Test ____ Minutes, After___ Pa ___Neg. Proof Press. (Overload Output: _____ _____)				Perf. Test __ Minutes, After ____ Pa ____Pos. Proof Press. (Overload Output: _____ _____)			
Pressure (Pa)	Theor. Output	Output (%VR)		Error (%VR)		Output (%VR)		Error (%VR)		Output (%VR)		Error (%VR)	
				+	−			+	−			+	−
				+	−			+	−			+	−
				+	−			+	−			+	−
		SEB.: +_____%, − _____ %VR				S.E.B.: +_____%, − _____ %VR				S.E.B.: +_____% − _____ %VR			

Full-Scale X-Y Plot: ______	Static Error Band (S.E.B.): (Calib.:) + ________%, − ________ %VR S.E.B.: (All Tests) +______%, −_____ %VR Allowed ± ____________ %VR

Tested By: ____________________ Date: ____________ Approved: ____________________

FIGURE 7.7

SPECIFICATIONS AND TEST FOR STRAIN GAGE FORCE TRANSDUCERS*

Instrument Society of America

TABLE OF CONTENTS

LIST OF FIGURES

*Reproduced by permission of Instrument Society of America.
ISA-S37.6, 1969.
ISA-S37.8, 1975.

1 PURPOSE

This standard establishes the following for strain-gage force transducers:

1.1 Uniform general specifications for design and performance characteristics.

1.2 Uniform acceptance and qualification test methods, including calibration techniques.

1.3 Uniform presentation of test data.

1.4 A drawing symbol for use in electrical schematics.

2 SCOPE

2.1 This Standard covers strain-gage force transducers, primarily those used in measurement systems.

2.2 Included among the specific versions of strain-gage force transducers, to which this Standard is applicable, are the following:

Tension Transducers
Compression Transducers
Universal (Combination Compression and Tension) Transducers

2.3 Terminology used in this document is defined either herein or in ISA Standard S37.1, Electrical Transducer Nomenclature and Terminology.

3 DRAWING SYMBOL

The drawing symbol for a strain-gage transducer is a square of dimensions 2x by 2x, with an added equilateral triangle, the base of which is the left side of the square. The triangle symbolizes the sensing element. The letter "F" in the triangle designates "force", and the additional sub-positioned letters denote the second modifier.

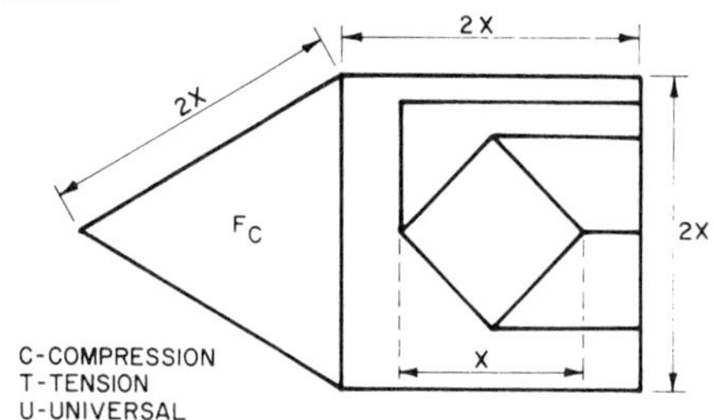

The strain-gage bridge is symbolized by a small square, with diagonals x by x, centered in the large square. The diagonals of the small square are drawn perpendicular to the sides of the large square. Lines from each apex of the small square projected to the right side of the large square represent the electrical leads.

4 CHARACTERISTICS

4.1 Design Characteristics

4.1.1 Basic Mechanical Design Characteristics

The following mechanical design characteristics shall be listed:

4.1.1.1 Type of Force Transducer — Tension, compression, or universal.

4.1.1.2 Physical Dimensions — Outline drawing to be provided with dimensions in millimeters (inches).

4.1.1.3 Force Connection — Force connections (both ends) shall be indicated on the outline drawing giving sufficient information regarding location, size, and tolerance of connection features, as well as any special considerations, to enable proper application of forces to the transducer.

4.1.1.4 Mountings and Mounting Dimensions — Outline drawing shall indicate method of mounting with dimensions in millimeters (inches).

4.1.1.5 Location of Electrical Connection — Indicate location and orientation of electrical connector or connecting wiring.

4.1.1.6 Overload Rating, Ultimate — Specify percentage of rated force that will not result in structural failure at environmental extremes.

4.1.1.7 Mounting Torque — Allowable mounting torque shall be specified if it will tend to effect transducer performance.

4.1.1.8 Weight — The weight of the transducer shall be specified in kilograms (pounds).

4.1.1.9 Identification — The following characteristics shall be given and preferably permanently affixed to the outside of the transducer case:

Nomenclature of transducer
(per ISA S37.1, Section 3)
Manufacturer's name, part number,
and serial number
Range
Excitation
Identification of electrical connections
Bridge identification, if more than one bridge provided.

Listing of the following characteristics is optional:

Sensitivity (usually millivolts
per volt at full scale load)
Customer's specification and/or part number
Maximum allowable force that will not
influence prescribed performance
Maximum operating temperature range

4.1.1.10 Temperature Range, Safe — Temperature range of environment in which the transducer may be used and which will not cause permanent calibration shift or permanent change in any of its characteristics shall be listed.

4.1.2 Supplemental Mechanical Design Characteristics

Listing of the following mechanical design characteristics is optional:

Case Material
Surface Finish
Type of Strain-Gage Used — Metallic;
bonded or unbonded, wire or foil;
Semiconductor; bonded or unbonded
Location of Strain-Gage — Mounted
directly on force sensing element or

mounted on auxiliary member activated by force sensing element
Number of Active Strain-Gage Bridge Arms (elements)
One, Two-arm active, Four-arm bridge
Number of Strain-Gage Bridges
Mounting Surface Requirements

4.1.3 Basic Electrical Design Characteristics

The following electrical design characteristics shall be listed. They are applicable at "ambient conditions" as specified in Section 4.2.

4.1.3.1* Excitation — Expressed as "____ volts dc" or "____, volts rms at ____ hertz," or, expressed as "____ milliamps dc" or "____ milliamps rms at ____ hertz." Preferred values of voltage 5, 10, 15, 20, and 28 volts.

4.1.3.2* Maximum Excitation — Expressed as "____ volts dc" or "____ volts rms at ____ hertz", or, expressed as "____ milliamps dc" or "____ milliamps rms at ____ hertz", and defined as the maximum value of excitation voltage which will not permanently damage the transducer.

4.1.3.3* Input Impedance — Expressed as "____ ± ____ ohms at ____ ± ____ hertz" and ____ °C (°F)." If impedance is resistive, specify "dc". Note: Output terminals are to be open-circuited for this measurement.

4.1.3.4* Output Impedance — Expressed as "____ ± ____ ohms at ____ ± ____ hertz" and ____ °C (°F)." If impedance is resistive, specify "dc". Note: If input terminals are to be short-circuited for this measurement, so specify.

4.1.3.5 Electrical Connections — Whether the electrical termination is by means of a connector or a cable, the pin designation or wire color code shall conform to the following:

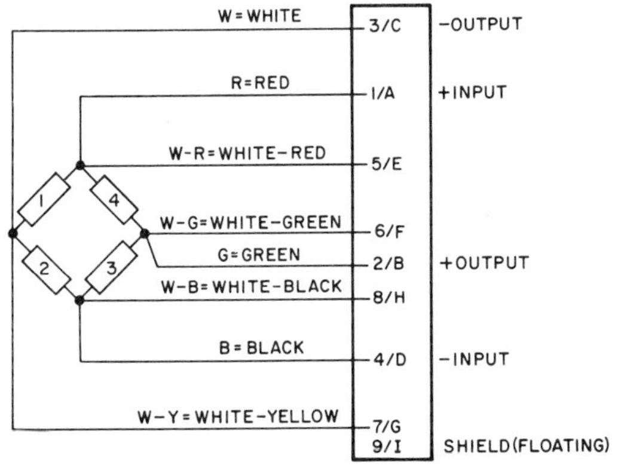

PRIMARY WIRING TERMINALS 1/A, 2/B, 3/C, 4/D
AUXILIARY WIRING TERMINALS 5/E, 6/F, 7/G, 8H (OPTIONAL)

Notes:

1. The output polarities indicated on the above wiring diagram apply when an *increasing* force (compression or tension) is applied to the transducer. For universal force transducers, the indicated polarities apply when the tension force is applied to the transducer; a compression force will produce a negative output.

2. For shielded transducers, pins 5, 7, and 9 shall be shield terminals for 4, 6, and 8 wire systems, respectively.

3. Type connection: Solder or weld.

4.1.3.6 Insulation Resistance — Expressed as "____ megohms at ____ volts dc at ____ °C (°F) between all terminals or leads connected in parallel, and the transducer case."

4.1.4 Supplemental Electrical Design Characteristics

Listing of the following design characteristics is optional:

4.1.4.1 Shunt Calibration Resistor(s) — Expressed as "____ ± ____ ohms for ____ % ± ____ % of full scale output at ____ °C(°F)."

Note: The terminals across which the resistor(s) is (are) to be placed shall be specified if the resistor(s) is (are) listed.

4.2 Performance Characteristics

The pertinent performance characteristics of strain-gage force transducers should be tabulated in the order shown. Unless otherwise specified, they apply at the following ambient conditions: Temperature 23 ±2°C (73.4°F) ±3.6°F; Relative Humidity 90% maximum; Barometric Pressure 98 ±kPa (29 ±3 inches of Hg).

4.2.1* Range — Usually expressed as "____ to ____ newtons (pounds force) compression or tension" or "____ to ____ newtons (pounds force) compression and ____ to ____ newtons (pounds force) tension".

Note: If 4.2.2 and 4.2.3 are used to specify performance characteristics, the tolerance in 4.2.3 may be omitted. Alternately, the following may be specified: 4.2.3 — 4.2.6.

4.2.2* End Points — Expressed as "____ ± ____ mV and ____ ± ____ mV open circuit per volt (milliamps) excitation", or "____ ± ____ mV and ____ ± ____ mV open circuit at ____ volts (milliamps) excitation".

4.2.3* Full Scale Output (FSO) — Expressed as "____ ± ____ mV open circuit per volt (milliamps) excitation," or "____ ± ____ mV open circuit at ____ volts (milliamps) excitation".

4.2.4 Zero-Measurand Output — Expressed as "± ____ % of full scale output". Determined at full rated excitation, with zero measurand applied to the force transducer.

4.2.5 Zero Drift — Expressed as "±____% of full scale output over a period of ____ (specify time) with no load applied".

4.2.6 Sensitivity Drift — Expressed as "±____% full ____ newtons (pounds force) applied".

4.2.7* Linearity — Expressed as "____ linearity within ±____% of full scale output in ____ (specify direction(s) of loading)".

Note: The type of linearity specified shall be one of the types defined in ISA S37.1; namely, end point, independent, least squares, terminal, or theoretical slope.

4.2.8 Hysteresis — Expressed as "____% of full scale output upon application of ascending and descending forces including rated force." Alternately, 4.2.7 and 4.2.8 may be combined as:

4.2.9 Hysteresis and Linearity — Expressed as "combined hysteresis and linearity within ±____% of full scale output upon application of ascending and descending forces including "rated force."

4.2.10* Repeatability — Expressed as "within ____% of full scale output over a period of ____ (specify time) and with ____ cycles of load application".

Alternately 4.2.7, 4.2.8, and 4.2.10 may be combined as:

4.2.11 Static Error Band — Expressed as "±____% of full scale output as referred to ____ straight line," (see 4.2.7).
Note: The static error band includes errors due to linearity, hysteresis and repeatability.

4.2.12 Creep at Load — Expressed as "± ____% of full scale output with the transducer subjected to rated force for a period of ____ (specify time)".

4.2.13 Creep Recovery — Expressed as "± ____% of full scale output measured at no load and over a period of ____ (specify time) immediately following removal of rated force, that force having been applied for an identical period of time as specified in 4.2.12".

4.2.14* Warm-up Period — Expressed as "____ minutes for subsequent drifts in sensitivity of zero-measurand balance not to exceed ____% of full scale output".

4.2.15 Static Spring Constant — Expressed in newtons per meter or (pounds force per inch), see Para. 6.3.

4.2.16 Equivalent Dynamic Masses — Expressed in kilograms (pounds mass), for both ends of transducer. See Para. 6.3.

4.2.17 Internal Mechanical Damping — Expressed in newtons per meter/second relative velocity (pounds force per inch/second relative velocity), between ends at a frequency of ____ Hz and a dynamic load of ±____ newtons (pounds force).

4.2.18 Overloading Rating, Safe — Expressed as "application of ____ newtons (pounds force) for ____ minutes will not cause permanent changes in transducer performance beyond specified static error band".

4.2.19 Rated Force — Expressed as "____ newtons (pounds force) either compression or tension." This is the maximum axial force the transducer is designed to measure within its specifications.

4.2.20* Thermal Sensitivity Shift — Expressed as "±____% of sensitivity____per °C (°F) temperature change over temperature range from ____ to____ °C (°F)".

4.2.21* Thermal Zero Shift — Expressed as "±____% of full scale output per____°C (°F) temperature change over temperature range from____to____°C (°F)."

4.2.22* Temperature Error Band — Expressed as "output values are within ±____% of full scale output from the straight line establishing static error band (as defined in 4.2.11) over temperature range from____to____°C (°F)."

4.2.23* Temperature Gradient Error — Expressed as "less than ± ____ % of full scale output while at zero load and subjected to a step function temperature change from____to____°C (°F) lasting for ____ minutes and applied to____ (specify particular part) of the transducer".

4.2.24 Cycling Life — Expressed as "____ full scale cycles over which transducer shall operate without change in characteristics beyond its specified tolerances".

4.2.25 Other Environmental Conditions — Other pertinent environmental conditions which should not change transducer performance beyond specified limits should be listed. Examples are:

Shock — Triaxial
High Level Acoustic Excitation
Humidity
Salt Spray
Electromagnetic Radiation
Magnetic Fields
Nuclear Radiation

4.2.26 Storage Life — Expressed as "Transducer can be exposed to specified environmental storage condition for ____ (days, months, years) without changing the following performance characteristics beyond their specified tolerances."

Note: Environmental storage conditions shall be described in detail. Pertinent performance characteristics (examples: sensitivity, zero drift) shall be specified.

4.2.27 Abnormal Loading Effects: (Refer to Figure I on page 8.)

4.2.27.1 Concentric Angular Load Effect — Expressed as "±____% of full scale output difference from true output (axially loaded output multiplied by cosine of angle) resulting from a load applied concentric with the primary axis at the point of application and at ______ degrees angle with respect to the primary axis".

4.2.27.2 Eccentric Angular Load Effect — Expressed as "±______% of full scale output difference from true output multiplied by cosine of angle) resulting from a load applied eccentric with the primary axis and at ______ degrees angle with respect to the primary axis".

4.2.27.3 Eccentric Load Effect — Expressed as "±______% of full scale output difference from axially loaded output resulting from a load parallel to but displaced ________ millimeters (inches) from concentricity with the primary axis".

4.3 Additional Terminology

Ambient Pressure Effects — The change in sensitivity and the change in zero-measurand output due to subjecting the transducer to a specified ambient pressure change.

Creep at Load — The change in output occurring with time under rated load and with all environmental conditions and other variables remaining constant.

Creep Recovery — The change in zero-measurand output occurring with time after removal of rated load, which had been applied for an identical period of time as employed in evaluating Creep at Load.

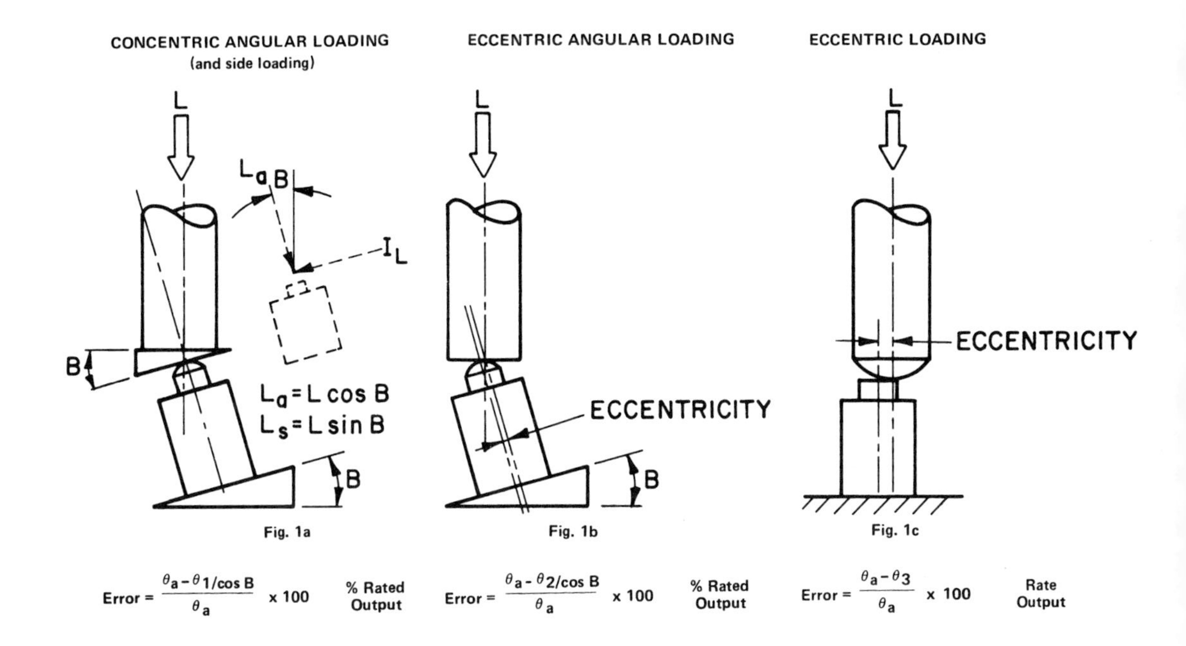

θ_a = Rated output at rated axial loading.

$\theta_1, \theta_2, \theta_3$ = Output under any unfavorably rated loading conditions.

L = Load
L_a = Axial Load
L_s = Side Load

FIGURE 1

4.4 TABULATED CHARACTERISTICS VERSUS TEST REQUIREMENTS

Characteristic	Paragraph	Design Characteristic		Verified During	
		Basic	Supp.	Acceptance	Qual.
Type of Force Transducer	4.1.1.1	X			
Physical Dimensions	4.1.1.2	X			
Force Connection	4.1.1.3	X			
Mounting Dimensions	4.1.1.4	X			
Electrical Connection Location	4.1.1.5	X			
Overload Rating, Ultimate	4.1.1.6	X			
Mounting Force or Torque	4.1.1.7	X			
Weight	4.1.1.8	X			
Identification	4.1.1.9	X			
Temperature Range	4.1.1.10	X			
Case Material	4.1.2		X		
Surface Finish	4.1.2		X		
Type of Strain-Gage Used	4.1.2		X		
Location of Strain-Gage	4.1.2		X		
Number of Active Strain-Gage Elements	4.1.2		X		
Number of Strain-Gage Bridges	4.1.2		X		
Excitation	4.1.3.1	X			
Maximum Excitation	4.1.3.2	X			
Input Impedance	4.1.3.3	X		5.2.9	
Output Impedance	4.1.3.4	X		5.2.9	
Electrical Connections	4.1.3.5	X			
Insulation Resistance	4.1.3.6	X		5.2.8	
Shunt Calibration Resistor	4.1.4.1		X		
Range	4.2.1	X			
End Point	4.2.2	X		5.2.3	
Full Scale Output	4.2.3	X		5.2.3	
Zero-Measurand Output	4.2.4	X		5.2.3	
Zero Drift	4.2.5	X		5.2.4	
Sensitivity Drift	4.2.6	X		5.2.4	
Linearity	4.2.7	X		5.2.3	
Hysteresis	4.2.8	X		5.2.3	
Hysteresis and Linearity	4.2.9	X		5.2.3	
Repeatability	4.2.10	X		5.2.3	
Static Error Band	4.2.11	X		5.2.3	
Creep	4.2.12	X		5.2.5	
Creep Recovery	4.2.13	X			
Warm-up Period	4.2.14	X		5.2.6	
Static Spring Constant	4.2.15	X			6.3
Equivalent Dynamic Masses	4.2.16	X			6.3
Internal Mechanical Damping	4.2.17	X			6.3
Overload Rating	4.2.18	X		5.2.7	
Rated Force	4.2.19	X			
Thermal Sensitivity Shift	4.2.20	X			6.1
Thermal Zero Shift	4.2.21	X			6.1
Temperature Error Band	4.2.22	X			6.1
Temperature Gradient Error	4.2.23	X			6.2
Cycling Life	4.2.24	X			6.4
Other Environmental Conditions	4.2.25		X		6.5
Storage Life	4.2.26	X			6.6
Abnormal Loading Effects	4.2.27	X			6.7

5 INDIVIDUAL ACCEPTANCE TESTS AND CALIBRATIONS

5.1 Basic equipment necessary to perform individual acceptance tests and calibrations of strain-gage force transducers.

The basic equipment for acceptance tests and calibration consists of a force calibrator, a source of electrical excitation for the strain-gages, and a device which measures the electrical output of the transducer. The errors or uncertainties of the measuring system comprising these three components should be less than one-third of the permissible tolerance of the transducer performance characteristic under evaluation. Traceability to the National Bureau of Standards should be established.

5.1.1 Force Calibrator — The maximum inaccuracy of the force calibrator, for ranges of 50 000 newtons and less, should be not more than one-fifth the permissible tolerance of the transducer performance characteristic under evaluation. Force ranges in excess of 50 000 newtons require force calibrator maximum inaccuracy not more than one-half the permissible tolerance of the transducer performance characteristic under evaluation. Range of the instrument supplying or monitoring the calibration force should be selected to provide the necessary accuracy to 125% of the full scale range of the transducer.

The force calibrator may be either continuously variable over the range of the instrument, or may vary in discrete steps provided that the steps can be programmed in such a manner that the transition from one force to the next during calibration is accomplished without creating a hysteresis error in the measurement due to over-shoot.

DEAD WEIGHT CALIBRATOR

Typical Ranges (Tension or Compression)

0-2000 newtons	Max. Error ±0.01% of Test Load
0-5000 newtons	Max. Error ±0.01% of Test Load
0-20 000 newtons	Max. Error ±0.01% of Test Load
0-50 000 newtons	Max. Error ±0.01% of Test Load
0-500 000 newtons	Max. Error ±0.01% of Test Load

PROVING RING CALIBRATOR

Typical Ranges (Tension or Compression)

0-200 newtons	Max. Error ±0.1% Full Scale
0-500 newtons	Max. Error ±0.1% Full Scale
0-2000 newtons	Max. Error ±0.1% Full Scale
0-5000 newtons	Max. Error ±0.1% Full Scale
0-20 000 newtons	Max. Error ±0.1% Full Scale
0-50 000 newtons	Max. Error ±0.1% Full Scale

HIGH ACCURACY REFERENCE FORCE TRANSDUCER

Typical Ranges (Tension or Compression)

0-500 newtons	Max. Error ±0.1% Full Scale
0-1000 newtons	Max. Error ±0.1% Full Scale
0-5000 newtons	Max. Error ±0.1% Full Scale
0-10 000 newtons	Max. Error ±0.1% Full Scale
0-50 000 newtons	Max. Error ±0.1% Full Scale
0-100 000 newtons	Max. Error ±0.1% Full Scale

5.1.2 Stable Source of Electrical Excitation of Accurately-Known Amplitude

For dc excitation, commonly used sources are line-powered, electronically regulated, power supplies.

For ac excitation, commonly used sources are the power line in association with a step-down transformer or an oscillator. Where radiometric measurement techniques are employed, input power should be within both instrument and force transducer specified range.

5.1.3 Readout Instrument

Examples of suitable devices are:

MANUALLY BALANCED POTENTIOMETER

Typical Ranges

0 to 16 millivolts, Maximum error ±0.015% of reading or ±1 microvolt, whichever is greater.
0 to 160 millivolts, Maximum error ±0.015% of reading or ±3 microvolts, whichever is greater.
0 to 1.6 volts, Maximum error ±0.015% of reading or ±30 microvolts, whichever is greater.

DIGITAL ELECTRONIC VOLTMETER WITH PREAMPLIFIER

Typical Ranges	Sensitivity	Maximum Error
0 to 10 millivolts	1 microvolt	±0.02% of reading, or ±2 microvolts, whichever is greater.
0 to 100 millivolts	10 microvolts	±0.01% of reading, or ±10 microvolts, whichever is greater.

Note: The input impedance of the readout instrument should be as high as possible. Unless otherwise stated, adjustments and compensation of the transducer apply to open circuit conditions on the output terminals.

5.2 Calibration and Test Procedures

Results obtained during the calibration and test procedures should be recorded on data sheets like the sample data sheets in Section 7 of this report. These procedures shall be performed under ambient conditions as defined in Paragraph 4.2.

5.2.1 The transducer is inspected visually for mechanical defects, poor finish, and improper identification markings.

5.2.2 The transducer shall be connected to the force calibrator with axial alignment as specified by the manufacturer. The excitation source and readout instrument shall be connected to the transducer and turned on. Adequate warm-up time for test equipment shall be allowed before tests are conducted. The force calibrator and connecting hardware shall have passed a prior test for proper operation. It may be desirable prior to calibration to exercise the force transducer by applying rated load and returning to zero load, if so, the number of cycles and time duration should be noted on the data sheet.

5.2.3 Two or more complete calibration cycles (dependent on desired statistical confidence levels) are run consecutively, including five to ten points in both ascending and descending directions. Excitation amplitude shall be monitored as required.

From the data obtained during these tests, the following characteristics should be determined:

End Points	4.2.2
Full Scale Output	4.2.3
Zero Balance	4.2.4
Linearity	4.2.7
Hysteresis	4.2.8
(or Hysteresis and Linearity)	4.2.9
Repeatability	4.2.10
(or Static Error Band)	4.2.11

5.2.4 Repeated calibration cycles over a specified period of time after warmup, establish the following characteristics for that period of time:

Zero Drift	4.2.5
Sensitivity Drift	4.2.6

Note: They may be abbreviated cycles with fewer data points than required in 5.2.3.

5.2.5 Application of rated force to the transducer during a specified short period of time and measurement of changes in output at constant excitation during this time should establish:

Creep at Load	4.2.12

Note: See Figure 2 (page 14). Rate of application of force to be as high as possible without resonant excitation of transducer.

5.2.6 By measuring zero-measurand output and sensitivity over a period of time (one hour should suffice), starting with the application of excitation to the transducer, the following characteristic should be determined:

Warmup Period	4.2.14

Note: It is desirable to test for these effects separately, establishing the warmup change of zero-measurand output first.

5.2.7 After application of the specified overload a specified number of times (and in the specified duration for compression or tension), at least one complete calibration cycle shall be performed to establish that the performance characteristics of the transducer are still within specifications.

Overload Rating, Safe	4.2.18

5.2.8 Measure the insulation resistance between all terminals, or leads connected in parallel, and the case of the transducer with a megohmmeter or similar acceptable device, using a potential of 50 volts, unless otherwise specified. Insulation resistance should be measured at room temperature.

Insulation Resistance	4.1.3.6

5.2.9 A Wheatstone bridge (for dc) or impedance bridge shall be used to measure:

Input Impedance	4.1.3.3
Output Impedance	4.1.3.4

6 QUALIFICATION TESTS

6.1 Steady State Temperature Effects

The transducer shall be placed in a suitable temperature chamber. After allowing adequate stabilization time at a specified temperature, one or more calibration cycles shall be performed within the chamber. This procedure shall be repeated at an adequate number of temperatures within the operating temperature range of the transducer. These tests should establish the following characteristics:

Thermal Sensitivity Shift	4.2.20
Thermal Zero Shift	4.2.21
Temperature Error Band	4.2.22

6.2 Temperature Gradient Error

The force transducer, at "room temperature," shall be subjected to a thermal transient by immersion in a fluid which is kept at a specified temperature above or below "room conditions." With no force applied, the output is observed over a specified period of time.

Note: The type of fluid and method of application shall be specified.

These tests should establish:

Temperature Transient Error	4.2.23

6.3 Dynamic Characteristics

The dynamic response of an installed force transducer depends on the stiffness and mass distribution throughout the entire system in which the force transducer is a part. In many cases, especially those which have many springs and masses, the dynamic response of the installed force transducer may best be determined experimentally by applying a suitable time-varying force to the complete system and measuring the output vs. time response of the transducer.

Alternatively, if the distribution of stiffness and mass throughout the system is known quantitatively, the response of the system to a time-varying force may be computed analytically. This approach, however, is apt to be very cumbersome unless (1) attention is limited to frequencies from zero to a little above the lowest natural frequency and (2) the stiffness and mass system is relatively simple.

The dynamic characteristics of the force transducer itself may be determined over a wide frequency range with suitable test fixtures and instrumentation either by applying sinusoidal forces or step function forces. Generally, such a determination yields very complex results. Over a limited low frequency range, however, the dynamic response of a force transducer may be defined adequately in terms of a simple equivalent model as shown in Figure 3.

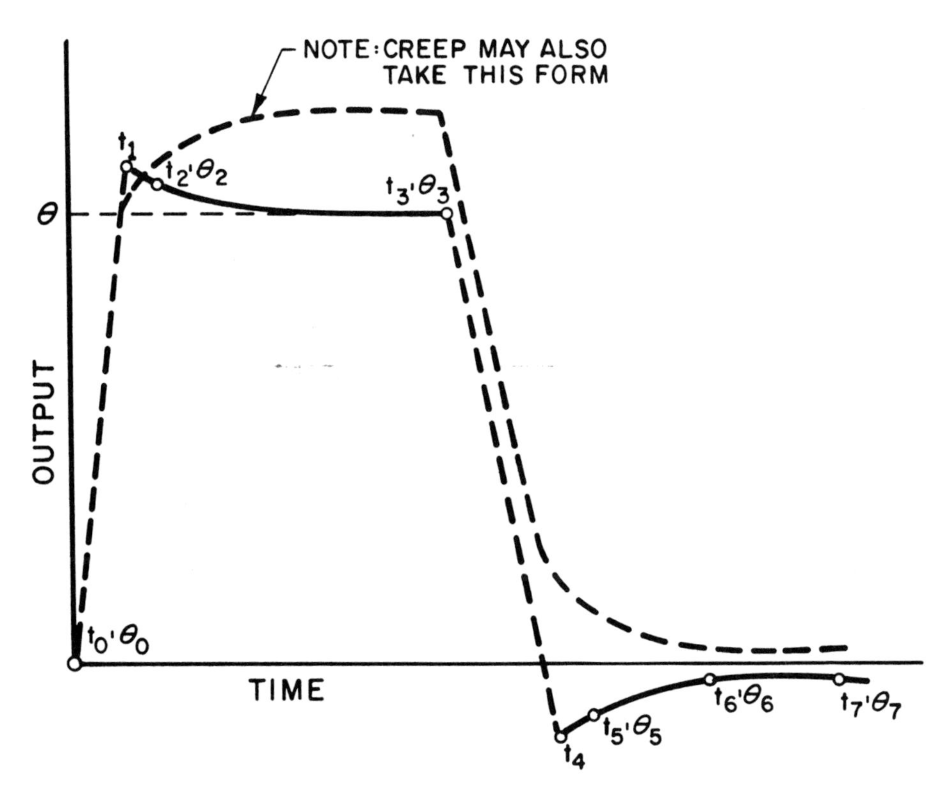

$t_1 - t_o$ = Load Application. Should be carefully considered in comparative creep measurement.

$t_2 - t_1$ = Should be as short as possible. (Suggested 5-10 s.)

$t_3 - t_2$ = Creep measurement period. Suggested 3 min., for short term and 30 min. for long term.

$\frac{\theta_2 - \theta_3}{\theta} \times 100$ = Creep in % rated output.

$t_4 - t_3$ = Load release period. (Should equal $t_1 - t_o$).

$t_5 - t_4$ = Should be as short as possible. (Suggested 5-10 s.)

$t_6 - t_5$ = Creep recovery period. Suggested 3 min. for short term and 30 min. for long term.

$\frac{\theta_5 - \theta_6}{\theta} \times 100$ = Creep recovery in % rated output.

t_7 = Time at which zero return is measured.

$\frac{\theta_7 - \theta_o}{\theta} \times 100$ = Zero return in % rated output.

FIGURE 2

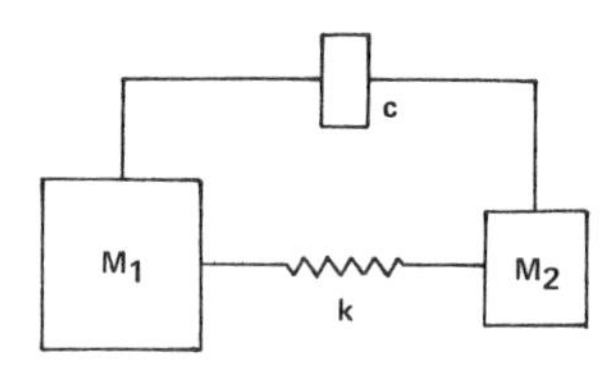

FIGURE 3

k = Static Spring Constant
c = Damping Force Parameter
M_1 = Effective Mass of "Base" of Transducer*
M_2 = Effective Mass of "Top" of Transducer*

*The "base" and "top" of the transducer are defined in the outline drawing.

M_1 and M_2 may be determined experimentally or analytically so that the calculated natural frequency of the M_1-k-M_2 model agrees with the actual first mode natural frequency of the transducer. The effective damping parameter generally varies with the force amplitude and possibly with frequency; consequently, the value of c must be determined for the particular load and frequency values of interest.

6.4 Life Test

After applying the specified number of full range excursions of force, at least one complete calibration cycle shall be performed to establish minimum value of:

Cycling Life 4.2.28

6.5 Effects of Other Environments

Expose transducer to other specified environmental conditions followed in each case by one complete calibration cycle to test ability of transducer to perform satisfactorily after such exposure.

6.6 Storage Life Test

After storing the transducer under specified conditions for the specified period of time, two complete calibration cycles shall be performed to establish:

Storage Life 4.2.30

6.7 Abnormal Loading Effects

6.7.1 For determination of the effects of concentric angular loading (and side loading), insert wedge blocks above and below the force transducer as illustrated in Figure 1a. The angle subtended by the two larger surface areas (B) of each block should be equivalent to the angle of interest or should result in the side load of interest.

6.7.1.1 Measure zero-measurand output.

6.7.1.2 Apply rated load and read output as soon as load has stabilized.

6.7.1.3 Remove load and record zero-measurand output after output has stabilized.

6.7.2 For determination of the effects of eccentric angular loading, remove the upper wedge block (Figure lb.) and repeat steps 6.7.1.1 through 6.7.1.3. If eccentricities other than that obtained in the foregoing are desired, a flat load button should be used and the amount of eccentricity adjusted through placement of the force transducer under a convex loading ram surface.

6.7.3 For determination of the effects of eccentric loading, remove the lower wedge block, use a flat load button, and adjust eccentricity through placement of the force transducer under the convex loading ram surface (see Figure 1c.). Repeat steps 6.7.1.1 through 6.7.1.3.

6.7.4 The effects of the various types of loading related to axial loading conditions can be determined in accordance with the expressions included in Figure 1.

7 TEST REPORT FORMS

7.1 The test report forms listed below are recommended for use during the testing of strain-gage force transducers.

7.2 When using the forms, all pertinent information shall be inserted in its proper place. On some forms, blank space has been provided for additional tests. Where the test is prolonged, more than one form may be required.

7.3 "Individual Acceptance Tests and Calibrations" (Figure 4) used during acceptance testing of Section 5 and may also be used during qualification testing of Section 6.

"Environmental Test Record" (Figure 5) used to record thermal sensitivity shift, thermal zero shift, temperature error band, temperature transient error, and other environmental tests.

8 BIBLIOGRAPHY

1. MIL-E-5272C (ASG) "*Environmental Testing, Aeronautical and Associated Equipment, General Specifications for*" (1970).

2. MIL-E-5400P, (ASG) "*Electronic Equipment, Aircraft, General Specification for*" (1974).

3. MIL-STD-810C, "*Environmental Test Methods for Aero-Space and Ground Equipment*" (1975).

4. ANSI MC6.1-1975 (ISA S37.1-75) "*Electrical Transducer Nomenclature and Terminology,*" September 10, 1975.

5. ANSI MC6.2-1975 (ISA S37.3-75) "*Specifications and Tests for Strain Gage Pressure Transducers,*" October 23, 1975.

6. "*Standard Load Cell Terminology and Definitions*", developed by the Industrial Instrument Section of the Scientific Apparatus Makers Association and published by the Scale Manufacturer's Association (1962).

7. "*Handbook of Transducers for Electronic Measuring Systems*", Norton, Harry N., Prentice-Hall, Inc. (1969).

VENDOR'S PART NO.	STRAIN GAGE FORCE TRANSDUCER INDIVIDUAL ACCEPTANCE TESTS & CALIBRATIONS	CUSTOMER'S PART NO.
TEST FACILITY		SERIAL NO. BRIDGE NO.
Ambient Conditions Temperature ______ °C Pressure ______ kPa Humidity ______ %	☐ Calibration ☐ ______ Overload ☐ Functional Test ☐ ______ Test	VENDOR
	☐ Concentric Angular, ___ ° ☐ Eccentric ______ mm ☐ Eccentric Angular, ___ ° and ______ mm	RANGE ____ TO ____ NEWTONS

1. Visual: Mechanical ☐ Finish ☐ Nameplate ☐ Electr. Conn. ☐
2. Electrical: Input Impedance ______ ohms, Output Impedance ______ ohms, Ins Res. ______ MΩ ______ Vdc
3. Calibration and Proof Load Test at ______ V at ______ Hz ______ Excitation after ______ minutes warm-up time
4. Load Impedance ______ ohms or ______ mA Excitation
5. Exercised ______ times, to ______ newtons.

Test Load Newtons	Theoretical Output (mV)	Output (mV) - Run 1		Output (mV) - Run 2		Overload Output (mV)	Output (mV) - Run 3		Maximum Error (mV)	
		Test Time	Minutes	Test Time	Minutes		Test Time	Minutes		
			Decrease	Increase	Decrease		Increase	Decrease	+	−
						NOT APPLICABLE				
Over / Rev.		NOT APPLICABLE				------		NOT APPLICABLE		

STATIC ERROR BAND + ______ % ______ % FSO (Allowed: ± ______ % FSO).
Referred to ______________________
______________________ Linearity

______ LINEARITY: + ______ − ______ % FSO (Allowed: ______ % FSO)
HYSTERESIS: ______ % FSO (Allowed ______ % FSO) REPEATABILITY: ______ % FSO (Allowed: ______ % FSO)
ZERO-MEASURAND OUTPUT: T ______ % FSO (Allowed: ______ % FSO) CREEP: ______ % FSO over ______ minutes
(Allowed: ______ % FSO)
ZERO SHIFT: ______ % FSO over a period of ______ (number) ______ (units) (Allowed: ______ % FSO)

SENSITIVITY SHIFT: ______ % over a period of ______ (number) ______ (units) (Allowed: ______ % sensitivity)

END POINTS (______): ______ and ______ mV (Allowed: ______ and ______ mV)
FULL SCALE OUTPUT (Run 2): ______ mV (Allowed: ______ mV) OVERLOAD (after Run 2) ______ %

Equipment Used:	Defects noted or Comments

BY: ______ DATE: ______ APPROVED BY: ______

Note:

FIGURE 4

VENDOR'S PART NO.	TEST FACILITY	CUSTOMER'S PART NO.
VENDOR		SERIAL NO. BRIDGE NO.
REPORT NO.	STRAIN GAGE FORCE TRANSDUCER	CUSTOMER
TYPE OF TEST	ENVIRONMENTAL TEST RECORD	RANGE ______ TO ______ NEWTONS

Maximum ______ volts
at ______ Hz.
or ______ mA
Warmup Time ______ Minutes

Type of Environment — Level
☐ Before
☐ During ______ ______
☐ After

For temperature tests, temperature measured at ______

Force	Theoretical	mV (Run 1)		mV (Run 2)		Overload	mV (Run 3)		Maximum Error	
Newtons	mV	Increase	Decrease	Increase	Decrease	mV	Increase	Decrease	+	-

Error Band + ______ % – ______ % FSO (Referred To ______) Allowed: ± ______ % FSO

Overload ______ % Rated Range for ______ Minutes

Ins. Resistance: ______ megohms At ______ Vdc

Zero Shift ______ % FSO

Sensitivity Shift ______ %

Comments: ______

Error Plot, Run 3

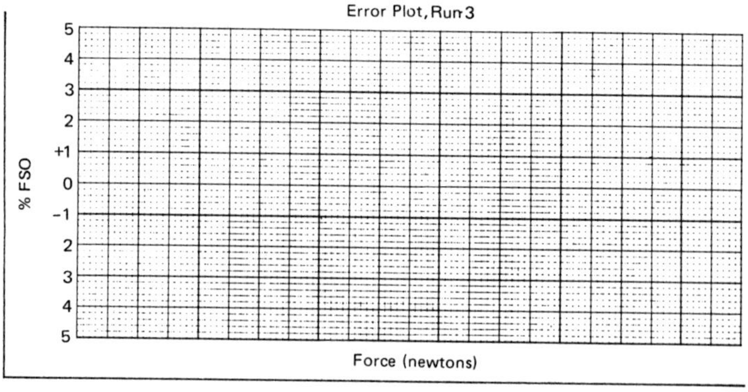

Tested By ______ Date Test Started ______ Date Test Finished ______

Approved By ______ Title

Approved By ______ Title

FIGURE 5

Appendix D
Engineering Data

1. METRIC-ENGLISH CONVERSION FACTORS
2. DATA SHEET FOR PRESSURE SWITCHES
3. CORROSION RESISTANCE OF VARIOUS MATERIALS
4. ALTITUDE-PRESSURE CHART
5. PRESSURE SWITCH COMPANY REFERENCE CHART

1. METRIC-ENGLISH CONVERSION FACTORS

Basic Unit of Measurement-Length (Meter)

The following charts give conversion factors for the basic unit of length and the units that are derived from this basic unit.

LENGTH

U.S. TO METRIC	METRIC TO U.S.
1 inch = 25.40 millimeters	1 millimeter = 0.03937 inch
1 inch = 2.540 centimeters	1 centimeter = 0.3937 inch
1 foot = 30.480 centimeters	1 meter = 39.37 inches
1 foot = 0.3048 meter	1 meter = 3.2808 feet
1 yard = 91.440 centimeters	1 meter = 1.0936 yards
1 yard = 0.9144 meter	1 kilometer = 0.62137 mile
1 mile = 1.609 kilometers	

To convert from	to	Multiply by
angstrom	meter (m)	$1.000\,000 \times 10^{-10}$
astronomical unit	meter (m)	$1.495\,98 \times 10^{11}$
caliber	meter (m)	$2.540\,000 \times 10^{-4}$
fathom	meter (m)	1.828 800
fermi (*fermtometer*)	meter (m)	$1.000\,000 \times 10^{-15}$
foot	meter (m)	$3.048\,000 \times 10^{-1}$
foot (*U.S. survey*)	meter (m)	1200/3937
foot (*U.S. survey*)	meter (m)	$3.048\,006 \times 10^{-1}$
inch	meter (m)	$2.540\,000 \times 10^{-2}$
league (*International nautical*)	meter (m)	$5.556\,000 \times 10^{3}$
league (*statute*)	meter (m)	$4.828\,032 \times 10^{3}$
league (*U.K. nautical*)	meter (m)	$5.559\,552 \times 10^{3}$
light year	meter (m)	$9.460\,55 \times 10^{15}$
micron	meter (m)	$1.000\,000 \times 10^{-6}$
mil	meter (m)	$2.540\,000 \times 10^{-5}$
mile (*international nautical*)	meter (m)	$1.852\,000 \times 10^{3}$
mile (*U.K. nautical*)	meter (m)	$1.853\,184 \times 10^{3}$
mile (*U.S. nautical*)	meter (m)	$1.852\,000 \times 10^{3}$
mile (*U.S. statute*)	meter (m)	$1.609\,344 \times 10^{3}$
parsec	meter (m)	$3.083\,74 \times 10^{16}$
pica (*printer's*)	meter (m)	$4.217\,518 \times 10^{-3}$
point (*printer's*)	meter (m)	$3.514\,598 \times 10^{-4}$
rod	meter (m)	5.029 200
statute mile (*U.S.*)	meter (m)	$1.609\,344 \times 10^{3}$
yard	meter (m)	$9.144\,000 \times 10^{-1}$

AREA

U.S. TO METRIC	METRIC TO U.S.
1 sq. inch = 645.16 sq. millimeters	1 sq. millimeter = 0.00155 sq. inch
1 sq. inch = 6.4516 sq. centimeters	1 sq. centimeter = 0.1550 sq. inch
1 sq. foot = 929.03 sq. centimeters	1 sq. meter = 10.7640 sq. feet
1 sq. foot = 0.0929 sq. meter	1 sq. meter = 1.196 sq. yards
1 sq. yard = 0.836 sq. meter	1 sq. hectometer = 2.471 acres
1 acre = 0.4047 sq. hectometer	1 hectare = 2.471 acres
1 acre = 0.4047 hectare	1 sq. kilometer = 0.386 sq. mile
1 sq. mile = 2.59 sq. kilometers	

Courtesy of *Design News Magazine*

Extracted from *Lyons' Enclyclopedia of Valves* (pages 250-267).

AREA (Continued)

To convert from	to	Multiply by
acre	meter² (m²)	$4.046\ 856 \times 10^{3}$
barn	meter² (m²)	$1.000\ 000 \times 10^{-28}$
circular mil	meter² (m²)	$5.067\ 075 \times 10^{-10}$
foot²	meter² (m²)	$9.290\ 304 \times 10^{-2}$
inch²	meter² (m²)	$6.451\ 600 \times 10^{-4}$
mile² (*U.S. statute*)	meter² (m²)	$2.589\ 988 \times 10^{6}$
section	meter² (m²)	$2.589\ 988 \times 10^{6}$
township	meter² (m²)	$9.323\ 957 \times 10^{7}$
yard²	meter² (m²)	$8.361\ 274 \times 10^{-1}$

VOLUME (Capacity)

U.S. TO METRIC	METRIC TO U.S.
1 fluid ounce = 2.957 centiliters = 29.57 cm³	1 centiliter = 10 cm³ = 0.338 fluid ounce
1 pint (liq.) = 4.732 deciliters = 473.2 cm³	1 deciliter = 100 cm³ = 0.0528 pint (liq.)
1 quart (liq.) = 0.9463 liter = 0.9463 dm³	1 liter = 1 dm³ = 1.0567 quarts (liq.)
1 gallon (liq.) = 3.7853 liters = 3.7853 dm³	1 liter = 1 dm³ = 0.26417 gallon (liq.)

To convert from	to	Multiply by
acre-foot	meter³ (m³)	$1.233\ 482 \times 10^{3}$
barrel (*oil, 42 gal*)	meter³ (m³)	$1.589\ 873 \times 10^{-1}$
board foot	meter³ (m³)	$2.359\ 737 \times 10^{-3}$
bushel (*U.S.*)	meter³ (m³)	$3.523\ 907 \times 10^{-2}$
cup	meter³ (m³)	$2.365\ 882 \times 10^{-4}$
fluid ounce (*U.S.*)	meter³ (m³)	$2.957\ 353 \times 10^{-5}$
foot³	meter³ (m³)	$2.831\ 685 \times 10^{-2}$
gallon (*Canadian liquid*)	meter³ (m³)	$4.546\ 122 \times 10^{-3}$
gallon (*U.K. liquid*)	meter³ (m³)	$4.546\ 087 \times 10^{-3}$
gallon (*U.S. dry*)	meter³ (m³)	$4.404\ 884 \times 10^{-3}$
gallon (*U.S. liquid*)	meter³ (m³)	$3.785\ 412 \times 10^{-3}$
gill (*U.K.*)	meter³ (m³)	$1.420\ 652 \times 10^{-4}$
gill (*U.S.*)	meter³ (m³)	$1.182\ 941 \times 10^{-4}$
inch³	meter³ (m³)	$1.638\ 706 \times 10^{-5}$
liter	meter³ (m³)	$1.000\ 000 \times 10^{-3}$
ounce (*U.K. fluid*)	meter³ (m³)	$2.841\ 305 \times 10^{-5}$
ounce (*U.S. fluid*)	meter³ (m³)	$2.957\ 353 \times 10^{-5}$
peck (*U.S.*)	meter³ (m³)	$8.809\ 768 \times 10^{-3}$
pint (*U.S. dry*)	meter³ (m³)	$5.506\ 105 \times 10^{-4}$
pint (*U.S. liquid*)	meter³ (m³)	$4.731\ 765 \times 10^{-4}$
quart (*U.S. dry*)	meter³ (m³)	$1.101\ 221 \times 10^{-3}$
quart (*U.S. liquid*)	meter³ (m³)	$9.463\ 529 \times 10^{-4}$
stere	meter³ (m³)	1.000 000
tablespoon	meter³ (m³)	$1.478\ 676 \times 10^{-5}$
teaspoon	meter³ (m³)	$4.928\ 922 \times 10^{-6}$
ton (*register*)	meter³ (m³)	2.831 685
yard³	meter³ (m³)	$7.645\ 549 \times 10^{-1}$

The following Nomogram provides a quick solution to problems involving inches and millimeters.

Example:
Convert 8 inches to millimeters

Solution:
Locate 8 inches on the Nomogram and read the answer of 203 mm.

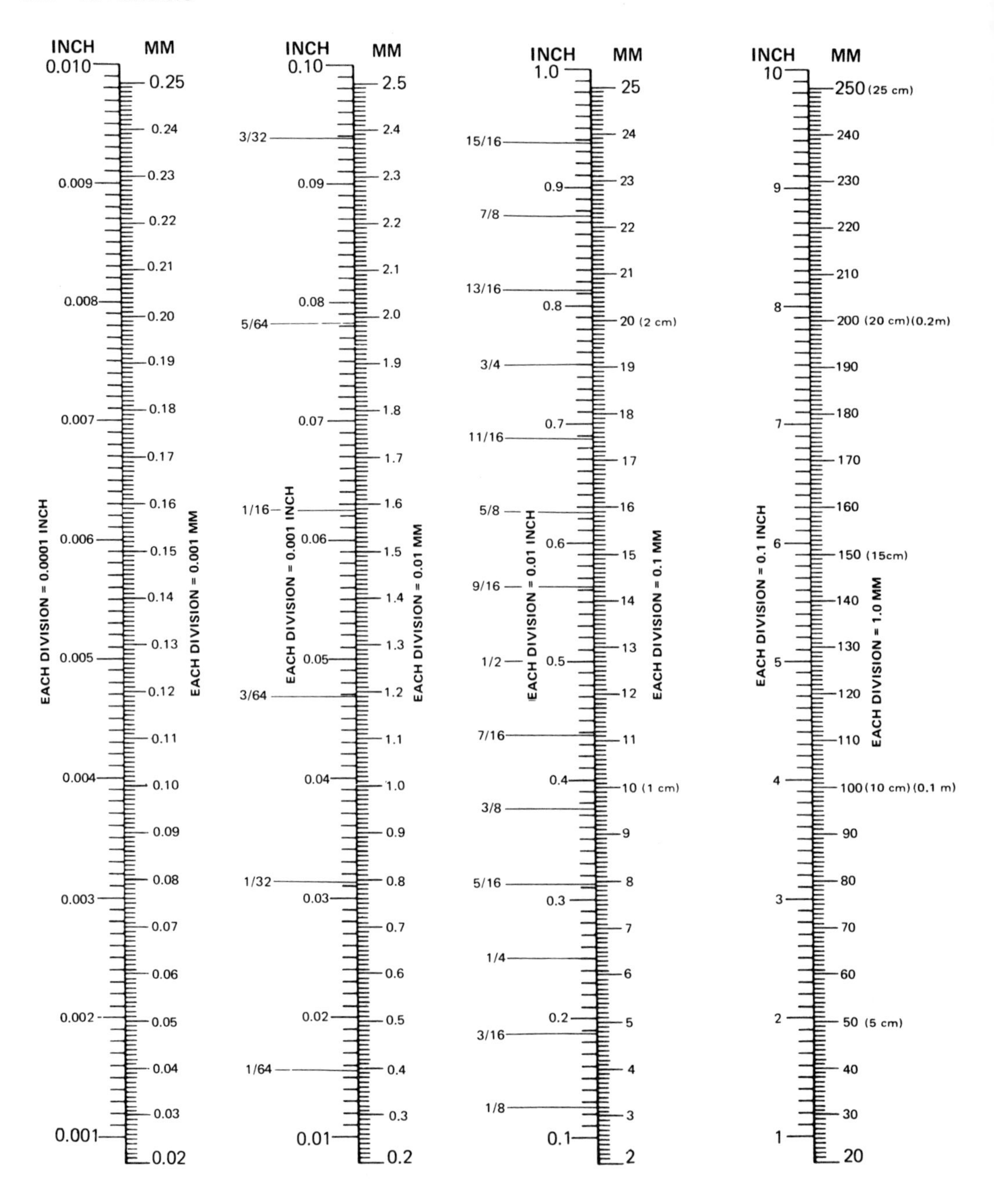
INCH
MM
EACH DIVISION = 0.0001 INCH
EACH DIVISION = 0.001 MM
EACH DIVISION = 0.001 INCH
EACH DIVISION = 0.01 MM
EACH DIVISION = 0.01 INCH
EACH DIVISION = 0.1 MM
EACH DIVISION = 0.1 INCH
EACH DIVISION = 1.0 MM
3/32
5/64
1/16
3/64
1/32
1/64
15/16
7/8
13/16
3/4
11/16
5/8
9/16
1/2
7/16
3/8
5/16
1/4
3/16
1/8
20 (2 cm)
10 (1 cm)
250 (25 cm)
200 (20 cm)(0.2m)
150 (15cm)
100 (10 cm) (0.1 m)
50 (5 cm)

Basic Unit of Measurement—Time (Second)

The following charts give conversion factors for the basic unit of time and the units that are derived from this basic unit.

TIME

To convert from	to	Multiply by
day (*mean solar*)	second (s)	$8.640\,000 \times 10^{4}$
day (*sidereal*)	second (s)	$8.616\,409 \times 10^{4}$
hour (*mean solar*)	second (s)	$3.600\,000 \times 10^{3}$
hour (*sidereal*)	second (s)	$3.590\,170 \times 10^{3}$
minute (*mean solar*)	second (s)	$6.000\,000 \times 10$
minute (*sidereal*)	second (s)	$5.983\,617 \times 10$
month (*mean calendar*)	second (s)	$2.628\,000 \times 10^{6}$
second (*sidereal*)	second (s)	$9.972\,696 \times 10^{-1}$
year (*calendar*)	second (s)	$3.153\,600 \times 10^{7}$
year (*sidereal*)	second (s)	$3.155\,815 \times 10^{7}$
year (*tropical*)	second (s)	$3.155\,693 \times 10^{7}$

ACCELERATION

To convert from	to	Multiply by
foot/second²	meter/second² (m/s²)	$3.048\,000 \times 10^{-1}$
free fall, standard	meter/second² (m/s²)	$9.806\,650$
gal (*galileo*)	meter/second² (m/s²)	$1.000\,000 \times 10^{-2}$
inch/second²	meter/second² (m/s²)	$2.540\,000 \times 10^{-2}$

VELOCITY (INCLUDES SPEED)

To convert from	to	Multiply by
foot/hour	meter/second (m/s)	$8.466\,667 \times 10^{-5}$
foot/minute	meter/second (m/s)	$5.080\,000 \times 10^{-3}$
foot/second	meter/second (m/s)	$3.048\,000 \times 10^{-1}$
inch/second	meter/second (m/s)	$2.540\,000 \times 10^{-2}$
kilometer/hour	meter/second (m/s)	$2.777\,778 \times 10^{-1}$
knot (*international*)	meter/second (m/s)	$5.144\,444 \times 10^{-1}$
mile/hour (*U.S. statute*)	meter/second (m/s)	$4.470\,400 \times 10^{-1}$
mile/minute (*U.S. statute*)	meter/second (m/s)	$2.682\,240 \times 10$
mile/second (*U.S. statute*)	meter/second (m/s)	$1.609\,344 \times 10^{3}$
mile/hour (*U.S. statute*)	kilometer/hour	$1.609\,344$

The following Nomogram provides a quick solution to problems involving the conversion from frequency to time units.

Example:

A component vibrates at 60 cps for 2 days. Determine the total number of cycles on the component.

Solution:

Construct a line from 60 on the F scale (cps) to 2 on the T scale (days). Where this line intersects the total number of cycles scale, read the answer of 10,000,000.

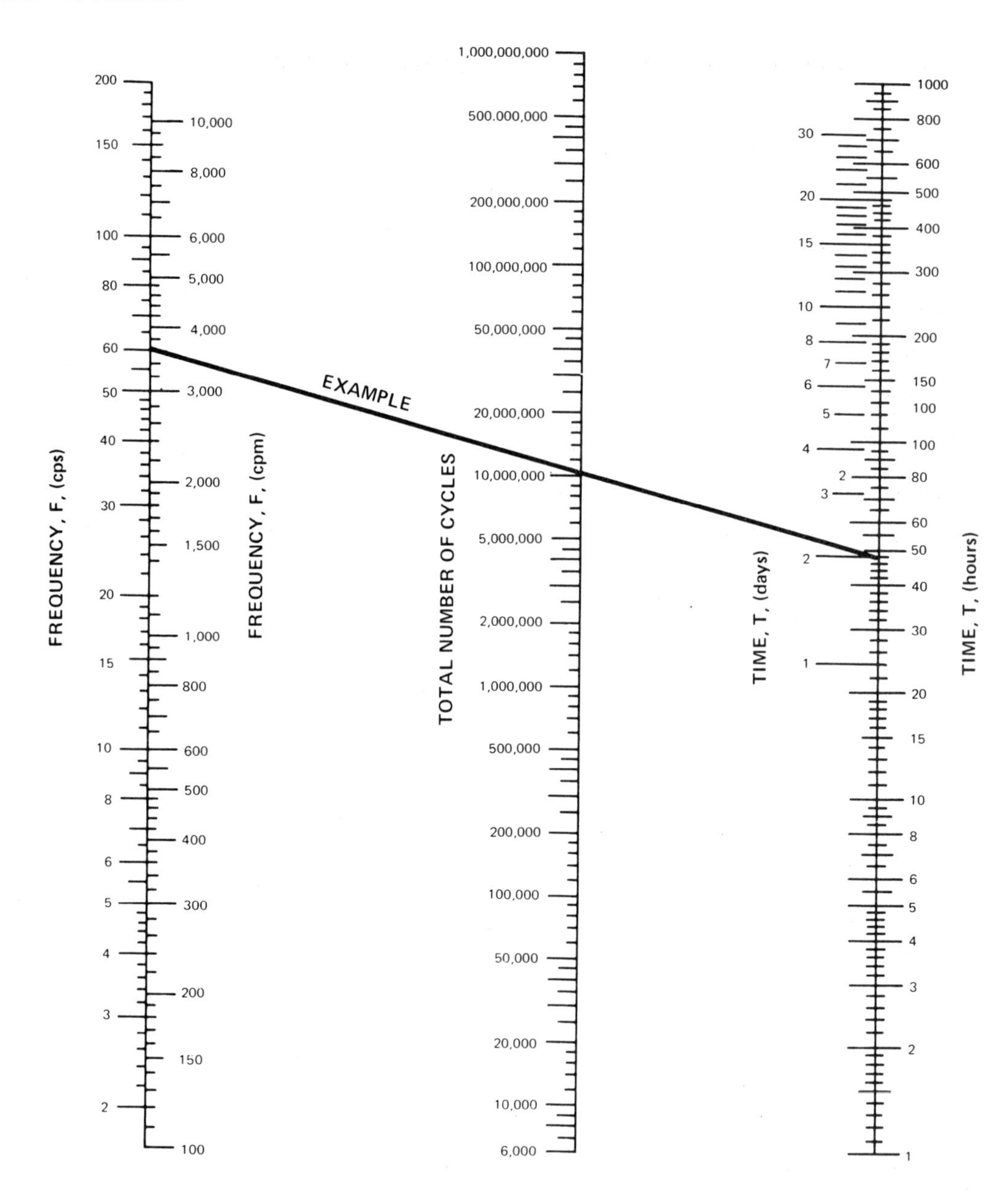
FREQUENCY, F, (cps)
200
150
100
80
60
50
40
30
20
15
10
8
6
5
4
3
2
FREQUENCY, F, (cpm)
10,000
8,000
6,000
5,000
4,000
3,000
2,000
1,500
1,000
800
600
500
400
300
200
150
100
EXAMPLE
TOTAL NUMBER OF CYCLES
1,000,000,000
500.000,000
200,000,000
100,000,000
50,000,000
20,000,000
10,000,000
5,000,000
2,000,000
1,000,000
500,000
200,000
100,000
50,000
20,000
10,000
6,000
TIME, T, (days)
30
20
15
10
8
7
6
5
4
2
3
2
1
TIME, T, (hours)
1000
800
600
500
400
300
200
150
100
100
80
60
50
40
30
20
15
10
8
6
5
4
3
2
1

Basic Unit of Measurement - Mass (Kilogram)

The following charts give conversion factors for the basic unit of mass and the units that are derived from this basic unit.

MASS (Weight)

U.S. TO METRIC		METRIC TO U.S.	
1 ounce (dry)	= 28.35 grams	1 gram	= 0.03527 ounce
1 pound	= 0.4536 kilogram	1 kilogram	= 2.2046 pounds
1 short ton (2000 lb.)	= 907.2 kilograms	1 metric ton	= 2 204.6 pounds
1 short ton (2000 lb.)	= 0.9072 metric ton	1 metric ton	= 1.102 tons (short)

MASS

To convert from	to	Multiply by
carat *(metric)*	kilogram (kg)	$2.000\ 000 \times 10^{-4}$
grain	kilogram (kg)	$6.479\ 891 \times 10^{-5}$
gram	kilogram (kg)	$1.000\ 000 \times 10^{-3}$
hundredweight *(long)*	kilogram (kg)	$5.080\ 235 \times 10$
hundredweight *(short)*	kilogram (kg)	$4.535\ 924 \times 10$
kilogram-force-second²/meter *(mass)*	kilogram (kg)	9.806 650
kilogram-mass	kilogram (kg)	1.000 000
ounce-mass *(avoirdupois)*	kilogram (kg)	$2.834\ 952 \times 10^{-2}$
ounce-mass *(troy or apothecary)*	kilogram (kg)	$3.110\ 348 \times 10^{-2}$
pennyweight	kilogram (kg)	$1.555\ 174 \times 10^{-3}$
pound-mass *(lbm avoirdupois)*	kilogram (kg)	$4.535\ 924 \times 10^{-1}$
pound-mass *(troy or apothecary)*	kilogram (kg)	$3.732\ 417 \times 10^{-1}$
slug	kilogram (kg)	$1.459\ 390 \times 10$
ton *(assay)*	kilogram (kg)	$2.916\ 667 \times 10^{-2}$
ton *(long, 2240 lbm)*	kilogram (kg)	$1.016\ 047 \times 10^{3}$
ton *(metric)*	kilogram (kg)	$1.000\ 000 \times 10^{3}$
ton *(short, 2000 lbm)*	kilogram (kg)	$9.071\ 847 \times 10^{2}$
tonne	kilogram (kg)	$1.000\ 000 \times 10^{3}$

ENERGY OR WORK

To convert from	to	Multiply by
British thermal unit *(International table)*	joule (J)	$1.055\ 056 \times 10^{3}$
British thermal unit *(mean)*	joule (J)	$1.055\ 87 \times 10^{3}$
British thermal unit *(thermochemical)*	joule (J)	$1.054\ 350 \times 10^{3}$
British thermal unit *(39 F)*	joule (J)	$1.059\ 67 \times 10^{3}$
British thermal unit *(60 F)*	joule (J)	$1.054\ 68 \times 10^{3}$
calorie *(International Table)*	joule (J)	4.186 800
calorie *(mean)*	joule (J)	4.190 02
calorie *(thermochemical)*	joule (J)	4.184 000
calorie *(15 C)*	joule (J)	4.185 80
calorie *(20 C)*	joule (J)	4.181 90
calorie *(kg, International Table)*	joule (J)	$4.186\ 800 \times 10^{3}$
calorie *(kg, mean)*	joule (J)	$4.190\ 02 \times 10^{3}$
calorie *(kg, thermochemical)*	joule (J)	$4.184\ 000 \times 10^{3}$
electron volt	joule (J)	$1.602\ 10 \times 10^{-19}$
erg	joule (J)	$1.000\ 000 \times 10^{-7}$
foot-pound-force	joule (J)	1.355 818
foot-poundal	joule (J)	$4.214\ 011 \times 10^{-2}$
joule *(International of 1948)*	joule (J)	1.000 165
kilocalorie *(International Table)*	joule (J)	$4.186\ 800 \times 10^{3}$
kilocalorie *(mean)*	joule (J)	$4.190\ 02 \times 10^{3}$
kilocalorie *(thermochemical)*	joule (J)	$4.184\ 000 \times 10^{3}$
kilowatt-hour	joule (J)	$3.600\ 000 \times 10^{6}$

kilowatt-hour *(international of 1948)*	joule (J)	3.600 59 × 10⁶
ton *(nuclear equivalent of TNT)*	joule (J)	4.20 × 10⁹
watt-hour	joule (J)	3.600 000 × 10³
watt-second	joule (J)	1.000 000

FORCE

To convert from	to	Multiply by
dyne	newton (N)	1.000 000 × 10⁻⁵
kilogram-force	newton (N)	9.806 650
kilopound-force	newton (N)	9.806 650
kip	newton (N)	4.448 222 × 10³
ounce-force *(avoirdupois)*	newton (N)	2.780 139 × 10⁻¹
pound-force *(lbf avoirdupois)*	newton (N)	4.448 222
pound-force *(lbf avoirdupois)*	kilogram-force	4.535 924 × 10⁻¹
poundal	newton (N)	1.382 550 × 10⁻¹

POWER

To convert from	to	Multiply by
Btu *(International Table)* /hour	watt (W)	2.930 711 × 10⁻¹
Btu *(thermochemical)* /second	watt(W)	1.054 350 × 10³
Btu *(thermochemical)* /minute	watt (W)	1.757 250 × 10
Btu *(thermochemical)* /hour	watt (W)	2.928 751 × 10⁻¹
calorie *(thermochemical)* /second	watt (W)	4.184 000
calorie *(thermochemical)* /minute	watt (W)	6.973 333 × 10⁻²
erg/second	watt (W)	1.000 000 × 10⁻⁷
foot-pound-force/hour	watt (W)	3.766 161 × 10⁻⁴
foot-pound-force/minute	watt (W)	2.259 697 × 10⁻²
foot-pound-force/second	watt (W)	1.355 818
horsepower *(550 ft. lbf/s)*	watt (W)	7.456 999 × 10²
horsepower *(boiler)*	watt (W)	9.809 50 × 10³
horsepower *(electric)*	watt (W)	7.460 000 × 10²
horsepower *(metric)*	watt (W)	7.354 99 × 10²
horsepower *(water)*	watt (W)	7.460 43 × 10²
horsepower *(U.K.)*	watt (W)	7.457 0 × 10²
kilocalorie *(thermochemical)* /minute	watt (W)	6.973 333 × 10
kilocalorie *(thermochemical)* /second	watt (W)	4.184 000 × 10³
watt *(international of 1948)*	watt (W)	1.000 165

PRESSURE OR STRESS (FORCE/AREA)

To convert from	to	Multiply by
atmosphere *(normal= 760 torr)*	newton/meter² (N/m²)	1.013 250 × 10⁵
atmosphere *(technical= 1 kgf/cm²)*	newton/meter² (N/m²)	9.806 650 × 10⁴
bar	newton/meter² (N/m²)	1.000 000 × 10⁵
centimeter of mercury *(0 C)*	newton/meter² (N/m²)	1.333 22 × 10³
centimeter of water *(4 C)*	newton/meter² (N/m²)	9.806 38 × 10
decibar	newton/meter² (N/m²)	1.000 000 × 10⁴
dyne/centimeter²	newton/meter² (N/m²)	1.000 000 × 10⁻¹
foot of water *(39.2F)*	newton/meter² (N/m²)	2.988 98 × 10³
gram-force/centimeter²	newton/meter² (N/m²)	9.806 650 × 10
inch of mercury *(32 F)*	newton/meter² (N/m²)	3.386 389 × 10³
inch of mercury *(60 F)*	newton/meter² (N/m²)	3.376 85 × 10³
inch of water *(39.2 F)*	newton/meter² (N/m²)	2.490 82 × 10²
inch of water *(60 F)*	newton/meter² (N/m²)	2.488 4 × 10²
kilogram-force/centimeter²	newton/meter² (N/m²)	9.806 650 × 10⁴
kilogram-force/meter²	newton/meter² (N/m²)	9.806 650
kilogram-force/millimeter²	newton/meter² (N/m²)	9.806 650 × 10⁶
kip/inch²	newton/meter² (N/m²)	6.894 757 × 10⁶
millibar	newton/meter² (N/m²)	1.000 000 × 10²
millimeter of mercury *(0 C)*	newton/meter² (N/m²)	1.333 224 × 10²
newton/meter²	pascal *(pa)*	1.000 000

pascal	newton/meter² (N/m^2)	1.000 000
poundal/foot²	newton/meter² (N/m^2)	1.488 164
pound-force/foot²	newton/meter² (N/m^2)	$4.788\ 026 \times 10$
pound-force/inch² *(psi)*	newton/meter² (N/m^2)	$6.894\ 757 \times 10^3$
pound-force/inch² *(psi)*	kilogram-force/mm²	$7.030\ 696 \times 10^{-4}$
psi	newton/meter² (N/m^2)	$6.894\ 757 \times 10^3$
torr *(mm Hg, 0 C)*	newton/meter² (N/m^2)	$1.333\ 22 \times 10^2$

The following Nomogram provides a quick solution to problems involving pounds and grams.

Example: Convert 500 grams to pounds.

Solution: Locate 500 grams on the Nomogram and read the answer of 1.1 lb.

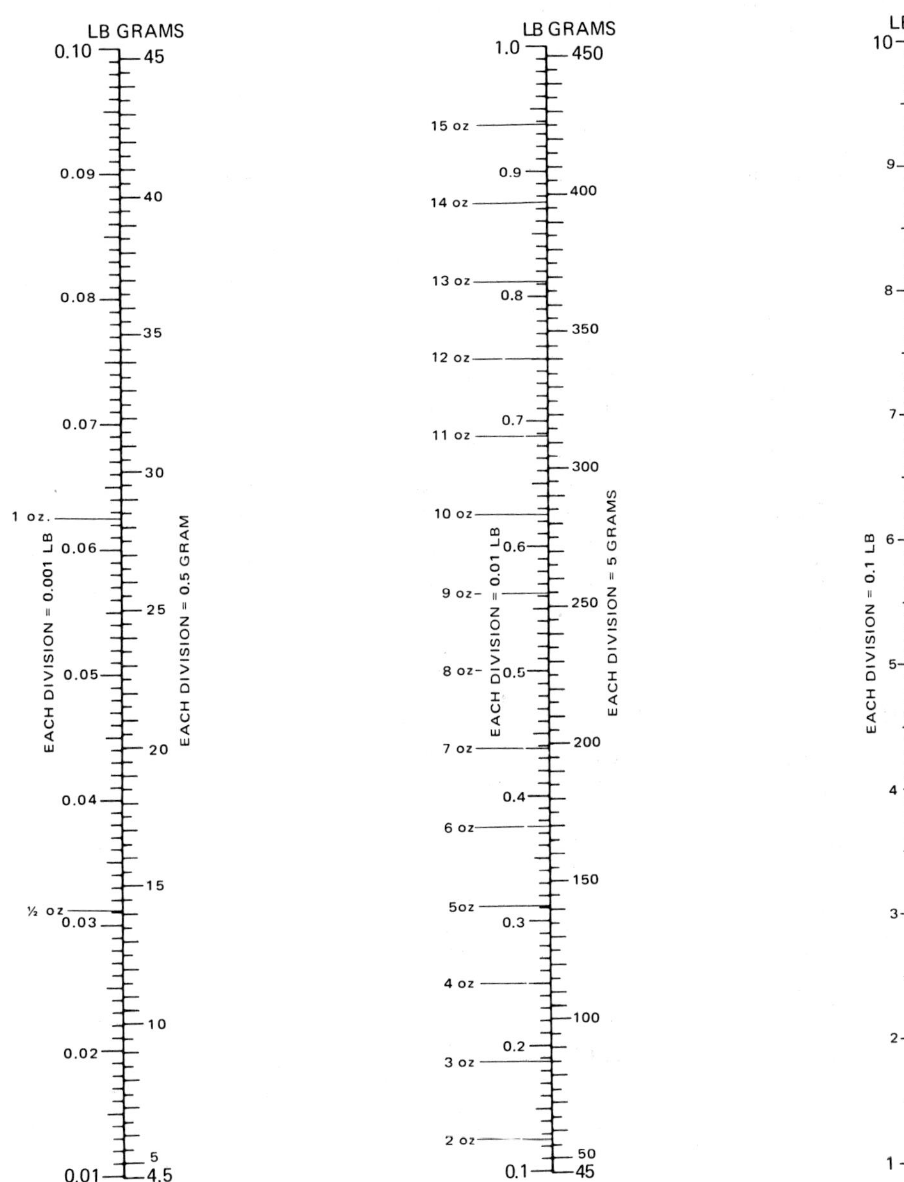

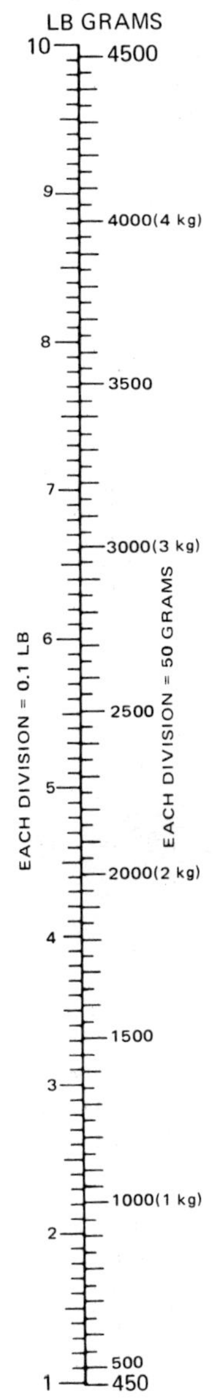

Basic Unit of Measurement—Temperature (Kelvin)

The following chart gives conversion factors for the basic unit of temperature.

TEMPERATURE

To convert from	to	Multiply by
degree Celsius	kelvin (K)	$t_K = t_C + 273.15$
degree Fahrenheit	kelvin (K)	$t_K = (t_F + 459.67)/1.8$
degree Rankine	kelvin (K)	$t_K = t_R/1.8$
degree Fahrenheit	degree Celsius	$t_C = (t_F - 32)/1.8$
kelvin	degree Celsius	$t_C = t_K - 273.15$

The following Nomograms provide a quick solution to problems involving the conversion from one temperature scale to another.

Example:

Given 32 deg F, determine its kelvin, celsius and rankine value.

Solution:

On chart 2, draw a straight line parallel to the bottom of the chart and read the corresponding values: kelvin = 273.15 deg, celsius = 0 deg and rankine = 491.69 deg.

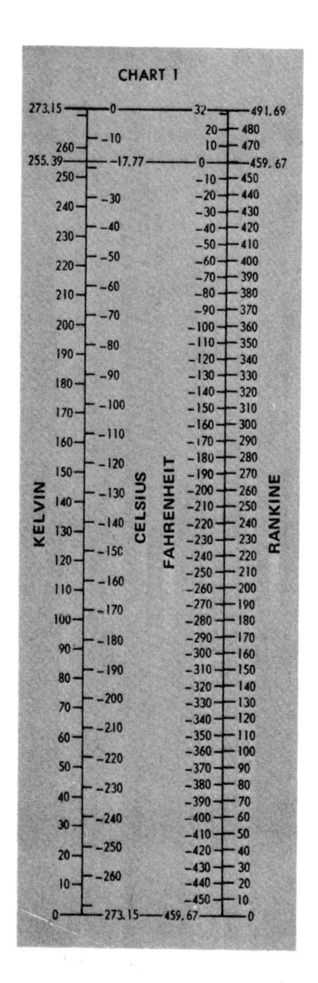

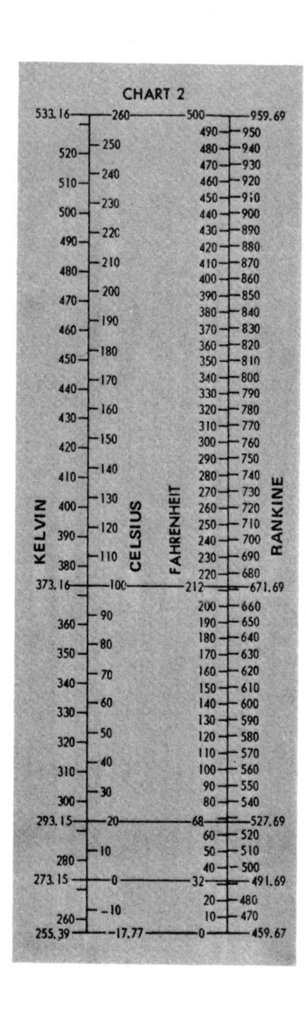

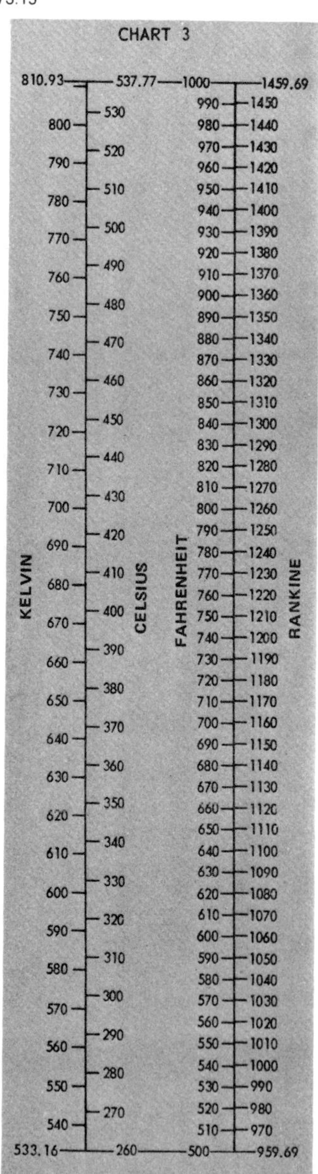

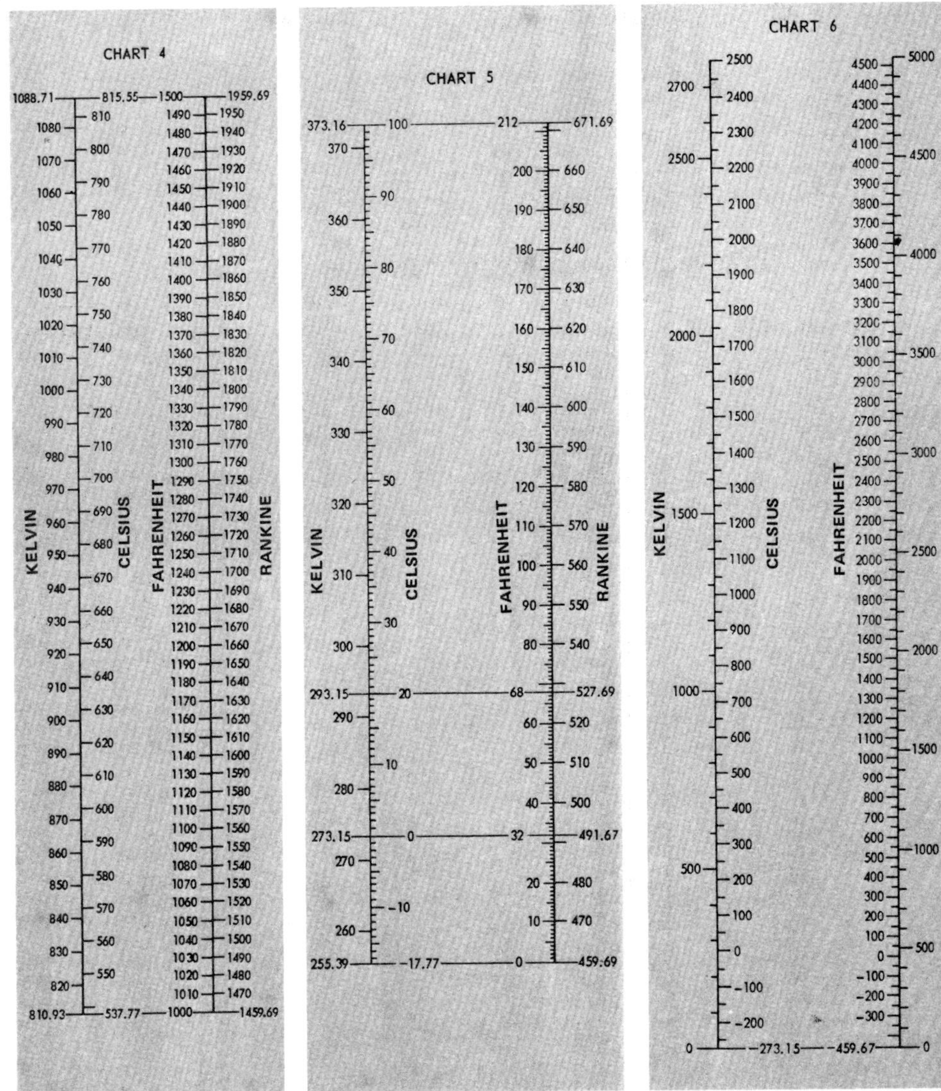
CHART 4
KELVIN
CELSIUS
FAHRENHEIT
RANKINE
1088.71
815.55
1500
1959.69
810.93
537.77
1000
1459.69
CHART 5
KELVIN
CELSIUS
FAHRENHEIT
RANKINE
373.16
100
212
671.69
293.15
20
68
527.69
273.15
0
32
491.67
255.39
-17.77
0
459.69
CHART 6
KELVIN
CELSIUS
FAHRENHEIT
RANKINE
0
-273.15
-459.67
0

Basic Unit of Measurement—Electric Current (Ampere)

The following chart gives conversion factors for the basic unit of electric current and the units that are derived from this basic unit.

ELECTRICITY AND MAGNETISM

To convert from	to	Multiply by
abampere	ampere (A)	$1.000\ 000 \times 10$
abcoulomb	coulomb (C)	$1.000\ 000 \times 10$
abfarad	farad (F)	$1.000\ 000 \times 10^{9}$
abhenry	henry (H)	$1.000\ 000 \times 10^{-9}$
abmho	mho	$1.000\ 000 \times 10^{9}$
abohm	ohm (Ω)	$1.000\ 000 \times 10^{-9}$
abvolt	volt (V)	$1.000\ 000 \times 10^{-8}$
ampere (*international of 1948*)	ampere (A)	$9.998\ 35 \times 10^{-1}$
ampere-hour	coulomb (C)	$3.600\ 000 \times 10^{3}$
coulomb (*international of 1948*)	coulomb (C)	$9.998\ 35 \times 10^{-1}$
EMU of capacitance	farad (F)	$1.000\ 000 \times 10^{9}$
EMU of current	ampere (A)	$1.000\ 000 \times 10$
EMU of electric potential	volt (V)	$1.000\ 000 \times 10^{-8}$
EMU of inductance	henry (H)	$1.000\ 000 \times 10^{-9}$
EMU of resistance	ohm (Ω)	$1.000\ 000 \times 10^{-9}$
ESU of capacitance	farad (F)	$1.112\ 6 \times 10^{-12}$
ESU of current	ampere (A)	$3.335\ 6 \times 10^{-10}$
ESU of electric potential	volt (V)	$2.997\ 9 \times 10^{2}$
ESU of inductance	henry (H)	$8.987\ 6 \times 10^{11}$
ESU of resistance	ohm (Ω)	$8.987\ 6 \times 10^{11}$
farad (*international of 1948*)	farad (F)	$9.995\ 05 \times 10^{-1}$
faraday (*based on carbon-12*)	coulomb (C)	$9.648\ 70 \times 10^{4}$
faraday (*chemical*)	coulomb (C)	$9.649\ 57 \times 10^{4}$
faraday (*physical*)	coulomb (C)	$9.652\ 19 \times 10^{4}$
gamma	tesla (T)	$1.000\ 000 \times 10^{-9}$
gauss	tesla (T)	$1.000\ 000 \times 10^{-4}$
gilbert	ampere-turn	$7.957\ 747 \times 10^{-1}$
henry (*international of 1948*)	henry (H)	$1.000\ 495$
maxwell	weber (Wb)	$1.000\ 000 \times 10^{-8}$
oersted	ampere/meter (A/m)	$7.957\ 747 \times 10$
ohm (*international of 1948*)	ohm (Ω)	$1.000\ 495$
ohm-centimeter	ohm-meter (Ω · m)	$1.000\ 000 \times 10^{-2}$
statampere	ampere (A)	$3.335\ 640 \times 10^{-10}$
statcoulomb	coulomb (C)	$3.335\ 640 \times 10^{-10}$
statfarad	farad (F)	$1.112\ 650 \times 10^{-12}$
stathenry	henry (H)	$8.987\ 554 \times 10^{11}$
statmho	mho	$1.112\ 650 \times 10^{-12}$
statohm	ohm (Ω)	$8.987\ 554 \times 10^{11}$
statvolt	volt (V)	$2.997\ 925 \times 10^{2}$
unit pole	weber (Wb)	$1.256\ 637 \times 10^{-7}$
volt (*international of 1948*)	volt (V)	$1.000\ 330$

Basic Unit of Measurement - Luminous Intensity (Candela)

The following chart gives conversion factors for the basic unit of luminous intensity and units that are derived from this basic unit.

LIGHT

To convert from	to	Multiply by
footcandle . . .	lumen/meter^2 (lm/m^2) . . .	1.076 391 × 10
footcandle . . .	lux (lx).	1.076 391 × 10
footlambert. . .	candela/meter^2 (cd/m^2). . .	3.426 3
lux.	lumen/meter^2 (lm/m^2) . . .	1.000 000

Almost anyone can select an incandescant lamp for a standard application. The lamp manufacturer gives nominal operating voltage, current and the light output you can expect. In addition, the lamps will be designed for some nominal life. Problems begin if you wish to use a lamp in a nonstandard application. For instance, what happens if you raise the operating voltage—what is your expected operating life then? How much more light can you expect from the lamp? Answers to these questions can be found by using the following nomogram.

Example: Given a small vacuum lamp, extend its life by a factor of 5.

Solution: The first step is to tabulate the manufacturer's data for the lamp in question. Write down the rated voltage, current and lamp life. Connect a line from 5 on the lamp life scale through the V life point on the E axis and extend it to the lamp operating voltage scale. Where this line intersects the lamp operating voltage scale, read the value of 0.89. This tells us that multiplying the lamp's design voltage by the factor 0.89 will extend the lamp's life by a factor of 5.

While the life may be extended, what does this do to the light output or luminous flux? Starting at the point just found, 0.89, connect a line to the V luminous flux point on the E axis and extend it to the luminous flux scale. Where this line intersects the luminous flux scale, read the value of 0.67. This means that under our new operating conditions, the light output will amount to only 67% of the normal or rated value. In the same manner, values for the other lamp parameters can be found.

Note that there are two points given for each factor along the E axis. G is for a gas-filled lamp while V is for a vacuum lamp. As a general rule, lamps having a rated current of less than 0.4 amp are vacuum, larger lamps are usually gas-filled.

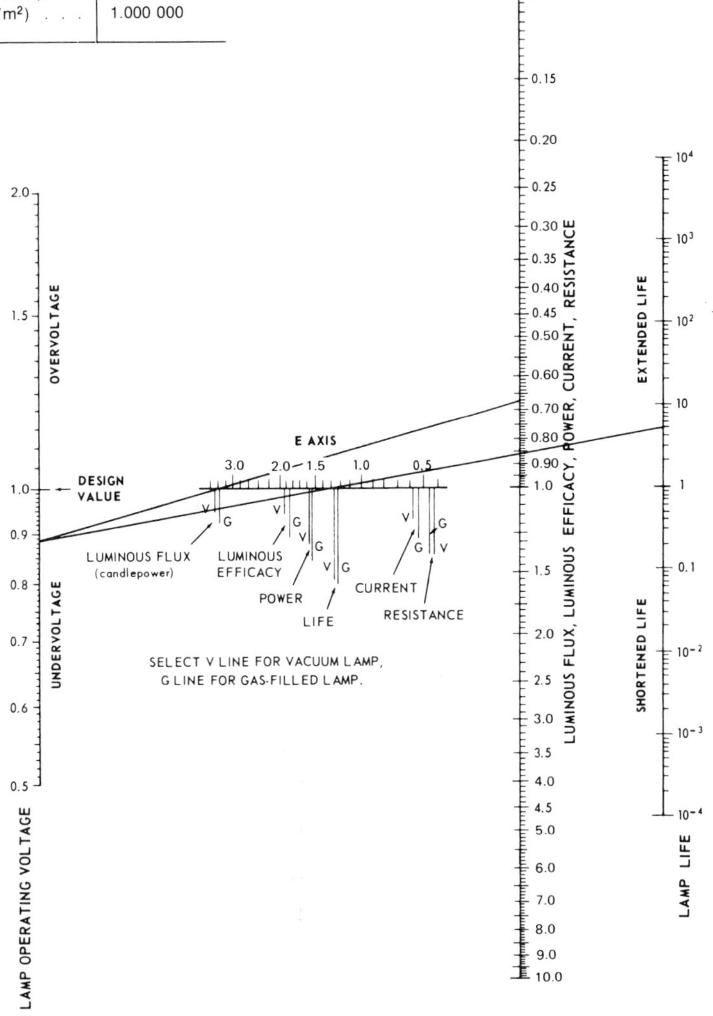

Miscellaneous Units of Measurement

The following charts give conversion factors for the miscellaneous metric units.

BENDING MOMENT OR TORQUE

To convert from	to	Multiply by
dyne-centimeter	newton-meter (N·m)	$1.000\ 000 \times 10^{-7}$
kilogram-force-meter	newton-meter (N·m)	9.806 650
ounce-force-inch	newton-meter (N·m)	$7.061\ 552 \times 10^{-3}$
pound-force-inch	newton-meter (N·m)	$1.129\ 848 \times 10^{-1}$
pound-force-foot	newton-meter (N·m)	1.355 818

BENDING MOMENT OR TORQUE/LENGTH

To convert from	to	Multiply by
pound-force-foot/inch	newton-meter/meter (N ·m/m)	$5.337\ 866 \times 10$
pound-force-inch/inch	newton-meter/meter (N ·m/n)	4.448 222

ENERGY/AREA TIME

To convert from	to	Multiply by
Btu (*thermochemical*)/foot²-second	watt/meter² (W/m²)	$1.134\ 983 \times 10^{4}$
Btu (*thermochemical*)/foot²-minute	watt/meter² (W/m²)	$1.891\ 489 \times 10^{2}$
Btu (*thermochemical*)/foot²-hour	watt/meter² (W/m²)	3.152 481
Btu (*thermochemical*)/inch²-second	watt/meter² (W/m²)	$1.634\ 246 \times 10^{6}$
calorie (*thermochemical*)/centimeter²-minute	watt/meter² (W/m²)	$6.973\ 333 \times 10^{2}$
erg-centimeter²-second	watt/meter² (W/m²)	$1.000\ 000 \times 10^{-3}$
watt/centimeter²	watt/meter² (W/m²)	$1.000\ 000 \times 10^{4}$

FORCE/LENGTH

To convert from	to	Multiply by
pound-force/inch	newton/meter (N/m)	$1.751\ 268 \times 10^{2}$
pound-force/foot	newton/meter (N/m)	$1.459\ 390 \times 10$

HEAT

To convert from	to	Multiply by
Btu (*thermochemical*)·in./s·ft.²·deg F (k, *thermal conductivity*)	watt/meter-kelvin (W/m·K)	$5.188\ 732 \times 10^{2}$
Btu (*International Table*)·in./s·ft.²·deg F (*k, thermal conductivity*)	watt/meter-kelvin (Wm/m·K)	$5.192\ 204 \times 10^{2}$
Btu (*thermochemical*)·in./h·ft.²·deg F (*k, thermal conductivity*)	watt/meter-kelvin (W/m·K)	$1.441\ 314 \times 10^{-1}$
Btu (*International Table*)·in./h·ft.²·deg F (*k, thermal conductivity*)	watt/meter-kelvin (W/m·K)	$1.442\ 279 \times 10^{-1}$
Btu (*International Table*)/ft.²	joule/meter² (J/m²)	$1.135\ 653 \times 10^{4}$
Btu (*thermochemical*)/ft.²	joule/meter² (J/m²)	$1.134\ 893 \times 10^{4}$
Btu (*International Table*)/h·ft.²·deg F (*C, thermal conductance*)	watt/meter²-kelvin (W/m²·K)	5.678 263
Btu (*thermochemical*)/h·ft.²·deg F (*C, thermal conductance*)	watt/meter²-kelvin (W/m²·K)	5.674 466
Btu (*International Table*)/pound-mass	joule/kilogram (J/kg)	$2.326\ 000 \times 10^{3}$
Btu (*thermochemical*)/pound-mass	joule/kilogram (J/kg)	$2.324\ 444 \times 10^{3}$
Btu (*International Table*)/lbm·deg F (*c, heat capacity*)	joule/kilogram-kelvin (J/kg·K)	$4.186\ 800 \times 10^{3}$
Btu (*thermochemical*)/lbm·deg F (*c, heat capacity*)	joule/kilogram-kelvin (J/kg·K)	$4.184\ 000 \times 10^{3}$
Btu (*International Table*)/s·ft.²·deg F	watt/meter²-kelvin (W/m²·K)	$2.044\ 175 \times 10^{4}$
Btu (*thermochemical*)/s·ft.²·deg F	watt/meter²-kelvin (W/m²·K)	$2.042\ 808 \times 10^{4}$
cal (*thermochemical*)/cm²	joule/meter² (J/m²)	$4.184\ 000 \times 10^{4}$
cal (*thermochemical*)/cm²·s	watt/meter² (W/m²)	$4.184\ 000 \times 10^{4}$
cal (*thermochemical*)/cm·s·deg C	watt/meter-kelvin (W/m·K)	$4.184\ 00 \times 10^{2}$
cal (*International Table*)/g	joule/kilogram (J/kg)	$4.186\ 800 \times 10^{3}$
cal (*International Table*)/g·deg C	joule/kilogram-kelvin (J/kg·K)	$4.186\ 800 \times 10^{3}$
cal (*thermochemical*)/g	joule/kilogram (J/kg)	$4.184\ 000 \times 10^{3}$
cal (*thermochemical*)/g·deg C	joule/kilogram-kelvin (J/kg·K)	$4.184\ 000 \times 10^{3}$
clo	kelvin-meter²/watt (K·m²/W)	$2.003\ 712 \times 10^{-1}$
deg F·h·ft.²/Btu (*thermochemical*) (*R, thermal resistance*)	kelvin-meter²/watt (K·m²/W)	$1.762\ 280 \times 10^{-1}$
deg F·h·ft.²/Btu (*International Table*) (*R, thermal resistance*)	kelvin-meter²/watt (K·m²/W)	$1.761\ 102 \times 10^{-1}$
ft.²/h (*thermal diffusivity*)	meter²/second (m²/s)	$2.580\ 640 \times 10^{-5}$

MASS/VOLUME (INCLUDES DENSITY AND MASS CAPACITY)

To convert from	to	Multiply by
grain (*lbm avoirdupois/7000*)/gallon (*U.S. liquid*)	kilogram/meter³ (kg/m³)	$1.711\ 806 \times 10^{-2}$
gram/centimeter³	kilogram/meter³ (kg/m³)	$1.000\ 000 \times 10^{3}$
ounce (*avoirdupois*)/gallon (*U.K. liquid*)	kilogram/meter³ (kg/m³)	6.236 027
ounce (*avoirdupois*)/gallon (*U.S. liquid*)	kilogram/meter³ (kg/m³)	7.489 152
ounce (*avoirdupois*) (*mass*)/inch³	kilogram/meter³ (kg/m³)	$1.729\ 994 \times 10^{3}$
pound-mass/foot³	kilogram/meter³ (kg/m³)	$1.601\ 846 \times 10$
pound-mass/inch³	kilogram/meter³ (kg/m³)	$2.767\ 990 \times 10^{4}$
pound-mass/gallon (*U.K. liquid*)	kilogram/meter³ (kg/m³)	$9.977\ 644 \times 10$
pound-mass/gallon (*U.S. liquid*)	kilogram/meter³ (kg/m³)	$1.198\ 264 \times 10^{2}$
slug/foot³	kilogram/meter³ (kg/m³)	$5.153\ 788 \times 10^{2}$
ton (*long, mass*)/yard³	kilogram/meter³ (kg/m³)	$1.328\ 939 \times 10^{3}$

MASS/AREA

To convert from	to	Multiply by
ounce-mass/yard^2	kilogram/meter^2 (kg/m^2)	$3.390\ 575 \times 10^{-2}$
pound-mass/foot^2	kilogram/meter^2 (kg/m^2)	4.882 428

MASS/TIME (INCLUDES FLOW)

To convert from	to	Multiply by
perm (*0 C*)	kilogram/newton-second (kg/N·s)	$5.721\ 35 \times 10^{-11}$
perm (*23 C*)	kilogram/newton-second (kg/N·s)	$5.745\ 25 \times 10^{-11}$
perm-inch (*0 C*)	kilogram-meter/newton-second (kg·m/N·s)	$1.453\ 22 \times 10^{-12}$
perm-inch (*23 C*)	kilogram-meter/newton-second (kg·m/N·s)	$1.459\ 29 \times 10^{-12}$
pound-mass/second	kilogram/second (kg/s)	$4.535\ 924 \times 10^{-1}$
pound-mass/minute	kilogram/second (kg/s)	$7.559\ 873 \times 10^{-3}$
ton (*short, mass*)/hour	kilogram/second (kg/s)	$2.519\ 958 \times 10^{-1}$

VISCOSITY

To convert from	to	Multiply by
centipoise	newton-second/meter^2 (N·s/m^2)	$1.000\ 000 \times 10^{-3}$
centistoke	meter^2/second (m^2/s)	$1.000\ 000 \times 10^{-6}$
foot^2/second	meter^2/second (m^2/s)	$9.290\ 304 \times 10^{-2}$
poise	newton-second/meter^2 (N·s/m^2)	$1.000\ 000 \times 10^{-1}$
poundal-second/foot^2	newton-second/meter^2 (N·s/m^2)	1.488 164
pound-mass/foot-second	newton-second/meter^2 (N·s/m^2)	1.488 164
pound-force-second/foot^2	newton-second/meter^2 (N·s/m^2)	$4.788\ 026 \times 10$
rhe	meter^2/newton-second (m^2/N·s)	$1.000\ 000 \times 10$
slug/foot-second	newton-second/meter^2 (N·s/m^2)	$4.788\ 026 \times 10$
stoke	meter^2/second (m^2/s)	$1.000\ 000 \times 10^{-4}$

VOLUME/TIME (INCLUDES FLOW)

To convert from	to	Multiply by
foot^3/minute	meter^3/second (m^3/s)	$4.719\ 474 \times 10^{-4}$
foot^3/second	meter^3/second (m^3/s)	$2.831\ 685 \times 10^{-2}$
inch^3/minute	meter^3/second (m^3/s)	$2.731\ 177 \times 10^{-7}$
yard^3/minute	meter^3/second (m^3/s)	$1.274\ 258 \times 10^{-2}$
gallon (*U.S. liquid*)/day	meter^3/second (m^3/s)	$4.381\ 264 \times 10^{-8}$
gallon (*U.S. liquid*)/minute	meter^3/second (m^3/s)	$6.309\ 020 \times 10^{-5}$

Miscellaneous Metric Nomograms

The following nomograms provide a quick solution to metric conversion problems.

Example:

Convert 30 psi to kg/sq cm.

Solution:

Locate 30 on the psi scale and read the answer of 2.1 kg/sq cm.

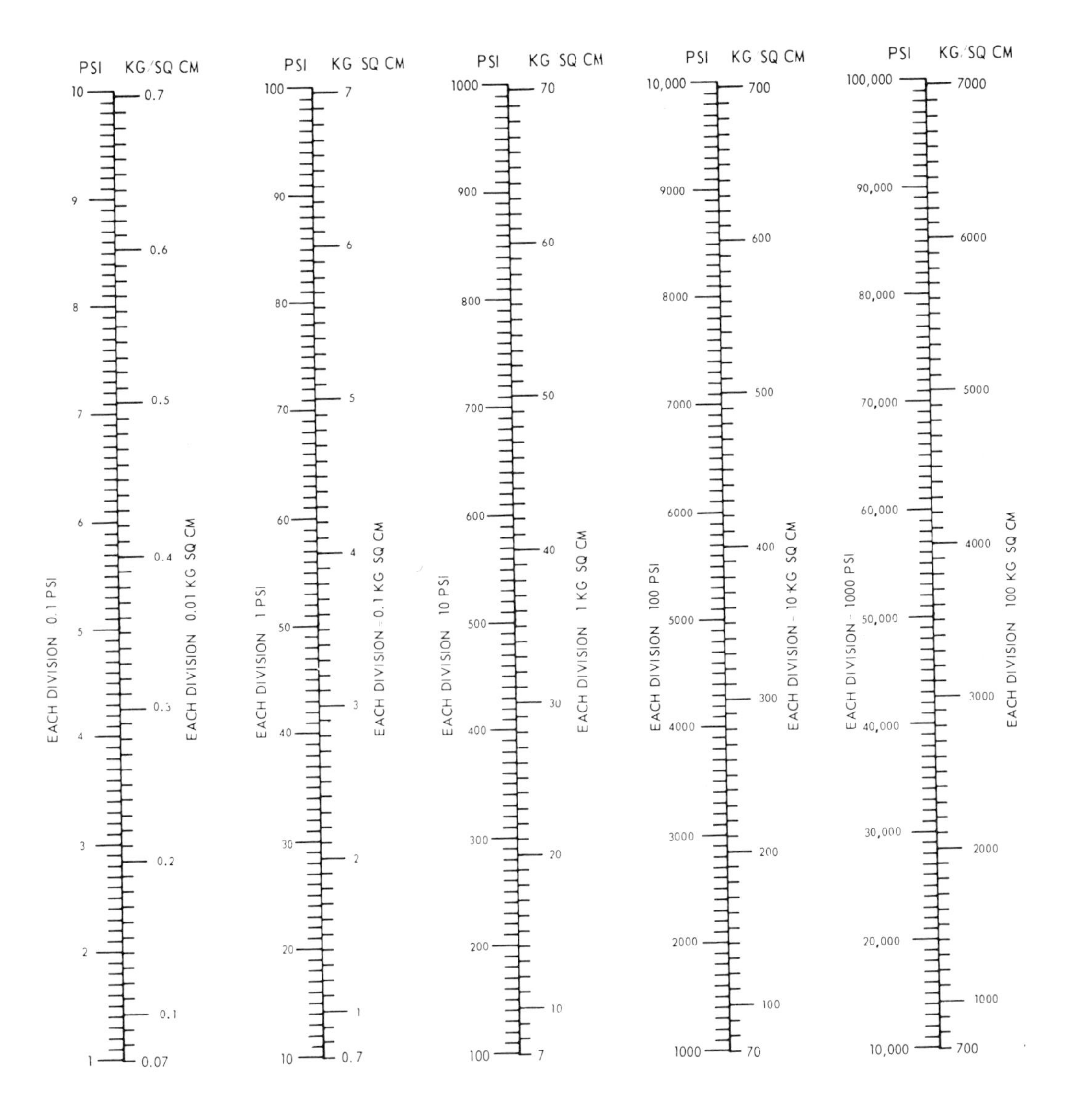

Example:
Convert 600 cu in to cc.

Solution:
Locate 600 on the cu in scale and read the answer of 9800 cc.

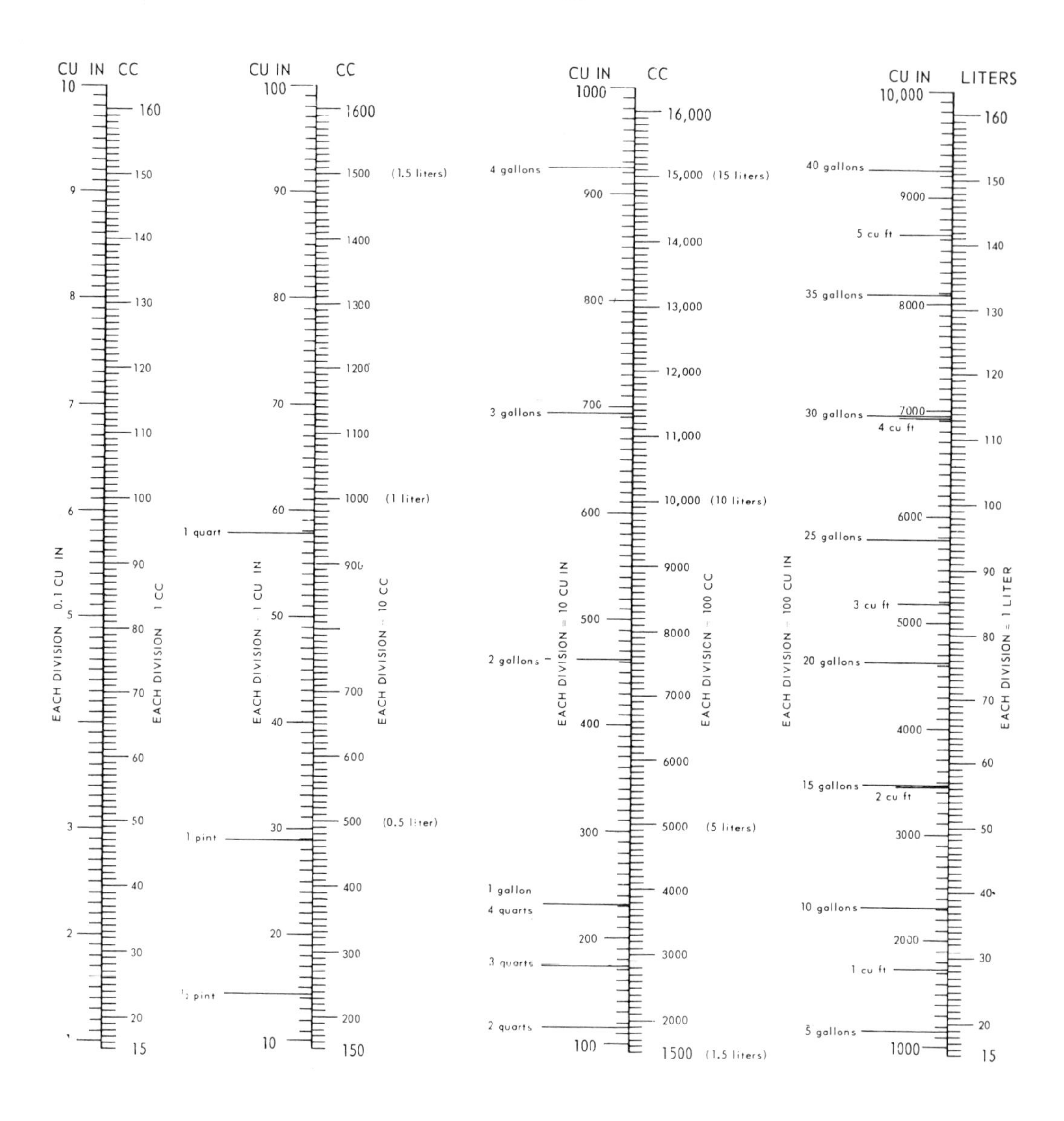

Example:

Convert 20 mph to knots, ft/min, ft/sec, km/hr, m/min and m/sec.

Solution:

Locate 20 mph on the mph scale and construct a horizontal line parallel to the base. Where this line intersects the various scales, read the answers of: 17.5 knots, 1750 ft/min, 30 ft/sec, 32 km/hr, 540 m/min and 9.0 m/sec.

The same procedure is used on the bottom nomogram.

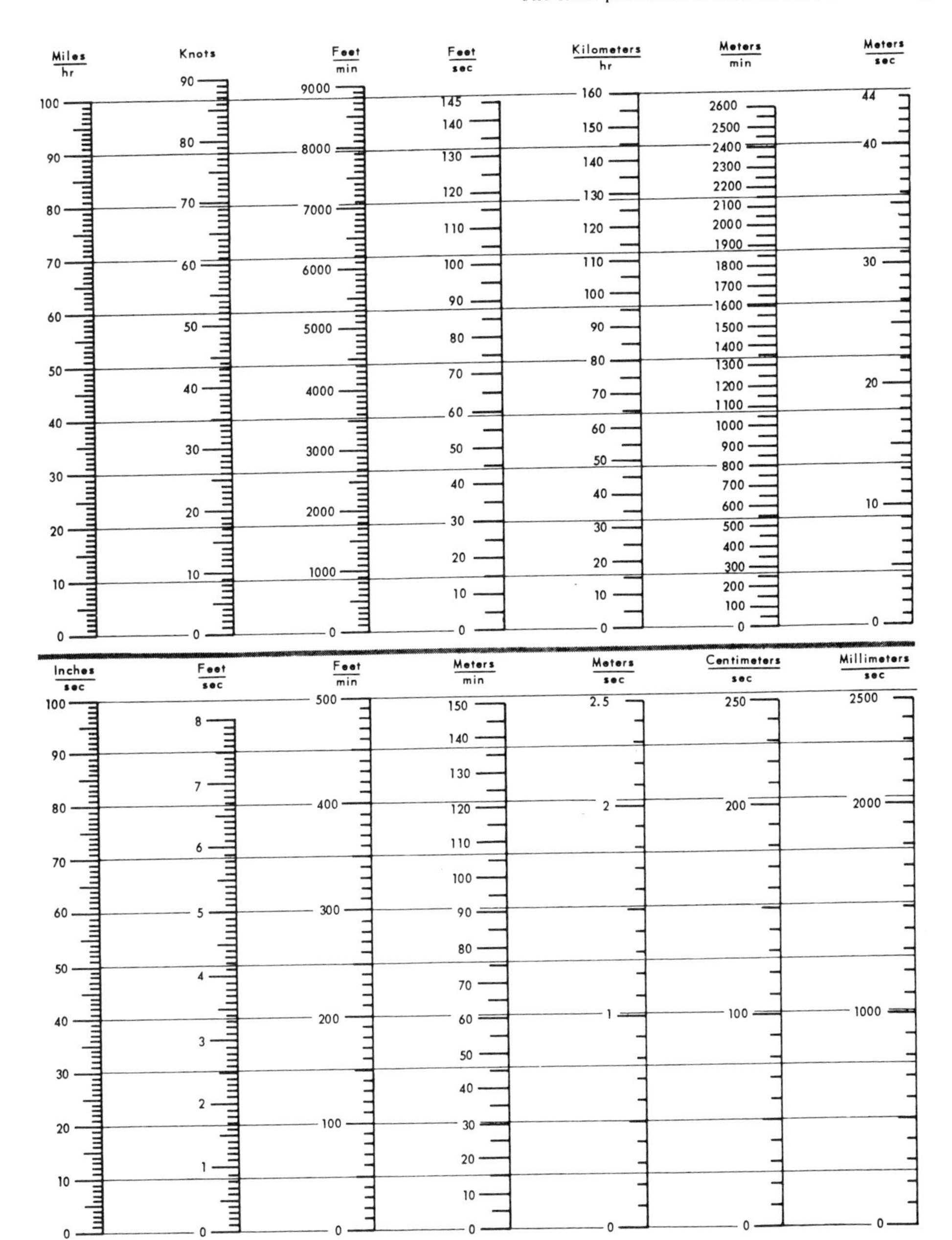

2. DATA SHEET FOR PRESSURE SWITCHES

PLEASE FILL OUT FORM COMPLETELY

Company ______________________________ Date: ____________

Address ______________________________ Name: ____________

______________________________ Title: ____________

______________________ ZIP ________ Tel. ( ) ____________

APPLICATION: ______________________________

____

ELECTRICAL: Load ______ Watts ______ Amps Inductive Resistive @ ______ Volts DC/AC ______________ Hertz

Cable/Connectors ______________________________

Switch Type: SPST ☐ SPDT ☐ Other ______________________

Fluid: ______________________ Contaminants: ______________________

ENVIRONMENTS: Temp. – Ambient ______ TO ______ °F. Max.

Operating ______ TO ______ °F. Max.

Vibration and Shock – Not Important ☐

Vibration __________ cps @ __________ G's Shock __________ G's for __________ Milliseconds

PRESSURE DATA:

System Normal Press. __________ Psi: Proof __________ Psi: Burst __________ Psi:

Switch Action Press. __________ Psi ± __________ Psi; ABS. ☐ Gauge ☐

contacts closing ☐ on increasing pressure ☐
opening ☐ decreasing ☐

Switch Set Point: Set & Sealed @ Factory ☐ or Field Adjustable ☐ over ______ to ______ Psia Range

Allowable "On to Off Spread" ("Reset Point" "Hysterisis") ______ Psi min. ☐ max. ☐

Pressure Port Fitting ______________________________

OTHER REQUIREMENTS:

Mounting and Envelope ______________________________

____

Life: ____________ cycles; Duty Cycle: ____________

SPECIAL REQUIREMENTS: None ☐ UL approval ☐ UL Component Listing ☐

MIL Specs ______________________ Other ______________________

PRODUCTION REQUIREMENTS:

Number of Prototype Units ______________ Delivery Dates ______________

Number of Production Units/Yr. ______________ Delivery Dates ______________

Standard Part No.

(Courtesy Whitman General Corporation, a Chapman Subsidiary.)

3. CORROSION RESISTANCE OF VARIOUS MATERIALS

Corrosion Resistance of Various Thermoplastics and Metals Used in Liquid Level Switches

Key:

E —Excellent to operating limit of material. See operating limits table.
G —Excellent to 80°F; good to max. oper. limit of material.
S —Satisfactory to 80°F.
L —Limited; may be used under certain conditions.
U —Unsatisfactory, not recommended.
- —Insufficient data available at this time.
PVC—Polyvinyl chloride.
PP —Polypropylene.
CU —Copper.

Thermoplastic Operating Limits in °F

	PRESSURE	NON-PRESSURE
PVC	150	160
Polypropylene	180	225
Polyurethane Foam	200	350

Temperature Derating Factor

Multiply room temperature strength by derating factor to obtain strength at elevated temperature.

TEMPERATURE (°F)	PVC	POLYPROPYLENE	TEMPERATURE (°F)	PVC	POLYPROPYLENE
73	1.00	1.00	130	.37	.52
80	.88	.93	140	.32	.46
90	.73	.84	150	.27	.40
100	.61	.74	160	-	.35
110	.51	.66	170	-	.31
120	.43	.58	180	-	.27

These data are based on tests that are believed to be reliable. They are given for information only, as we cannot accept responsibility for operations not under our direct control.

	PVC	PP	CU
Acetaldehyde	U	L	-
Acetate solvents, crude	U	S	G
Acetate solvents, pure	U	S	E
Acetic acid 20%	E	S	G
Acetic acid 80%	S	S	G
Acetic acid glacial	S	S	-
Acetic anhydride	U	U	G
Acetone	U	E	E
Adipic acid	E	S	-
Allyl alcohol	S	G	E
Allyl chloride	U	G	-
Alum	E	E	-
Aluminum chloride	E	E	-
Aluminum fluoride	E	E	-
Aluminum hydroxide	E	E	-
Aluminum oxychloride	E	E	-
Aluminum nitrate	E	E	-
Aluminum sulfate	E	E	G
Ammonia—dry gas	E	E	L
Ammonium bifluoride	E	E	-
Ammonium carbonate	E	E	-
Ammonium chloride	E	E	G
Ammonium fluoride, 25%	S	S	-
Ammonium hydroxide	E	E	U
Ammonium metaphosphate	E	E	U
Ammonium nitrate	E	E	L
Ammonium oxalate	-	-	-
Ammonium persulfate	E	E	-
Ammonium phosphate	E	E	L
Ammonium sulfate	E	E	L
Ammonium sulfide	E	E	-
Ammonium thiocyanate	E	E	-
Amyl alcohol	E	E	-
Amyl chloride	U	U	-
Aniline	U	L	-
Aniline chlorohydrate	U	U	-
Aniline hydrochloride	U	U	-
Anthraquinone	E	L	-
Anthraquinone sulfonic acid	E	L	-
Antimony trichloride	E	E	-
Aqua regia	G	L	-
Arsenic acid	G	E	-
Arylsulfonic acid	E	-	-
Asphalt	E	G	E
Barium carbonate	E	E	-
Barium chloride	E	E	-
Barium hydroxide	E	E	-
Barium sulfate	E	E	-
Barium sulfide	E	E	-
Beer	E	E	G
Beet sugar liquors	E	E	E
Benzaldehyde	U	L	-
Benzene, benzol	U	U	E
Benzenesulfonic acid 10%	E	-	-
Benzoic acid	E	-	-
Bismuth carbonate	E	E	-
Black liquor	E	E	-
Bleach 12.5%, active 01_2	G	S	-
Borax	E	E	G
Boric acid	E	E	G
Boron trifluoride	E	E	-
Breeder pellets, fish	E	E	-
Brine	E	E	-
Bromic acid	E	E	-
Bromine—liquid	U	U	-
Bromine water	E	L	-
Butadiene	E	-	E
Butane	E	S	E
Butanol, primary	E	-	-
Butanol, secondary	S	-	-
Butyl acetate	U	S	-
Butyl alcohol	G	E	-
Butyl phenol	S	-	-
Butylene	S	-	E
Butyric acid	S	-	-
Calcium bisulfite	E	E	G
Calcium carbonate	E	E	-
Calcium chlorate	E	E	-
Calcium chloride	E	E	-
Calcium hydroxide	E	E	-
Calcium hypochlorite	E	E	L
Calcium nitrate	E	E	-
Calcium sulfate	E	E	-
Cane sugar liquors	E	E	E
Carbon bisulfide	U	U	U
Carbon dioxide	E	E	E
Carbon monoxide	E	E	-
Carbon tetrachloride	L	U	E
Carbon acid	E	E	-
Carbonated beverages	-	-	-
Casein	E	E	-
Castor oil	E	E	-
Caustic potash	E	S	-
Caustic soda	E	S	-
Cellosolve	S	G	-
Chloracetic acid	E	-	-
Chloral hydrate	E	U	-
Chloric acid, 20%	E	-	-
Chlorine gas—dry	S	S	G
Chlorine gas—wet	S	S	L
Chlorine water	E	G	-
Chlorobenzene	U	U	-
Chloroform	U	L	-
Chlorosulfonic acid	L	L	-
Chrome alum	E	E	-
Chromic acid, 10%	E	E	U
Chromic acid, 50%	S	G	-
Citric acid	E	E	G

	PVC	PP	CU		PVC	PP	CU
Coconut oil	E	S	–	Gelatin	E	E	E
Coffee	–	–	–	Glucose	E	E	E
Copper chloride	E	E	–	Glycerine (Glycerol)	E	E	E
Copper cyanide	E	E	–	Glycol	E	E	–
Copper fluoride	E	E	–	Glycolic acid	E	E	–
Copper nitrate	E	E	–	Green liquor	E	E	–
Copper sulfate	E	E	L	Heptane	G	U	–
Core oils	E	E	E	Hexane	S	L	–
Cottonseed oil	E	E	E	Hexanol, tertiary	E	G	–
Cresol	U	U	G	Hydrobromic acid, 20%	E	E	–
Cresylic acid, 50%	E	U	–	Hydrochloric acid, 0–25%	G	E	L
Croton aldehyde	U	L	–	Hydrochloric acid, 25–40%	E	E	–
Crude oil	E	L	–	Hydrocyanic acid	E	E	U
Cyclohexanol	U	S	–	Hydrofluoric acid, 10%	S	E	L
Cyclohexanone	U	U	–	Hydrofluoric acid, 30%	S	E	–
Demineralized water	E	E	–	Hydrofluoric acid, 60%	S	–	–
Dextrin	E	E	–	Hydrofluosilicic acid	E	–	–
Dextrose	E	E	–	Hydrogen	E	E	–
Diazo salts	E	E	–	Hydrogen cyanide	E	E	–
Diglycolic acid	G	E	–	Hydrogen fluoride	–	–	G
Dimethylamine	U	U	–	Hydrogen peroxide, 30%	E	G	–
Diotylphthalate	U	U	–	Hydrogen peroxide, 50%	E	U	–
Disodium phosphate	E	E	–	Hydrogen peroxide, 90%	E	U	–
Ethers	U	U	E	Hydrogen phosphide	E	E	–
Ethyl acetate	U	E	–	Hydrogen sulfide, aq. sol.	E	E	U
Ethyl acrylate	U	S	–	Hydrogen sulfide, dry	E	E	U
Ethyl alcohol	E	E	–	Hydroquinone	E	E	–
Ethyl chloride	U	U	–	Hydroxylamine sulfate	E	S	–
Ethyl ether	U	L	–	Hypochlorous acid	E	E	–
Ethylene bromide	U	U	–	Iodine (in alcohol)	U	S	–
Ethylene chlorohydrin	U	U	–	Isopropyl alcohol	E	E	–
Ethylene dichloride	U	U	–	Kerosene	E	L	–
Ethylene glycol	E	E	E	Kraft liquor	E	E	–
Ethylene oxide	U	L	–	Lacquer thinners	L	L	E
Fatty acids	E	G	–	Lactic acid, 25%	E	E	–
Ferric chloride	E	E	U	Lard oil	E	–	–
Ferric nitrate	E	E	–	Lauric acid	E	S	–
Ferric sulfate	E	E	L	Lauryl chloride	E	L	–
Ferrous chloride	E	E	–	Lauryl sulfate	E	L	–
Ferrous sulfate	E	E	–	Lead acetate	E	E	–
Fish solubles	E	E	–	Lime sulfur	E	E	–
Fluoboric acid	E	E	–	Lineolic acid	E	S	–
Fluorine gas	L	L	–	Linseed oil	E	E	–
Fluosilicic acid	E	E	–	Liquors, liqueurs	E	E	–
Formaldehyde	G	E	G	Lubricating oils	E	L	–
Formic acid	S	E	G	Lysol	–	–	–
Freon-12	G	G	E	Magnesium carbonate	E	E	–
Frutose	E	E	–	Magnesium chloride	E	E	G
Fruit juices, pulp	E	E	–	Magnesium hydroxide	E	E	E
Fuel oil	E	L	–	Magnesium nitrate	E	E	–
Furfural	U	U	–	Magnesium sulfate	E	E	E
Gallic acid	E	E	–	Maleic acid	E	E	–
Gas—coke oven	E	S	L	Malic acid	E	E	–
Gas—manufactured	U	S	–	Mercuric chloride	E	E	–
Gas—natural	E	S	–	Mercuric cyanide	E	E	–
Gasoline—refined	S	U	E	Mercurous nitrate	E	E	–
Gasoline—sour	E	U	U	Mercury	E	E	U

	PVC	PP	CU
Methyl alcohol	E	E	-
Methyl chloride	U	U	-
Methyl ethyl ketone	U	U	-
Methyl sulfate	S	L	-
Methyl sulfuric acid	E	E	-
Methylene chloride	U	S	-
Milk	E	S	L
Mineral oils	E	L	-
Molasses	E	E	-
Naphtha	E	U	-
Naphthalene	U	S	-
Nickel acetate	E	E	-
Nickel chloride	E	E	L
Nickel nitrate	E	E	-
Nickel sulfate	E	E	G
Nicotine	E	E	-
Nicotinic acid	E	E	-
Nitric acid, 10%	E	E	U
Nitric acid, 30%	E	L	-
Nitric acid, 60%	E	L	-
Nitric acid, 68%	E	U	-
Nitric acid, anhydrous	U	L	-
Nitrobenzene	U	E	-
Nitropropane	-	L	-
Nitrous acid (conc)	-	-	-
Nitrous oxide	E	L	-
Ocenol	E	S	-
Oils and fats	E	E	-
Oleic acid	E	S	G
Oleum	U	U	-
Oxalic acid	E	G	G
Oxygen	E	L	E
Ozone	S	L	-
Palmitic acid, 10%	E	S	G
Palmitic acid, 70%	S	S	-
Peracetic acid, 40%	S	L	-
Paraffin (molten)	-	-	-
Perchloric acid, 10%	E	L	-
Perchloric acid, 70%	S	L	-
Petroleum oils (refined)	-	-	E
Petroleum oils (sour)	-	-	U
Phenol	U	L	-
Phenylhydrazine	U	-	-
Phenylhydrazine hydrochloride	S	L	-
Phosgene—gas	G	L	-
Phosgene—liquid	U	L	-
Phosphoric acid, 0-50%	E	E	G
Phosphoric acid, 50-75%	E	E	G
Phosphorous, yellow	G	L	-
Phosphorous, red	E	-	-
Phosphorous pentoxide	S	S	-
Phosphorus trichloride	S	L	-
Photographic solutions	E	E	-
Picric acid	U	S	U
Pine tar oil	-	-	-
Plating solution, brass	E	E	-
Plating sol., cadmium	E	E	-
Plating sol., chromium	G	E	-
Plating sol., copper	E	E	-
Plating sol., gold	E	E	-
Plating sol., lead	E	E	-
Plating sol., nickel	E	E	-
Plating sol., silver	E	E	-
Plating sol., tin	E	E	-
Plating sol., zinc	E	E	-
Potassium acid sulfate	E	E	-
Potassium antimonate	E	-	-
Potassium bicarbonate	E	E	-
Potassium bichromate	E	E	-
Potassium bisulfite	E	-	-
Potassium borate	E	E	-
Potassium bromate	E	E	-
Potassium bromide	E	E	-
Potassium carbonate	E	E	-
Potassium chlorate	E	E	-
Potassium chloride	E	E	G
Potassium chromate	E	E	-
Potassium cuprocyanide	E	E	-
Potassium cyanide	E	E	-
Potassium dichromate	E	E	-
Potassium ferricyanide	E	E	-
Potassium fluoride	E	E	-
Potassium hydroxide, 0-20%	E	E	L
Potassium hydroxide, 35%	E	-	-
Potassium hypochlorite	G	-	-
Potassium nitrate	E	E	-
Potassium oxalate	-	-	-
Potassium perborate	E	E	-
Potassium perchlorite	E	E	-
Potassium permanganate	E	E	-
Potassium persulfate	E	E	-
Potassium sulfate	E	E	E
Potassium sulfide	E	E	-
Propane	E	L	E
Propargyl alcohol	E	L	E
Propyl alcohol	E	E	E
Propylene dichloride	U	L	-
Quinine bisulfate	-	-	-
Quinine sulfate	-	-	-
Rayon coagulating bath	E	E	-
Salicylic acid	E	G	-
Silver bromide	-	-	-
Silver cyanide	E	E	-
Silver nitrate	E	E	-
Soaps	E	E	E
Sodium acetate	E	E	-
Sodium acid sulfate	E	E	-
Sodium antimonate	E	E	-
Sodium arsenite	E	E	-
Sodium benzoate	E	E	-
Sodium bicarbonate	E	E	-
Sodium bisulfate	E	E	-
Sodium bisulfite	E	E	-
Sodium borate	-	-	-

	PVC	PP	CU
Sodium bromide	E	E	-
Sodium carbonate	E	E	E
Sodium chlorate	G	E	-
Sodium chloride	E	E	G
Sodium cyanide	E	E	U
Sodium dichromate	E	E	-
Sodium ferricyanide	E	E	-
Sodium fluoride	E	E	-
Sodium hydroxide, 0-70%	E	E	L
Sodium hypochlorite	E	G	L
Sodium lactate	-	-	-
Sodium nitrate	E	E	G
Sodium nitrite	E	E	-
Sodium perborate	-	-	G
Sodium peroxide	-	-	G
Sodium phosphate acid	E	E	G
Sodium silicate	E	E	G
Sodium sulfate	E	E	E
Sodium sulfide	E	E	U
Sodium thiosulfate (hypo)	-	-	U
Soybean oil	-	-	-
Stannic chloride	E	E	-
Stannous chloride	E	E	-
Stearic acid	E	E	G
Stoddard's solvent	E	E	-
Sulfur	E	E	L
Sulfur chloride	-	-	U
Sulfur dioxide, dry	E	S	E
Sulfur dioxide, wet	S	S	G
Sulfuric acid, 0-10%	E	E	G
Sulfuric acid, 10-50%	E	E	L
Sulfuric acid, 50-75%	E	E	L
Sulfuric acid, 75-90%	E	S	L
Sulfuric acid, 95%	G	S	U
Sulfur dioxide—liquid	S	L	-
Sulfur trioxide	E	L	-
Syrup	-	-	-
Tannic acid	E	E	-
Tar	-	-	E
Tartaric acid	E	E	G
Tetraethyl lead	G	S	-
Tetrahydrofurate	U	L	-
Thionyl chloride	U	S	-
Titanium tetrachloride	S	S	-
Toluol or toluene	U	U	E
Tributyl phosphate	U	U	-
Trichloroethylene	U	U	E
Tricesylphosphate	G	S	-
Triethanolamine	G	S	-
Triethylamine	G	L	-
Trimethylpropane	E	L	-
Trisodium phosphate	E	E	-
Turpentine	E	U	E
Urea and urine	E	E	-
Varnish	-	-	G
Vegetable oils	-	-	G
Vinegar	E	E	G
Vinyl acetate	U	S	-
Water—demineralized, distilled, fresh, salt, sewage, acid mine	E	E	G
Whiskey, wine	E	E	G
White liquor	E	E	-
X-Ray dev. sol.	-	-	-
Xylene or xylor	U	U	E
Zinc chloride	E	E	G
Zinc chromate	E	E	-
Zinc cyanide	E	E	-
Zinc nitrate	E	E	-
Zinc sulfate	E	E	G

Valox^T Polyester Type Thermoplastic

Valox[T] resins are characterized by their excellent resistance to many environments. In particular, they are resistant to a variety of chemicals including aliphatic hydrocarbons, gasoline, trichlorethylene, carbon tetrachloride, perchloroethylene, oils and fats, alcohols, glycols, ethers, high molecular weight esters and ketones, dilute acids and bases, detergents, and most aqueous salt solutions. They are attacked by strong acids and bases. The table shows the long-term effects of some of these chemicals on Valox[T] resins. With this resistance, Valox[T] resins can be used effectively under the hood in automobiles and in such applications as pumps.

Chemical Resistance (% Weight Change)

CHEMICAL	% WEIGHT CHANGE
Acetic acid (5%)	0.2
Acetone	1.7
Brake fluid	0.0
Carbon tetrachloride	0.3
1, 2 Dichlorethylene	6.3
Ethyl acetate	1.2
Gasoline (high test)	0.1
Heptane	0.1
Hydrochloric acid (10%)	−1.0
Ivory brand soap (1%)	0.1
Kerosene	0.1
Methanol	0.9
Motor oil (detergent)	0.1
Nitric acid (10%)	0.1
Perclean	0.6
Phenol (5%)	6.1
Sodium chloride (10%)	0.1
Sodium carbonate (20%)	0.1
Sodium hydroxide (10%)	−6.1
Sulfuric acid (5%)	0.1
Sulfuric acid (30%)	0.1
Sulfuric acid (30%)	0.1
Toluene	0.7
Transformer 0.1	0.0
Water—6 mo. exp. @ 73° F	0.4

General Electric Company has no control over the usage of this material and therefore does not guarantee results as above. Tests must be made to prove suitability for specific uses. Each user of the material should make his own tests to determine the material's suitability for his own particular use. Statements concerning possible or suggested uses of the materials described herein are not to be construed as constituting a license under any General Electric patent covering such use or as recommendations for use of such materials in the infringement of any patent.

T = Trademark of General Electric Company.

Polyurethane Foam (Closed Cell with Solid Outer Skin, 20 lb/ft^3)*

This material is satisfactory for use with most petroleum products, e.g., gasoline, lubricating oils, hydraulic oils, JP-4, JP-5, skydrol, etc. It can also be used with common weak acids and bases (sulfuric acid excepted), water, and alcohols. It is attacked by acetone, mek, carbon tetrachloride, xylene, and related chemicals.

Corrosion Resistance of Some Valve Materials*

*Courtesy of Harwil Company, Santa Monica, California.

*Extracted from *Lyons' Enclyclopedia of Valves* (pages 110–118). *Courtesy of Van Nostrand Reinhold Company and *Design News Magazine* pages 244–267.

Table 1 Corrosion Resistance of Some Typical Valve Materials. *(Courtesy of Jordan Valve, Division of Richards Industries, Inc.)*

CHEMICAL 1	MATERIAL																		CHEMICAL 2
	Aluminum	Asbestos	Bronze	Carpenter 20	Cast Iron	Carbon Steel	Ductile Iron	Hastelloy B	Jordanite	Monel	Neoprene	Nickel	Plastisol	Polyethylene	303 S.S.	304 S.S.	316 S.S.	Teflon	
Acetaldehyde	1	2	1 2		1	2		1	1 2	1 2	1 2	1	1 2	1 2	1 2	1 2	1 2	1 2	Zinc Sulfate (acid)
Acetate Solvents (crude)	1 2		1 2		1 2	2	1 2	2	1 2	1 2	2	1	2	2	1 2	1 2	1 2		Zinc Sulfate
Acetate Solvents (pure)	1		1		1		1			1		1		1	1	1	1		Zinc Plating Solution
Acetic Acid (crude)	1		1	1	1	1	1	1	1	1	2	1	1	1	1 2	1	1 2	2	Zinc Hydrosulfite
Acetic Acid (pure)	1 2		1 2	1 2	1	1 2	1 2	1 2	1 2	1 2	2	1 2	1 2	1 2	1 2	1 2	1 2	1 2	Zinc Chloride
Acetic Acid Vapors	1		1	1 2	1	1 2	1	1	1	1		1			1	1 2	1		Zinc Cyanide Solution
Acetic Anhydride	1		1	1 2		1	1	1	1	1	1	1	1		1	1	1	1	Zinc Carbonate
Acetone	1	1	1	1	1	1	1	1	1	1	2	1	1	1 2	1	1	1	1	Zinc Ammonium Chloride
Acetylene	1	1	1		1	1	1		1	1 2	2				1	1	1 2		Zinc Acetate
Alcohol – Amyl	1		1	1	1	1	1	1	1	1 2	2	1	1		1	1	1	1	Zeolite
Alcohol – Butyl	1 2		1 2	1	2	1 2	1	1	1	1 2	2	1 2	1	1 2	1 2	1 2	1 2	1 2	Xylene
Alcohol – Ethyl	1 2	2	1 2	1	1 2	1 2	1 2	1	1	1 2	2	1	1 2	1 2	1 2	1 2	1 2	1 2	Whiskey & Wines
Alcohol – Isopropyl	1		1 2	1		1	1	1	1	1		1	1	2	1 2	1	1 2	1	White Liquor
Alcohol – Methyl	1		1		1	1 2	1	1	1	1 2	2		1		1	1 2	1 2	1	Wax - molten
Alcohol – Diacetone			2	2	1 2	2		2		1 2	1 2	2				2	1 2		Water - sea
Alcohol – Oleyl	2	2	2	2	2	2	2	2		1 2	2	2	2		2	2	2	2	Water - fresh
Alkaform						2		1	1							1 2	1		Vinyl Chloride
Alum	1	1	1 2	1		2	1		1	1	1		1		1	1 2	1 2	1	Vinyl Acetate
Alumina	2	2	1 2	2	2	2	2	2	1	1 2	1 2	2	1 2	2	2	2	2	1 2	Vinegar
Aluminum Acetate	1 2		2	1 2		2	2	1 2		1 2		2	2	2	2	1 2	1 2	2	Vegetable Oils
Aluminum Chloride	1 2	1	1 2	1	1 2	1 2	2	1 2	1	1 2	1	1 2	1	1	1 2	1 2	1 2	1 2	Varnish
Aluminum Fluoride	1		1	1 2	1			1 2	1	1	1	1 2	1		1	1 2	1 2	1	Uric Acid
Aluminum Hydroxide	2	2	2	1 2	1 2	2	2	2	1 2	1 2	1	1 2	2		1 2	1 2	1 2	1 2	Turpentine
Aluminum Sulfate 10% Boil	1		1 2	1	1	1	1		1	12	2	1	1 2	1 2	1 2	1	1 2	1	Trisodium Phosphate
Aluminum Sulfate Saturate room	1		1	1 2	1	1	1		1	1			1	1	1	1 2	1 2	1	Triodium Phosphate
Aluminum Sulfate Saturate Boil	1		1	1 2	1	1	1	1	1	1			1	1	1	1 2	1 2	1	Tripotassium Phosphate
Amines	1		1	2		1			1	1		1			1	1 2	1 2	1	Triphenylphosphite
Ammonia, Dry	1	1		1	1	1		1	1	1 2	1	1	1	1	1	1	1		Triethylamine
Ammonia (gas of liquid)	1	1	1 2		1	1	1		1	1 2	2		1	1	1 2	1	1 2	1	Triethanolamine
Ammonium – Bicarbonate	1		1	1 2	1				1	1	1	1			1	1 2	1 2	1	Trichlorotrifluroethane
Ammonium Carbonate	1		1	1 2	1			1 2	1	1	1	1	1		1	1 2	1 2	1	Trichloropropane
Ammonium Chloride	1	1	1 2	1 2	1	1	1	1	1	1	1	1	1	1	1	1 2	1 2	1	Trichloromonofluoroethane
Ammonium Dophospate	2						2		1 2	1	1 2			2	2	1 2	1 2	1	Trichloroethylene (moist)
Ammonium Hydroxide	1 2	1	1 2	1	1	1 2	1 2	1	1 2	1 2	2	1 2	1	1 2	1 2	1 2	1 2	1 2	Trichloroethylene (dry)
Ammonium Hydrozide					1				1		1 2						1		Trichlorobenzene
Ammonium Monosulfate	2			2	1			1	1		1 2					1 2	1 2	1	Trichloroacetic Acid
Ammonium Nitrate	1		1	1	1	1	1	1	1	1 2	1	1	1	1	1	1	1	1	Tretolite

KEY: **1 - Chemical in column 1 may be used** ● 1 - Chemical in column 1 may not be used
2 - Chemical in column 2 may be used ● 2 - Chemical in column 2 may not be used
( ) **Conditional or no information**

Table 1 (*Continued*)

CHEMICAL 1	Aluminum	Asbestos	Bronze	Carpenter 20	Cast Iron	Carbon Steel	Ductile Iron	Hastelloy B	Jordanite	Monel	Neoprene	Nickel	Plastisol	Polyethylene	303 S.S.	304 S.S.	316 S.S.	Teflon	CHEMICAL 2
Ammonium Oxalate	1		2	1	1 2	1		1		1 2						1	1 2		Transmission Oil
Ammonium Persulfate	1			1		1		1		1 2		1	1	1		1	1		Toxaphene
Ammonium Phosphate (mono)	1 2		1	1 2	1	1	1 2	1 2		1 2		2	1			1 2	1 2		Tomato Juice
Ammonium Phosphate (di)	1		1	1		1	1	1 2		1			1	1	1	1	1	1	Toluene Sulfonic Acid
Ammonium Phosphate (tri)	1 2	2	1 2	1	1 2	1 2	1 2	1	2	1 2	1 2		1	2	1 2	1 2	1 2	2	Toluene or Toluol
Ammonium Sulfate	1 2	1	1	1 2	1	1 2	1	2	1	1 2	1	1	1	1	1	1 2	1	1	Titanium Tetrachloride
Ammonium Thiocyanate									1		1							1	Tin Plating Solution
Amyl Acetate	1	1		1		1		1	1	1 2	1	1		1		1	1	1	Thiophene
Amyl Chloride	1			1		1		1	1	1	1	1				1	1	1 2	Thiamine Hydrochloride
Aniline	1	1	1	1	1	1		1	1	2	1 2	1		1	1	1	1	1 2	Tetraphosphoglucosate
Aniline Sulfite	1			1 2				1 2		1							1	1	Tetraphosphoric Acid
Aniline Dyes		1							1 2	1							1	1	Tetramine
Aniline Oils		1	1		1				1 2	1	1				1		1 2	1	Tetrachloroethane
Aniline Hydrochloride	1					2		1		1		1				1 2	1	1	Terpene Monocyclic
Antimony Chloride								1	1	1 2								1	Tennox
Antimony Trichloride:	2		2	1		1 2	2	1 2	2	1 2	2	1 2	1 2	1	1	1 2	1 2	2	Tartaric Acid
Antioxidants										1 2	1							1	Tar Acids
Arochlor	2	2	2		2	1 2	2			2					2	1 2	2		Tar
Arsenic Acid			1	1	1	1			1	1 2	1 2		1		2	1	1 2	1	Tannin
Arsenic Trichloride			2						1	1	1		2	2				1	Tanning Liquor
Asphalt	1		1 2	2	1	1 2	1	2	1	1 2	2	1 2	1 2	2	1 2	1 2	1 2	1 2	Tannic Acid
Barium Carbonate	1			1				1	1	1 2	1	1		2	1	1	1	1	Tallow, Molten
Barium Chloride		1	1	1 2	1	1		1 2	1	1	1	1	1	1	1	1	1 2	1	Tall Oil
Barium Hydrate	1			1				1		1 2	2	1				1	1		Talc Slurry
Barium Hydroxide	2	1 2	1 2	1 2	1 2	1 2	2	2	1	1 2	1	2	1	1 2	1 2	1 2	1 2	1 2	Sulfurous Acid
Barium Nitrate				1	1	1				2			1	1		1	1		Sulfuric Acid (spent)
Barium Sulfate	2		2	1 2	2		2	2	1 2	1 2	1 2	1		2	1 2	1 2	1 2	1	Sulfuric Acid (95-100%)
Barium Sulfide	2		2	2	2	1 2	2	2	1 2	1 2	1 2		1	1 2	2	1 2	1 2	1	Sulfuric Acid (90-95%)
Beer	1 2	1	1 2	2	1 2	1 2	1 2	1 2	1 2	1 2	1	1	1	1 2	1 2	1 2	1 2	1	Sulfuric Acid (75-90%)
Beet Sugar Liquor	1 2	1	1 2	2	1	1 2	1 2	2	2	1 2	1	1	1 2	2	1 2	1 2	1 2	1	Sulfuric Acid (10-75%)
Benzaldehyde	2	2	2	2	2	1 2	2	2	2	2			2	2	2	1 2	2		Sulfuric Acid (0-10%)
Benzene (Benzol)	1 2		1 2	1	1	1 2	1 2	1	1	1 2	1 2	1	2	1	1 2	1 2	1 2	1	Sulfur Trioxide (dry)
Benzenesulfonic Acid	1 2	2	2	1	2	2	2	2	1	1 2	2		2	2	2	2	2	1	Sulfur Dioxide (dry)
Benzine	1 2		1 2	2	1 2	1 2	1 2	2	1 2	1 2					1	1 2	1 2	1 2	Sulfur Chloride
Benzoic Acid	1 2	2	2	1 2	1 2	1 2	2	1 2	1	1 2		1 2	1	1	1 2	1 2	1 2	1 2	Sulfur (molten)
Black Liquor		1				1					1			1		1	1 2	1	Sulphonyl Chloride
Blast Furnace Gas						2					1		2				2	2	Sulfite Liquor

KEY: **1 - Chemical in column 1 may be used** • 1 - Chemical in column 1 may not be used
2 - Chemical in column 2 may be used • 2 - Chemical in column 2 may not be used
( ) **Conditional or no information**

Table 1 *(Continued)*

CHEMICAL 1	Aluminum	Asbestos	Bronze	Carpenter 20	Cast Iron	Carbon Steel	Ductile Iron	Hastelloy B	Jordanite	Monel	Neoprene	Nickel	Plastisol	Polyethylene	303 S.S.	304 S.S.	316 S.S.	Teflon	CHEMICAL 2
Bleaching Powder, Wet	1			1		1		1 2	1	1	1	1		1		1	1	1	Sulfate Oils
Boiler Compounds (pH 8.0)	2		2		2		2			1 2	1				2	2	2		Sulfate Liquor
Boiler Acid - phosphate type	2	2	2	2		2		2		1 2	1 2	2			2	2	2		Sugar Solution
Borax	1	1	1		1 2	2	1	1		1 2		1	1	1	1	1 2	1 2	1	Styrene
Bordeaux Mixture	2		2		2	2				1 2	1					2	2		Stoddard Solvent
Boric Acid	1 2	2	1 2	1 2	1	1 2	1 2	1 2	1 2	1 2	1	1 2	1 2	1 2	1 2	1 2	1 2	1	Stearic Acid
Boron Trichloride						1					2					1			Steam Condensate
Boron Trifluoride			2	2	2	1 2			2	2		2	2		2	1 2	1 2	2	Steam
Brine			1 2			1 2				1 2	1 2			2	2	1 2	1 2		Starch
Bromine (Wet)			1 2	1 2	1	2		1	1		1			1	1	2	1 2		Stannous Chloride
Bromine (Dry)	1		1	1		1		1 2	1	1		1	1	1	1	1	1	1	Stannous Bisulfate
Butadiene	1 2	2	1 2	2	1 2	2	1	2	2	1 2	2	2	2	2	1	1 2	1 2	2	Stannic Chloride
Butane	1 2	1 2	1 2	2	1 2		1	2		1 2	1 2				1	1 2	1 2		Soybean Oil
Buttermilk	1		1		1	1 2		1		1	1	1				1 2	1	1	Sorbitol
Butyl Acetate	1	1	1	1	1 2	1		1		1 2	2				2	1	1 2		Sodium Triphosphate
Butyl Catechol Tert										1 2	2								Sodium Tetraphosphate
Butyl Cellosolve	1		1		1 2				1	1 2	2				2		1 2		Sodium Tetraborate
Butyl Chloride	2		2	2	2	1		2		2	2		2	2	2	1 2	2	2	Sodium Sulfite
Butyl-p-Aminopheno	2	2	2	2	2	2	2	2		1 2	2	2	2	2	2	2	2	2	Sodium Sulfide
Butyl – Stearate	1 2	2	1 2		1 2	2	2	2		2	2	2	2	2	2	2	1 2	2	Sodium Sulfate
Butylene	1		1		1		1			1 2					1	1	1		Sodium Silicofluoride
Butyric Acid	1 2	2	1 2	1	1 2	1 2	2	1 2		1 2	1 2	1 2	2		2	1 2	1 2	1 2	Sodium Silicate
Borax					2					1 2	1 2						1 2		Sodium Salts
Cadmium Sulfate				1		2		1		1						2	1		Sodium Salicylate
Calcium Acetate	1		1			2				1 2						2	1		Sodium Resinate
Calcium Bisulfite	1		1			1	1			1 2	1 2		1	1	1	1	1		Sodium Pyrophosphate
Calcium Carbonate	1			1	2			1		1 2	1 2	1			1 2	1	1 2		Sodium Polyphosphate
Calcium Chlorate		2		1		2		2		1		1			2	1 2	1 2		Sodium Phosphate (neutral)
Calcium Chloride	1 2	1	1 2	1 2	1	1 2	1	1 2		1 2	1	1	1	1	1 2	1 2	1 2		Sodium Phosphate (acid)
Calcium Nitrate 40%	1									1 2							1		Sodium Plumbite
Calcium Hydroxide	1 2	1	1 2	1 2	1	1 2	2	1 2		1 2	1	1 2	1	1	1 2	1 2	1 2	1 2	Sodium Peroxide
Calcium Hypochlorite	1	1	1 2	1	1	1	1 2	1		1 2	2	1 2	1 2	1	2	1 2	1 2	1 2	Sodium Perborate
Calcium Sulfate	1		1	1	1					1 2	1 2					1	1		Sodium Orthosilicate
Calgon	2		2	2	1					1	1 2				1 2	2	1 2		Sodium Nitrite
Camphene	2		2		2	1 2	2	2	2	2	2	2	2	2	2	1 2	2	2	Sodium Nitrate
Camphor	1			1				1		2		1		1		1	1		Sodium Naphthsulfonate
Cane Sugar Liquor	1	1	1 2		1 2	1	1			1 2	1				1	1	1 2		Sodium Triphosphate

KEY: 1 - Chemical in column 1 may be used • 1 - Chemical in column 1 may not be used
2 - Chemical in column 2 may be used • 2 - Chemical in column 2 may not be used
( ) Conditional or no information

Table 1 (*Continued*)

CHEMICAL 1	Aluminum	Asbestos	Bronze	Carpenter 20	Cast Iron	Carbon Steel	Ductile Iron	Hastelloy B	Jordanite	Monel	Neoprene	Nickel	Plastisol	Polyethylene	303 S.S.	304 S.S.	316 S.S.	Teflon	CHEMICAL 2
	MATERIAL																		
Carbolic Acid (phenol)	1		1		2	1	1	1		1 2		1			1	1	1 2	1	Sodium Diphosphate
Carbon Monoxide		1	2		2			1		1 2	2	1				1	1 2		Sodium Monophosphate
Carbon Dioxide (wet)	1	1	1	1			1			1 2					1	1	1		Sodium Methylate
Carbon Dioxide (dry)	1	1	1		1	1	1			1 2	2	1	1	1	1	1	1	1	Sodium-M-Silicate
Carbon Disulfide	1 2	2	1 2		1 2	1	1 2		1	1 2	1 2	1		1	1 2	1 2	1 2	1	Sodium-M-Phosphate
Carbon Tetrachloride			1	1		1	1	1	1	1 2	1 2	1			1	1	1 2	1	Sodium Diphosphate
Carbonated Beverages	1		1	1		2	1	1		1		1				1 2	1		Sodium Oleate
Carbonated Water	1		1			2	1			1	1				1	1 2	1		Sodium Lactate
Carbonic Acid	1 2			2	1	1 2	1	1 2	1	1 2	2	1 2	1	1		1 2	1 2	1	Sodium Hyposulfite
Castor Oil	2	1 2	1 2	2	1 2	2	2	2		1 2	1		2	2	1	2	1 2	2	Sodium Hypochlorite
Catechol	2	2	2		2	2	2	2	2	1 2	2	2			2	2	2	2	Sodium Hydroxide 20% or Hot
Caustic Soda	1 2	2	1 2	1	1 2	2	2	1 2	2	1 2	2	1 2	2	2	2	1 2	1 2	1 2	Sodium Hydroxide 0-20%
Cellosolve (butyl or ethyl)	2		1	2	1			2		1 2		2					1		Sodium Hydrosulfite
Cellosolve (methyl)	1		1		1	2				1						2	1		Sodium Glutamate
China Wood Oil (tung)	1 2	1	2	1 2		1 2		2		1 2		1 2	2	2	1	2	1 2	1 2	Sodium Fluoride
Chloric Acid	2			1 2		1		2		2			2		1	1			Sodium Ferrocyanide
Chlorinated Water			1	1		1 2		1		1		1			1	1 2	1	1	Sodium Ethylate
Chlorine (wet)	1 2		1 2			2	1	1	1	1			1	1	1 2	1 2	1 2	1	Sodium Dichromate
Chlorine (dry)	1 2		1 2		1 2	1 2	1 2	1	1	1 2	2	1	2	1 2	1 2	1 2	1 2	2	Sodium Cyanide
Chloroacetic Acid	1 2		1	1 2	1	1		1 2		1 2	1	1 2	1		1	1 2	1 2	1	Sodium Citrate
Chlorobenzene			1 2	1 2	2	1		2		1	1	1			1 2	1 2	1 2		Sodium Chromate
Chlorobromomethane	2	2	1 2		2	2	2	2	2	1 2	2	2	2	2			1 2	2	Sodium Chloride
Chloroethane	2	2	2	2	2	1 2	2	2		2	2	2	2	2	2	1 2	2	2	Sodium Carbonate (soda ash)
Chloroethylbenzene	2			2		2		2	1	2		2	2	2	2	2	2	2	Sodium Bromide
Chloroform	1 2		1 2	1 2	1	1		1 2	1	1 2	1 2	1			1	1 2	1 2		Sodium Borate
Chlorex	1		1		2		1			1 2	2		2	2		1	1 2		Sodium Bisulfite
Chlorosulfonic Acid	1 2	2	2	1 2		1 2	2	1 2		1 2	2	1 2	1 2	1 2	2	1 2	1 2	1 2	Sodium Bisulfate
Chlorox	2		1	2	1			2		1 2	1					2	2		Sodium Bichromate
Chromic Acid (free of SO_3)	1 2	2	1 2	1		1	1 2	2	1	1 2	1 2	1	1 2	1 2	2	1 2	1 2	1	Sodium Bicarbonate
Chromic Acid (contains SO_3)	1 2		1	1		1	1	1 2	1	1 2	1		1	1 2		1	1	1	Sodium Benzoate
Chrome Plating Sol.			1 2	1		2			1	1 2	2	1			1 2	1 2	1 2		Sodium Aluminate
Chromium Sulfate				1							2					1	1		Sodium Acid Sulfate
Cider	1 2		1	1 2		2		1 2		1 2	1	1 2	2	1 2	2	1 2	1 2	2	Sodium Acetate
Citric Acid	1 2		1	2	1	1 2	1	1 2	1 2	1 2	1 2	1 2	1 2	1 2	1 2	1 2	1 2	1	Soap (molten)
Clay Slurries	2		2		2	1	2			2						1	2		Sludge Acid
Coal Tar (creosote)	1	1	1	1	1	1	1		1	1 2					1	1	1	1	Sizing, Alkaline
Coca Cola Syrup						1				2						1	1		Sizing, Acid

KEY: 1 - Chemical in column 1 may be used • 1 - Chemical in column 1 may not be used
2 - Chemical in column 2 may be used • 2 - Chemical in column 2 may not be used
( ) Conditional or no information

Table 1 *(Continued)*

CHEMICAL 1	MATERIAL: Aluminum	Asbestos	Bronze	Carpenter 20	Cast Iron	Carbon Steel	Ductile Iron	Hastelloy B	Jordanite	Monel	Neoprene	Nickel	Plastisol	Polyethylene	303 S.S.	304 S.S.	316 S.S.	Teflon	CHEMICAL 2
Coconut Oil			1 2		1				1	1			2				2		Silver Plating Sol.
Coffee	1 2		2	1 2		1		1 2		1	1 2	1	2	2	2	1 2	1 2	2	Silver Nitrate
Coke Oven Gas	1 2	1	1	2	1		1	2		1 2	1				1	1 2	1 2		Silver Cyanide
Cod Liver Oil				2	1	2		2		1 2	1					2	1 2	2	Silver Chloride
Copal Varnish	1 2			1 2		2		1 2		1 2		1			2	1 2	1 2	2	Silver Bromide
Copper Acetate	1			1				1 2								1	1		Silicon Tetraiodide
Copper Carbonate	1			1		2		1		2						1 2	1		Silicon Tetrachloride
Copper Chloride	1 2	1	1 2		1		2	1		1 2	1	1			1 2	1 2	1 2		Shellac (bleached)
Copper Cyanide	1 2		2	1	1 2		2	1		1 2		1			2	1 2	1 2		Shellac Orange
Copper Nitrate	1		1	1		1		1		1 2	1	1	1	1	1	1	1	1	Santosite
Copper Sulfate	1	1	1	1	1	1	1	1		1 2	1	1	1	1	1	1	1	1	Santophen
Cupric Chloride	1			1		1		1	1	1 2		1	1	1		1	1	1	Santomerse
Cupric Nitrate	1			1				1		1 2		1				1	1		Santobrite
Core Oil	1 2		1	2	1	2	1	2	2	1 2	2	2			1	1 2	1 2	2	Salicylic Acid
Cornstarch Slurries	2		2	2	2	2	2	2	2	1 2	1 2	2	2	2	2	2	2	2	Sal Ammoniac
Cottonseed Oil	1	1	1		1	1	1			1 2	1				1	1	1		Rustang
Cream of Tartar	1 2		2	1 2		2	2	2		2		2			2	1 2	1 2	2	Rosin (light)
Cresylic Acid	2		2	2	1	1 2	2	2		1 2	1	1 2			1 2	1 2	1 2	2	Rosin (dark)
Cyanogen Chloride						1				2						1			Resorcinol
Cyanohydrin				2		1 2		2		2						1 2	2		Quinine Sulfate
Cyclohexane	1		1	2		2		2		1 2						2	1 2		Quinine Bisulfate
Cyclohexylamine										1 2	2								Querbracho
DDT	1		1							2		1					1	1	Quasol 80
Detergents	1		1			2				1						2	1		Pyroligneous Acid
Developing Solutions	2			1 2	2	1 2		1 2		2		2		1		1 2	1 2	2	Pyrogallic Acid
Dextrose			1		2	2			1						1 2	2	1 2		Pyridine
Diacetone										1 2									Pyrethrum Sol.
Diamylamine						2				1						2	2	2	Propylene Oxide
Dichloroethane			2		2	1 2				2						1 2			Propylene Glycol
Dichloropentane						2				1 2	1 2					2			Propylene Dichloride
Diesel Oil (light)			2							1	2				2		2		Propyl Alcohol
Diethanolamine										1 2	1								Propene, Liquefied
Diethylbenzene	2				2					2	1 2	2	2		2	2	2	2	Propane, Liquefied
Diethyl Sulfate	2	2	2		2	2	2			1 2		2	2		2	2	2	2	Propane Gas
Diethylene Glycol		2			1	1				1 2	1 2				1	1	1 2		Producer Gas
Dimethyl Phthalate										2	1								Prestone
Dinitrochlorobenzene					2	1				2	2				2	1	2		Potassium Triphosphate

KEY: **1 - Chemical in column 1 may be used** • 1 - Chemical in column 1 may not be used
2 - Chemical in column 2 may be used • 2 - Chemical in column 2 may not be used
( ) **Conditional or no information**

Table 1 (*Continued*)

CHEMICAL 1	Aluminum	Asbestos	Bronze	Carpenter 20	Cast Iron	Carbon Steel	Ductile Iron	Hastelloy B	Jordanite	Monel	Neoprene	Nickel	Plastisol	Polyethylene	303 S.S.	304 S.S.	316 S.S.	Teflon	CHEMICAL 2
Dioctyl Phthalate				2		1 2						2		2		1 2	2	2	Potassium Sulfide
Dioxane	2		2		2	2	2	2	1	1 2	2	2	2	2	2	2	2	2	Potassium Sulfate
Dipentene						2				1 2	2					2			Potassium Phosphate (alkaline)
Diphenyl						1 2				1 2	2					1 2	1		Potassium Phosphate (acid)
Diphenyloxide			1	2	1					1						2	2		Potassium Peroxide
Distilled Water	2		1	2	1 2	2		2		1 2	1 2	2	2	2	2	2	1 2	2	Potassium Permanganate
Distillery Wort	1 2		1	2				2				2				1 2	1 2		Potassium Oxalate
Doctor Sol.	1 2		1 2	2			1	2		1 2	1	2		2		1 2	1 2		Potassium Nitrate
Dowtherm					2	1			1	1 2	2				2	1	1 2		Potassium Monophosphates
Dyewood Liquor	2			2		1 2		2		2	2	2				1 2	1 2		Potassium Iodide
Embalming Fluid	2			2		1 2		2		2		2	2			1 2	2	2	Potassium Hypochlorite
Enamel	2		2			2	2	2		1 2	2	2	2	2	2	2	2	2	Potassium Hydroxide
Ethanolamine	1 2			2		1		2		1 2	1	2				1 2	1 2		Potassium Hydrate
Ether, Diethyl	1 2	1	1	1 2		1 2	1	1 2		1 2		1 2	2	1 2	1	1 2	1 2	2	Potassium Ferrocyanide
Ether, Dibutyl	1	1	1	1 2		1 2	1	1 2		1 2		1 2	2	2	1 2	1 2	1 2	2	Potassium Ferricyanide
Ether, Petroleum	1	1	1	1	2	1	1	1		1 2	2	1			1	1	1 2		Potassium Diphosphate
Ethyl Acetate	1 2	1	1	1 2	1	1 2		1 2	1	1 2	1	1	2	1 2	1	1 2	1 2	1 2	Potassium Dichromate
Ethyl Acrylate	2		2	2	2	1 2		2		2	2	2	2	2	2	1 2	2	2	Potassium Cyanide
Ethylbenzene			2	2	2			2		1	1					2	2		Potassium Chromate
Ethyl Cellulose	2		2		2	1 2	2	2		2	2	2	2	2		1	2	2	Potassium Chloride
Ethyl Chloride	1 2	1	1	1 2	2	1 2		1 2		1 2	2	1 2	2	2	2	1 2	1 2	1 2	Potassium Chlorate
Ethyl Mercaptan					2	1				1					2	1	1 2		Potassium Bicarbonate
Ethyl Sulfate	2			2	2	1 2		2		1 2	1 2	2	2	2	2	1 2	1 2	2	Potassium Carbonate
Ethylene (liquefied)	1 2		1	2	1	2		2		1 2	2	2				2	1 2	2	Potassium Bromide
Ethylene Chloride	1			1						1 2	2	1				1	1		Potassium Bisulfite
Ethylene Chlorohydrin	2			2		1		2								1 2	2		Potassium Bichromate
Ethylene Dibromide				1							2						1		Potassium Antimonate
Ethylene Dichloride			1		1	1			1	1	1 2			1		1	1		Potassium Alum
Ethylene Glycol	1	1	1		1	1	1			1 2			1 2	1 2	1	1	1	1	Plating Solution
Ethylene Oxide			1			1				1 2	1					1	1	1	Pitch
Esters	2		2	2				2		1 2	2					2	2		Pine Oil
Fatty Acids	1 2		1 2	1		2	2	1 2		1 2		1	1		1 2	1 2	1 2	1 2	Picric Acid, Aqueous Sol.
Ferric Chloride	1 2		1 2	1 2	1 2	2	1 2	1 2	1	1 2	1	1	1	1	1 2	1 2	1 2	1 2	Picric Acid, Molten
Ferric Hydroxide	1		2	1	2	2		1		1 2	2	1				1 2	2		Phthalic Anhydride
Ferric Nitrate	1			1		1 2		1		1	1	1	1	1	1	1 2	1	1	Phthalic Acid
Ferric Sulfate	1 2		1	1 2	1	1	1	1 2	2	1	1	2	1	1	1	1 2	1 2	1	Phosphorous Trichloride
Ferrous Ammonium Citrate	1					2		1		2						2			Phosphorous Molten

KEY: 1 - Chemical in column 1 may be used • 1 - Chemical in column 1 may not be used
2 - Chemical in column 2 may be used • 2 - Chemical in column 2 may not be used
( ) Conditional or no information – contact: Jordan Valve for assistance

Table 1 (*Continued*)

CHEMICAL 1	MATERIAL																		CHEMICAL 2
	Aluminum	Asbestos	Bronze	Carpenter 20	Cast Iron	Carbon Steel	Ductile Iron	Hastelloy B	Jordanite	Monel	Neoprene	Nickel	Plastisol	Polyethylene	303 S.S.	304 S.S.	316 S.S.	Teflon	
Ferrous Chloride	1		1	1 2	1	2		1	1	1	1	1	1		1 2	2	2		Phosphoric Anhydride
Ferrous Sulfate	1 2	2	1 2	1 2	1 2	2	2	1 2	1	1 2	1 2	1	1 2	1 2	1 2	1 2	1 2	1 2	Phosphoric Acid 45%
Filter Aid	2	2	2	2	2	2	2	2	2	1 2	1 2	2	2	2	2	2	2	2	Phosphoric Acid 0-45%
Fish Oil	2	2	2	2	2	1 2	2	2	2	1 2	2	2	2	2		1 2	1 2	2	Phosphoric Acid, Crude
Flue Gases	1			1		2		1								2	1		Phosgene
Fluoboric Acid										1 2	1		1 2	2			1		Phoscaloid
Fluorine	1			1		2			1	1		1				1 2			Phenolic Sulfonate
Fluosilicic Acid								2		1	1		1	1			2		Phenosulfonic Acid
Formaldehyde	1 2		1	1 2		1	1	1 2		1	1	1 2	1	1	1	1 2	1 2	1	Phenolic Resins
Formalin	1 2		2	2	2	2		2	2	2		2	2	2	2	2	1 2		Phenol
Formic Acid	1 2		1 2		1 2	1 2	1 2	1	1	1 2	1	1		1	1 2	1 2	1 2	1	Petroleum Oils (refined)
Freon (liquefied)	1 2	2	1 2		1 2	1	1 2	2	1	1 2	1	1	1		1 2	1 2	1 2	1	Petroleum Oils (sour)
Freon (dry)	1		1		1	2	1		1	1					1	1 2	1		Perfume
Fruit Juices	1			1	1	1		1		1 2	1 2	1	1			1	1		Pentane
Fuel Oil	1	1	1	1	1	1 2		1		1 2	1 2	1				1 2	1	1	Penicillin, Sol.
Fumeric Acid										2	1								Pelargonic Acid
Furfural	1		1	1	1	1 2	1	1	1	1					1	1 2	1	1	Pectin
Gallic Acid	1		2	1	2	1		1 2		1	1	1			1	1	1 2		Peanut Oil
Gasoline (refined)	1		1	1	1	1 2	1	1	1	1		1	1	1	1 2	1	1 2	1	Parez 607
Gasoline (sour)	1		1	1	1	2	1	1 2	1	1		1	1	1	1	1 2	1 2		Paregoric Compound
Gasoline (antioxident)									2	1 2									Paraldehyde
Ginger Ale						1				2						1	1		Para-formaldehyde
Gelatin	1	1	1		1	1	1			1 2	1 2				1	1	1 2		Paraffin Oil
Glauber's Salt	1 2			1 2		2		1 2		1 2	1 2	1				1 2	1 2		Paraffin
Glucose	1 2	1 2	1 2		1	1 2	1 2			1 2	1		1		1	1 2	1 2		Palmitic Acid
Glue	1	1	1	1	1	1	1	1		1 2	1				1	1	1		Palmic Acid
Glycerine	1	1	1 2	1	1 2	1	1	1	1	1 2	1 2	1	1	1	1	1	1 2	1	Palm Oil
Glycerol	1	1	1				1			1 2	1		1		1	1	1		Paint Vehicles (except soya)
Glutamic Acid				1				1		2							1		Paint
Grease			1		1	2				1						2	1		Ozone
Green Sulfate Liquor	2	1 2	2		1 2	2	2			1 2	1 2				1 2	2	1 2		Oxygen
Gypsum	2	2	2	1 2		2	2		2	2	2	2	2	2	2	1 2	1 2	2	Oxalic Acid
Hagan Solution										1 2									Organic Esters
Heptane (liquefied)			1 2		1 2					1 2	1 2				1	1	1 2		Olive Oil
Hexamine	2			1	2	2		1		2	2					1 2	1 2	2	Oleum
Hexane	1 2	2	1 2	2	1	2	2	2	2	1 2		2	2	2	1 2	2	1 2		Oleic Acid
Hydrazine Hydrate									2	1 2	2						1		Octyl Alcohol

KEY: **1 - Chemical in column 1 may be used** • 1 - Chemical in column 1 may not be used
2 - Chemical in column 2 may be used • 2 - Chemical in column 2 may not be used
( ) **Conditional or no information**

Table 1 (*Continued*)

CHEMICAL 1	Aluminum	Asbestos	Bronze	Carpenter 20	Cast Iron	Carbon Steel	Ductile Iron	Hastelloy B	Jordanite	Monel	Neoprene	Nickel	Plastisol	Polyethylene	303 S.S.	304 S.S.	316 S.S.	Teflon	CHEMICAL 2
Lead Sulfamate					2	2				2	1	2			2	2	2	2	Methyl Chloride (dry)
Lime Slurry						1				1 2	1 2					1	1		Methyl Cellosolve
Lime Sulfur	1 2		1 2		1 2		1			1 2	1				1	1	1		Methyl Benzene
L.P.G.	1									2							2		Methyl Acrylate
Levulinic Acid								1		2	2								Methyl Acetate
Linoleic Acid	1 2				2					1 2							1 2		Methane
Linseed Oil	1	1	1	1	1	1 2		1		1	1	1	1		1	1 2	1		Mesityl Oxide
Lithium Chloride						1					2				2	1	1 2		Mercury Salts
Lithium Hydroxide	2		2	1 2	2	2	2	1 2		2	2	2	2	2	2	1 2	1 2	2	Mercury
Lubricating Oils	2	1	1	2	1	1 2		2		1				1		1 2	1 2		Mercurous Nitrate
Magnesium Carbonate	1 2			1 2		2		1 2		1		1 2	2	2		1 2	1 2		Mercuric Cyanide
Magnesium Chloride	1 2	1	1 2	1	1 2	1 2	1 2	1 2		1 2	1	1	1 2	1 2	1 2	1 2	1 2	1 2	Mercuric Chloride
Magnesium Hydroxide	1 2	1	1		1	1	1	1 2		1	1	1	1	1	1	1 2	1 2	1	Mercuric Bichloride
Magnesium Nitrate	1		1	1				1		2					1	1	1		Mercaptobenzothiazole
Magnesium Oxide						2				1	1					2			Mercaptans
Magnesium Oxychloride			2	1	2	1				2					2	1	1 2		Melamine Resins
Magnesium Sulfate	1 2		1 2		1 2	1 2	1	1 2		1 2	1 2	1 2	1	1	1	1 2	1 2	1	Mayonnaise
Maleic Acid	2		2		1	1		2			1				1	1 2	1 2		Mash
Maleic Anhydride									2	1									Manganese Sulfate
Malic Acid	1			1 2	1			1		1 2	1	1 2	1			1 2	1 2		Manganese Chloride
Malt Beverages				2						1 2	1	2				2	2		Manganese Carbonate

KEY: 1 - Chemical in column 1 may be used
2 - Chemical in column 2 may be used
( ) Conditional or no information
● 1 - Chemical in column 1 may not be used
● 2 - Chemical in column 2 may not be used

Table 1 *(Continued)*

CHEMICAL 1	MATERIAL																		CHEMICAL 2
	Aluminum	Asbestos	Bronze	Carpenter 20	Cast Iron	Carbon Steel	Ductile Iron	Hastelloy B	Jordanite	Monel	Neoprene	Nickel	Plastisol	Polyethylene	303 S.S.	304 S.S.	316 S.S.	Teflon	
Hydraulic Oil	1		1		1					1 2	1						1		Oakite
Hydrobromic Acid	1		1	1	1			1	1	1 2	1	1	1	1	1	1			Nordihydroguaraetic Acid
Hydrocarbons (chlorinated)	2		1	2		2		1 2		2	1	2				2			Nitrous Oxide
Hydrocarbons (alkylated)						2		1		1						2			Nitrous Acid
Hydrocarbons (H_2SO_4)						2		1								2			Nitropropane
Hydrochloric Acid, Cold	1		1		1	1 2	1	1	1	1		1	1	1	1	1 2	1	1	Nitroethane
Hydrocyanic Acid	1		1	1	2	1 2	1	1		1 2	1 2		1	1	1 2	1 2	1 2	1	Nitrobenzene
Hydrofluoric Acid 5%	1 2	1 2	1 2	2	1 2	1 2	1 2	1 2	1 2	1 2	1 2	1 2	1	1 2	1	1	1 2	1 2	Nitric Acid 50-100%
Hydrofluoric Acid 50%	1 2	1	1 2	2	1 2	1 2	1 2	1 2	1 2	1 2	1	2	1 2	1 2	1 2	1 2	1 2	1 2	Nitric Acid 40%
Hydrofluoric Acid 60%	1 2	1	1 2	2	1 2	1 2	1 2	1 2	1 2	1 2	1	2	1 2	1 2	1 2	1 2	1 2	1 2	Nitric Acid 20%
Hydrofluosilicic Acid	2	1	1 2	1 2	1 2	1 2	2	1 2	2	1 2	1	1 2	1 2	1 2	2	1 2	1 2	1 2	Nitric Acid 5%
Hydrogen Chloride (gas)	2		2	2	2	1 2	2	1 2	2	1 2	2	1 2	1 2	1 2	2	1 2	1 2	1 2	Nitric Acid (crude)
Hydrogen Gas	1 2	1	1 2	2	1 2	1 2	1 2	2	2	1 2	2	2	2	2	1 2	1 2	1 2	2	Nickel Sulfate
Hydrogen Fluoride	2		2			1				2					2	1 2	2		Nickel Plating Sol.
Hydrogen Peroxide	1 2		1 2	1 2	1	1	1	1 2		1 2	1 2	1 2	1	1	1	1 2	1 2	1	Nickel Nitrate
Hydrogen Sulfide	1 2		1 2	1 2	1 2	1 2	1 2	1 2	1 2	1 2	2	1 2	1 2	1 2	1 2	1 2	1 2	1 2	Nickel Chloride
Hydrogen Sulfide (Wet)	1		1				1			1	1 2	1	1	1	1	1	1	1	Nickel Acetate
Hydroquinone	2	2	2		2	1 2	2			1 2	2				2	1 2	1 2		Natural Gas
"HYPO" (hyposulfite soda)				2	1	2				1		1				1 2	1 2		Naphthalenic Acid
Ink	1 2		1 2	1	1 2	1 2		1	2	1 2	2	1			1 2	1 2	1 2		Naphthalene
Iodine	1 2		2	1 2	2	1 2		1 2	1 2	1 2		1 2	1 2	1	1 2	1 2	1 2	1 2	Naphtha
Iodoform	1					1				2	1 2				1	1	1		Nalco Solution
Isobutane			2		1 2	2				1					1	2	1 2		Mustard
Isobutyl Acetate	1		1						2	1							1	1	Monoethanolamine
Isoctane				2						1	1					2	2		Monochlorodifluromethane
Isopropyl Acetate					1				1 2		1 2						1	1	Monochlorobenzene
Isopropyl Ether									1	1	1 2								Monochloroacetic Acid
Jet Fuel	1 2		1 2		1 2	2	2	2		1 2	2	2	2	2	2	2	1 2	2	Molasses
Kerosene	1	1 2	1 2	1	1 2	1		1	1	1 2	2	1	1	2	1	1	1 2	1	Mineral Oil U.S.P.
Ketchup				1 2	1	1 2		1 2		1 2	1 2	2			1	1 2	1 2		Mine Water
Ketones	2		1 2		2	1 2	2	2	2	2	2	2	2	2	1 2	1 2	1 2	2	Milk
Lacquers and Lacquer Solvents	1 2		1 2	2	2	1	1	2		1 2	1	2			1 2	1 2	1 2		Methylene Chloride
Lactic Acid	1	1	1	1	1	1 2		1	1	1	1	1	1	1	1	1 2	1	1	Methyl Methacrylate
Lard	1		1	1	1			1	2		1					1	1		Methyl Ketone
Latex			1			1				2	2				1	1	1 2		Methyl Isobutyl Ketone
Lead Acetate	1			1		1		1		1 2	1	1	1	1		1	1	1	Methyl Formate
Lead Nitrate	2		2		2	1				2	1 2					1	2	2	Methyl Ethyl Ketone

KEY: **1 - Chemical in column 1 may be used** • 1 - Chemical in column 1 may not be used
2 - Chemical in column 2 may be used • 2 - Chemical in column 2 may not be used
( ) **Conditional or no information**

4. ALTITUDE-PRESSURE CHART

ALTITUDE—PRESSURE CHART

ALTITUDE (Feet)	PRESSURE In. Hg.	Mm. Hg.	P.S.I.
—1,000	31.02	787.9	15.25
— 500	30.47	773.8	14.94
Sea Level	29.92	760.0	14.70
500	29.38	746.4	14.43
1,000	28.86	732.9	14.18
1,500	28.33	719.7	13.90
2,000	27.82	706.6	13.67
2,500	27.31	693.8	13.41
3,000	26.81	681.1	13.19
3,500	26.32	668.6	12.92
4,000	25.84	656.3	12.70
4,500	25.36	644.2	12.45
5,000	24.89	632.3	12.23
5,500	24.43	620.6	12.00
6,000	23.98	609.0	11.77
6,500	23.53	597.6	11.56
7,000	23.09	586.4	11.34
7,500	22.65	575.3	11.12
8,000	22.22	564.4	10.90
8,500	21.30	553.7	10.70
9,000	21.38	543.2	10.50
9,500	20.98	532.8	10.30
10,000	20.58	522.6	10.10
10,500	20.18	512.5	9.91
11,000	19.79	502.6	9.73
11,500	19.40	492.8	9.53
12,000	19.03	483.3	9.35
12,500	18.65	473.8	9.15
13,000	18.29	464.5	8.97
13,500	17.93	455.4	8.81
14,000	17.57	446.4	8.63
14,500	17.22	437.5	8.46
15,000	16.88	428.8	8.28
15,500	16.54	420.2	8.13
16,000	16.21	411.8	7.96
16,500	15.89	403.5	7.81
17,000	15.56	395.3	7.64
17,500	15.25	387.3	7.49
18,000	14.94	379.4	7.34
18,500	14.63	371.7	7.19
19,000	14.33	364.0	7.04
19,500	14.04	356.5	6.90
20,000	13.75	349.1	6.75
20,500	13.46	341.9	6.61
21,000	13.18	334.7	6.48
21,500	12.90	327.7	6.34
22,000	12.63	320.8	6.21
22,500	12.36	314.1	6.08
23,000	12.10	307.4	5.94
23,500	11.84	300.9	5.82
24,000	11.59	294.4	5.70
24,500	11.34	288.1	5.58
25,000	11.10	281.9	5.45
25,500	10.86	275.8	5.33
26,000	10.62	269.8	5.22
26,500	10.39	263.9	5.11
27,000	10.16	258.1	4.99
27,500	9.94	252.5	4.88
28,000	9.72	246.9	4.78
28,500	9.50	241.4	4.67
29,000	9.29	236.0	4.56
29,500	9.08	230.7	4.46
30,000	8.88	225.6	4.36
30,500	8.68	220.5	4.27
31,000	8.48	215.5	4.17
31,500	8.29	210.6	4.07
32,000	8.10	205.8	3.98
32,500	7.91	201.0	3.89
33,000	7.73	196.4	3.80
33,500	7.55	191.8	3.71
34,000	7.38	187.4	3.63
34,500	7.20	183.0	3.54
35,000	7.04	178.7	3.46
35,500	6.87	174.5	3.375
36,000	6.71	170.4	3.296
36,500	6.55	166.4	3.220
37,000	6.39	162.4	3.140
37,500	6.24	158.6	3.067
38,000	6.10	154.9	2.994
38,500	5.95	151.2	2.925
39,000	5.81	147.6	2.852
39,500	5.68	144.1	2.798
40,000	5.54	140.7	2.720
40,500	5.41	137.4	2.660
41,000	5.28	134.2	2.595
41,500	5.16	131.0	2.535
42,000	5.04	127.9	2.470
42,500	4.92	124.9	2.415
43,000	4.80	122.0	2.360
43,500	4.69	119.1	2.304
44,000	4.58	116.3	2.250
44,500	4.47	113.5	2.195
45,000	4.36	110.8	2.140
45,500	4.26	108.2	2.094
46,000	4.16	105.7	2.042
46,500	4.06	103.2	1.997
47,000	3.97	100.7	1.948
47,500	3.873	98.38	1.900
48,000	3.781	96.05	1.858
48,500	3.693	93.79	1.813
49,000	3.605	91.57	1.772
49,500	3.520	89.41	1.729
50,000	3.436	87.30	1.689
51,000	3.276	83.22	1.610
52,000	3.124	79.34	1.533
53,000	2.978	75.64	1.463
54,000	2.839	72.12	1.395
55,000	2.707	68.76	1.330
56,000	2.581	65.55	1.269
57,000	2.460	62.49	1.208
58,000	2.346	59.58	1.152
59,000	2.236	56.80	1.098
60,000	2.132	54.15	1.048
61,000	2.033	51.63	1.000
62,000	1.938	49.22	0.952
63,000	1.847	46.92	0.906
64,000	1.761	44.73	0.865
65,000	1.679	42.65	0.825
66,000	1.601	40.66	0.786
67,000	1.526	38.76	0.748
68,000	1.455	36.95	0.714
69,000	1.387	35.23	0.681
70,000	1.322	33.59	0.649
71,000	1.261	32.02	0.619
72,000	1.202	30.53	0.590
73,000	1.146	29.10	0.562
74,000	1.093	27.75	0.536
75,000	1.041	26.45	0.512
76,000	0.993	25.22	0.488
77,000	0.946	24.04	0.465
78,000	0.902	22.92	0.443
79,000	0.860	21.85	0.423
80,000	0.820	20.83	0.403
85,000	0.646	16.41	0.317
90,000	0.508	12.91	0.249
95,000	0.401	10.18	0.197
100,000	0.327	8.30	0.161
110,000	0.213	5.40	0.105
120,000	0.138	3.50	0.068
130,000	0.095	2.40	0.047
140,000	0.063	1.60	0.031
150,000	0.043	1.10	0.021
160,000	0.030	0.76	0.0147
170,000	0.021	0.53	0.0103
180,000	0.015	0.37	0.0074
190,000	0.098	0.25	0.0048
200,000	0.067	0.17	0.0033

*Courtesy of Consolidated Controls Corp., Bethel, Connecticut.

5. PRESSURE SWITCH COMPANY REFERENCE CHART

switches, pressure & temperature, hydraulic	PRESSURE													TEMPERATURE					
	SENSING ELEMENT				FEATURES				PRESSURE SETTING -psi		DIFFERENTIAL -psi (dead band)			FEATURES					
	DIAPHRAGM	BELLOWS	PISTON	BOURDON TUBE	MANUAL RESET	ADJUSTABLE DIFFERENTIAL	EXTERNAL SETTING	DUAL PRESSURES	MINIMUM	MAXIMUM	MINIMUM	MAXIMUM	MAXIMUM PROOF PRESSURE-psi	SINGLE SETTING	DUAL SETTING	ADJUSTABLE DIFFERENTIAL	TEMPERATURE RANGE-F	MAXIMUM ALLOWABLE PROOF PRESSURE-psi	METRIC DIMENSIONS AVAILABLE
▪ Automatic Switch Co.	Diaphragm-piston				•	•	•	•	30'' Vac.	3,000	1'' H_2O	3,000	18,000	•	•	•	-60 to 510	1500	
▪ Barksdale Controls Div., Delaval Turbine, Inc.	•		•	•			•		30'' Vac.	18,000	.018	800	24,000	•	•		-65 to 165		
Consolidated Controls Corp.	•	•	•				•		10	5,000	2	1,000	15,000	•			-65 to 500	15,000	
Cook Electric Co.	•	•				•			Vac.	3000	20%	20%	4500						
▪ Custom Component Switches, Inc.	•		•				•	•	.8'' H_2O	5,000	.7'' H_2O	800	7500	•			-44 to 630	1500	
Delta Power Hydraulic Co.			•				•		500	3000			3750						
Duff-Norton Co., Inc.			•				•		700	10,000	200	600						15,000	
▪ Dynamco Inc.			•		•	•	•	•	135	3,000			10,000						•
▪ ENERPAC			•				•		500	10,000	300		15,000						
▪ Hydra-Electric Co.	•		•		•				.03	5000	.05	1000	7500	•			-85 to 400	6000	•
Hydro-Stack Mfg. Co.			•				•	•	50	5000			10,000	•	•			10,000	•
Kratos				•					300	5000	150	400	7500						•
Mid-West Instrument	•	•	•	•			•		1'' H_2O	5000	1/2%	1/2%	12,000						
▪ Miller Fluid Power Corp.			•		•	•	•	•	75	3000									
Minnesota Automotive, Inc.			•			•			1000	2000	50	100				•		3000	•
Neo-Dyn, Inc	•		•				•		0	10,000	5%	100%	15,000	•			-60 to 500	2250	
▪ Oildyne Inc.	•		•				•	•	50	5000	20	400	10,000	•	•		-65 to 250	10,000	
▪ Paul-Munroe Hydraulics, Inc			•				•	•	300	3000	150	300							
Pneu-Hydro Products Inc.			•			•			50	1500	10	20	2300	•					
▪ Rexroth Corp.			•	•		•	•	•	22	7500	7	5700		•	•	•	-4 to 160		•
▪ SOR, Inc.	•		•		•	•	•	•											•
▪ Sperry Vickers			•			•	•	•	100	5000	75								
▪ Square D Co.	•	•	•		•	•	•	•	0	15,000	.2	2400	25,000	•		•	470	480	
▪ Teledyne Sprague Engineering	•					•	•		10	150			7500						
Tridon, Inc. Hyflo Products			•			•	•	•	10	10,000	20	350	20,000						•
United Electric Controls Co.	•	•	•		•	•	•	•	1'' WC	6000	.07'' WC	300	10,000	•	•	•	-40 to 160	10,000	•
Whitman General Corp.	•	•	•			•	•	•	.5	8000	.3'' H_2O	2500	15,000						
▪ Zinga Industries, Inc.	•								4'' Vac.	25									

switches, pressure & temperature, pneumatic

	PRESSURE													TEMPERATURE					
	SENSING ELEMENT				FEATURES				PRESSURE SETTING -psi		DIFFERENTIAL -psi (dead band)			FEATURES					
	DIAPHRAGM	BELLOWS	PISTON	BOURDON TUBE	MANUAL RESET	ADJUSTABLE DIFFERENTIAL	EXTERNAL SETTING	DUAL PRESSURES	MINIMUM	MAXIMUM	MINIMUM	MAXIMUM	MAXIMUM ALLOWABLE PROOF PRESSURE-psi	SINGLE SETTING	DUAL SETTING	ADJUSTABLE DIFFERENTIAL	TEMPERATURE RANGE-F	MAXIMUM ALLOWABLE PROOF PRESSURE-psi	METRIC DIMENSIONS AVAILABLE
Allen-Bradley Co.	•	•	•		•	•	•		30″	5,000	0.2	3,000	10,000	•		•	-100 to 570		
▪ Automatic Switch Co.	Diaphragm-piston				•	•	•	•	30″ Vac.	1500	1″ H_2O	1500	18,000	•	•	•	-60 to 510	1500	
Bachman Valve Corp.	•						•						100	•					
▪ Barksdale Controls Div. Delaval-Turbine Inc.	•		•	•			•		30″ Vac.	18,000	.018	800	24,000	•	•		-65 to 165		
▪ Clippard Instrument Laboratory, Inc.	•								2.8″ H_2O	3.2″ H_2O		.4″ H_2O	10						
Consolidated Controls Corp	•	•	•				•		.25″ H_2O	5000	.2″ H_2O	1000	15,000	•			-65 to 500	15,000	
Cook Electric	•	•				•			.5	500	20%	20%	1000						
▪ Custom Component Switches, Inc.	•		•				•	•	.8″ H_2O	5,000	.7″ H_2O	800	7500	•			-44 to 630	1500	
Cutler Controls Inc.	•		•			•	•		0.1	100	0.05	4	150						
Dwyer Instruments, Inc.	•					•	•	•		20			100	•	•	•	120	100	
▪ Dynamco Inc			•		•	•	•	•	1	115			255						•
Fairchild Camera & Instrument, Ind. Prod. Div	•					•			.02	5.0	0	1	15						
Fairchild Industrial Products Div.	•					•			0.4	40	.01	.03							
Festo Corp.	•		•				•		10	60			150				160	150	•
Gagne Associates Inc.		•					•		0.01	100	10%	25%	150						
▪ Hydra-Electric Co	•				•				.03	5000	.05	1000	7500	•			-85 to 750	6000	•
Kratos				•					300	5000	150	400	7500						•
▪ Lang Pneumatic Div. Sperry Vickers	•								5	150		- 1%	160						•
Mecman Inc.	•		•			•	•		15	150			150						•
Micro Pneumatic Logic, Inc.	•						•		.002	11			25						•
Mid-West Instrument	•	•	•	•			•		1″ H_2O	5000	1/2%	1/2%	12,000						
▪ Miller Fluid Power Corp	•		•			•			4.7	14			150						
Neo-Dyn, Inc.	•		•				•		0	3500	2%	100%	7500	•			-65 to 500	2250	
Pneu-Hydro Products Inc			•			•			50	1500	10	20	2300	•					
Pneumation Systems, Inc.	Belleville spring				•	•	•	•	30″ Vac.	15,000	1%	100%	30,000	•					
▪ SOR, Inc.	•		•		•	•	•	•											•
Spectrum Inc.	•	•	•			•	•	•	2	500	As required		10,000	•	•	•	-65 to 500	6000	•
▪ Square D Co	•	•			•	•	•	•	.25	2250	.2	350	2500	•		•	470	480	
▪ Teledyne Sprague Engineering	•					•	•		10	150			7500						
Tridon, Inc. Hyflo Products			•			•	•	•	10	10,000	20	350	20,000						
United Electric Controls Co.	•	•	•		•	•	•	•	1″ WC	6,000	.07″ wc	300	10,000	•	•	•	-40 to 160	10,000	•
Whitman General Corp.	•	•	•			•	•	•	.3	8000	.3″ H_2O	2500	15,000						

switches, level

switches, level	TYPE						TEMPERATURE-F		PRESSURE-psi		INDICATOR					MOUNTING CONFIGURATION		MATERIAL					SWITCH TYPE			FEATURES		APPROVAL		
	FLOAT	CONDUCTIVITY	PRESSURE	ULTRASONIC	DISPLACER	RADAR	MINIMUM	MAXIMUM	MINIMUM	MAXIMUM	SINGLE-POINT	MULTI-POINT	CONTINUOUS	ADJUSTABLE	CUSTOM BUILT	MINIMUM	MAXIMUM	BRASS	STAINLESS STEEL	CARBON STEEL	PLASTIC	ALUMINUM	SPST	SPDT	DPDT	HERMETICALLY SEALED SWITCH	SHOCK/VIBRATION RESISTANT	MEET MIL SPECS	UL	CSA
▪ Barksdale Controls, Delaval Turbine			•						.22	150	•	•		•					•					•			•		•	•
Consolidated Controls Corp.	•						-65	275		15	•				•			•	•			•	•			•	•	•		
Cutler Controls Inc.			•				-40	900	0.1	100		•	•	•	•				•				•		•					
Dwyer Instruments, Inc.			•				32	110		100	•	•		•								•		•	•				•	•
Fairchild Industrial Products			•				-40	150	.02	8	•										•	•	•				•		•	
Formsprag Co.			•						0.1	20	•										•			•		•			•	
Gagne Associates, Inc.		Air bubbler					32	1000											•					•			•			
Gems Sensors Div. Delaval Turbine Inc.	•	•					-32	350	0	4,000	•	•	•	•	•	1/4	5	•	•	•	•		•	•		•	•	•	•	•
Great Lakes Hydraulics Inc.	•						-40	220	0	300	•							•			•				•	•				
Kratos		•					30	150			•												•				•			
Mid-West Instrument			•					250	300	6000	•		•						•		•	•		•						
▪ Oil-Rite Corp.	•							220		200									•				•			•	•			
United Electric Controls Co.			•				-40	160	1''wc	6000	•		•					•	•	•	•	•		•		•	•	•	•	•

transducers, pressure

transducers, pressure	TYPE			TRANSDUCTION METHOD						NO. OF PRESSURE RANGES	MINIMUM PRESSURE RATING-psi	MAXIMUM PRESSURE RATING-psi	TOTAL ERROR at 77° F	REPEATABILITY - %	HYSTERESIS - %	LINEARITY - %	SENSITIVITY - mv/v	THRESHOLD and/or RESOLUTION	TEMPERATURE SENSITIVITY - %/°F
	GAGE	ABSOLUTE	DIFFERENTIAL	STRAIN GAGE	PIEZOELECTRIC	POTENTIOMETRIC	CAPACITIVE	LVDT	VARIABLE RELUCTANCE				ACCURACY - %						
Atos Oleodinamica S.p.A.				•						3			±0.5	0.1					
BLH Electronics	•	•	•	•						20	20	200,000	0.25	0.15	0.15	0.25			.0025
Consolidated Controls Corp.	•	•	•	•				•			0.3	6000	0.25	<.10	<.10	.15	3	Inf.	.01
▪ Gould Inc., Measurement Systems Div.	•	•	•	•								10,000	.15	.1	.1	.15	Var.	Inf.	Var.
S. Himmelstein & Co.	•	•	•	•						49	5	75,000	±.1	±.05	±.05	±.1	3		.005
Kavlico Corp.	•	•	•				•	•		30	0.5	2,500	1	0.5	<1	<.1		Inf.	0.02
Kratos	•	•		•			•		•	16	50	10,000	1						
Mensor Corp.	•	•	•						•	37	1	2500	.04	.01		.01		.01%	.001
Schaevitz Engineering	•	•	•	•				•		15	0.2	10,000	0.20	0.1	0.15	0.15	2.5	Inf	±0.008
Spectrum Inc.	•	•	•			•				4	5	3000	5	2	10	5	As required		
Standard Controls, Inc.	•	•	•	•						40	5	100,000	.15	.05	.05	.07	2.5	Inf.	.005
Transducers Inc.	•	•	•	•						15	25	12,500	0.3	0.05	0.15	0.25	3	Inf.	0.005
▪ Tyco Instrument Div.										Contact manufacturer directly for specification data									

switches, flow

	TYPES							SET POINT		ACCURACY -%		PRESSURE -psi		LINE SIZE -in.		SWITCH TYPE			MATERIAL					FEATURES			APPROVAL	
switches, flow	CUSTOM	PADDLE	SHUTTLE	DIAPHRAGM	BELLOWS	ULTRASONIC	ORIFICE	MINIMUM-cc/min	MAXIMUM-gpm	REPEATABILITY	SET POINT	MINIMUM	MAXIMUM	MINIMUM	MAXIMUM	SPST	SPDT	DPDT	BRASS	STAINLESS STEEL	CARBON STEEL	PLASTIC	ALUMINUM	HERMETICALLY SEALED SWITCH	LOW PRESSURE DROP	SHOCK/VIBRATION RESISTANT	UL LISTED	CSA
Custom Component Switches, Inc.							•	1900	8.8	1	2	0	300		1/2'' Pipe		•		•						•	•		
Dwyer Instruments, Inc.				•						2	2		100				•	•					•				•	•
Gems Sensors Div. Delaval Turbine Co.	•	•	•					60	100	1	±10	5	2,000	1/4	3	•	•		•	•		•		•	•	•	•	•
■ Hydra-Electric	•			•			•	380	50	±2.5	±7	.5	800	1/4	1.5	•	•	•		•			•	•	•	•		
Mid-West Instrument				•	•			No Limit		1/2	1/2	300	6000				•			•		•	•					

switches, limit & proximity

	TYPE					ACTUATOR				OPERATION							SPECIAL ENCLOSURE						MOUNTING					
switches, limit & proximity	ELECTROMECHANICAL	RF INDUCTIVE	PERMANENT MAGNET	HALL EFFECT	PHOTOELECTRIC	FORK	ROLLER LEVER	PUSH ROD	WOBBLE STICK	OVERCENTER SNAP ACTION	LATCH SNAP ACTION	NEUTRAL POSITION	TWO-STEP	SLOW-ACTION	MOMENTARY CONTACT	MAINTAINED CONTACT	WATER AND DUST TIGHT	OIL TIGHT	EXPLOSION PROOF	HERMETICALLY SEALED	OPERATING FORCE - oz.	SENSING DISTANCE-in.	SURFACE	CARTRIDGE	PLUG-IN	MANIFOLD	INTEGRAL TO CYLINDER	METRIC DIMENSIONS AVAILABLE
Allen-Bradley Co.	•	•				•	•	•	•	•		•			•	•	•	•	•	•	12.5	3-	•		•	•		
Allis-Chalmers, Industrial Controls Div.	•						•	•		•		•		•	•		•	•					•		•			•
Bettis Corp.	•						•			•		•			•		•	•	•	•	6		•					•
■ Carter Controls, Inc.	•							•		•						•	•	•	•	•							•	
W. C. Dillon & Co., Inc.	•							•				•	•			•	•	•	•	•		.024	•				•	
■ ENERPAC	•						•								•													
Festo Corp.							•	•	•	•	•	•	•		•	•	•	•	•	•	8	4	•		•	•	•	•
Gagne Associates, Inc.			Fluidic				Nozzle									•	•				≈0	1.2	•		•			
Kavlico Corp.							Ferrous Mat'l.					Momentary, Non-contact					•	•		•	0	1			•			•
Micro Pneumatic Logic, Inc.							•								•						1/3							•
Micro Switch							•	•									•											
Mid-West Instrument			•		•						•						•		•									
■ Miller Fluid Power Corp.	•						•	•	•							•	•				7.2		•					•
Namco Controls Div. Acme Cleve. Corp.	•	•	•			•	•	•	•	•	•	•				•	•	•	•	•			•		•			
PHD, Inc.			•							•							•	•		•							•	•
The S-P Mfg. Corp.			•														•	•	•	•		.025		•			•	
■ Sheffer Corp.	•		•												•		•						•				•	
■ Square D Co.	•	•				•	•	•	•	•	•	•	•	•	•	•	•	•	•				•		•	•		•
■ Tann Controls Co.			•			Metal Target									•		•	•	•	•		1/2	•				•	
■ Versa Product Co., Inc.						•	•	•	•			•	•	•	•	•		•	•		7	Var.	•			•		•

Index